# RF Circuits and Applications for Practicing Engineers

For a listing of recent titles in the
*Artech House Microwave Series*,
turn to the back of this book.

# RF Circuits and Applications for Practicing Engineers

Mouqun Dong

**ARTECH HOUSE**
BOSTON | LONDON
artechhouse.com

Library of Congress Cataloging-in-Publication Data
A catalog record for this book is available from the U.S. Library of Congress

British Library Cataloguing in Publication Data
A catalog record for this book is available from the British Library.

ISBN-13: 978-1-63081-631-5

Cover design by Mark Bergeron, Publishers' Design and Production Services, Inc.

© 2021 Artech House
685 Canton St.
Norwood, MA

All rights reserved. Printed and bound in the United States of America. No part of this book may be reproduced or utilized in any form or by any means, electronic or mechanical, including photocopying, recording, or by any information storage and retrieval system, without permission in writing from the publisher.

All terms mentioned in this book that are known to be trademarks or service marks have been appropriately capitalized. Artech House cannot attest to the accuracy of this information. Use of a term in this book should not be regarded as affecting the validity of any trademark or service mark.

10 9 8 7 6 5 4 3 2 1

*To my parents,*
*Dong Yizhong and Hua Tongwen*

# Contents

| | Preface | xv |
|---|---|---|
| **1** | **Essentials of Radio Frequency (RF) Circuits and S-Parameters** | **1** |
| 1.1 | Introduction | 1 |
| 1.2 | Transmission Lines | 6 |
| 1.2.1 | Characteristic Impedance | 7 |
| 1.2.2 | Electrical Length | 8 |
| 1.2.3 | Attenuation | 10 |
| 1.3 | Power Waves and the Reflection Coefficient | 13 |
| 1.3.1 | Power Waves | 13 |
| 1.3.2 | Reflection Coefficient | 14 |
| 1.3.3 | Impedance Transformation | 17 |
| 1.4 | RF Source | 20 |
| 1.5 | Scattering Parameters (S-Parameters) | 27 |

| 1.5.1 | Two-Port S-Parameters | 27 |
| 1.5.2 | N-Port Networks and the SnP File Format | 33 |
| | References | 34 |

| **2** | **Circuit Analysis Using S-Parameters** | **37** |
|---|---|---|
| 2.1 | Power Gains | 37 |
| 2.1.1 | $\Gamma_{out}$ and $\Gamma_{in}$ | 38 |
| 2.1.2 | $P_{av\_out}$ and $P_{L\_in}$ | 39 |
| 2.2 | Stability of Amplifiers | 44 |
| 2.2.1 | Analysis in the $\Gamma_S$ and $\Gamma_L$ Planes | 45 |
| 2.2.2 | Stability Circles | 47 |
| 2.2.3 | Stability Criteria of $K$ and $\mu$ Parameters | 50 |
| 2.2.4 | Additional Comments | 52 |
| 2.3 | Conjugate Matching and Constant Circles of Gains and Mismatch Factors | 55 |
| 2.3.1 | Simultaneous Conjugate Matching | 55 |
| 2.3.2 | Constant Circles of Gains and Mismatch Factors | 58 |
| 2.3.3 | Linear Circuit Design Using the Constant Circle Method | 63 |
| 2.4 | Invariance of the Mismatch Factors of Lossless Networks | 68 |
| | References | 72 |

| **3** | **Matching Networks and the Smith Chart** | **73** |
|---|---|---|
| 3.1 | Circuit Realization of Matching Networks | 73 |
| 3.1.1 | Case 1: Resistor Allowed for the Matching Circuit | 74 |
| 3.1.2 | Case 2: Lossless Matching Networks | 75 |
| 3.2 | Construction of the Smith Chart | 76 |
| 3.3 | The Smith Chart Representation of Networks | 82 |
| 3.4 | Bandwidth of Matching Networks and $Q$ Lines on the Smith Chart | 86 |
| 3.4.1 | Review of $Q$ Factors | 86 |

| 3.4.2 | Bandwidth Expressed in $Q$ | 89 |
| 3.4.3 | $Q$ Lines on the Smith Chart | 95 |
| 3.5 | Matching Network Design Using the Smith Chart | 97 |
| 3.5.1 | Lumped Elements and Transmission Lines on the Smith Chart | 97 |
| 3.5.2 | Computerized Smith Chart | 101 |
| 3.5.3 | Design Considerations of Matching Networks | 103 |
| | References | 111 |

## 4 Noise and Its Characterization in RF Systems — 113

| 4.1 | Noise Sources | 114 |
| 4.2 | Thermal Noise | 115 |
| 4.3 | Equivalent Noise Circuits | 118 |
| 4.4 | The Noise Figure of Linear Two-Port Networks | 122 |
| 4.5 | The Noise Figure in Practical Applications | 129 |
| 4.5.1 | Noise Figure and Source Impedance | 129 |
| 4.5.2 | Temperature Effects on Noise Performance and Characterization | 135 |
| | References | 140 |

## 5 RF Amplifier Designs in Practice — 141

| 5.1 | Circuit Specifications and Transistor Selection | 141 |
| 5.1.1 | Circuit Specifications of Linear Amplifiers | 141 |
| 5.1.2 | Transistors for RF Applications | 143 |
| 5.1.3 | Datasheet Specifications for RF Transistors | 145 |
| 5.2 | Stability Considerations | 152 |
| 5.2.1 | Conditions for Oscillation | 152 |
| 5.2.2 | Stability Analysis Using Stability Parameters | 155 |
| 5.2.3 | Some Common Unintended Oscillations and Solutions | 159 |
| 5.3 | Low-Noise RF Amplifier Design with the Constant-Circle Method | 169 |

| 5.3.1 | Estimate of $RL_{in}$ When $F = F_{min}$ and $RL_{out}$ Is Perfect | 171 |
| 5.3.2 | Trade-Offs Among $F$, $RL_{in}$, and $RL_{out}$ | 172 |
| 5.4 | Configurations for RF Amplifiers | 175 |
| 5.4.1 | Balanced Amplifiers | 175 |
| 5.4.2 | Broadband Amplifiers with Negative Feedback | 177 |
| 5.4.3 | LC Balun for Differential and Single-Ended Circuits | 179 |
| | References | 181 |

| 6 | **RF Power Amplifiers** | **183** |
|---|---|---|
| 6.1 | DC Characteristics of RF Transistors | 184 |
| 6.2 | Load-Line and Matching for Power | 187 |
| 6.3 | Nonlinearity Specifications | 194 |
| | References | 202 |

| 7 | **Efficiency of RF Power Amplifiers** | **203** |
|---|---|---|
| 7.1 | RF Cooling and Efficiencies of RF Power Amplifiers | 203 |
| 7.2 | Conduction Angle and Class A, AB, B, and C Power Amplifiers | 206 |
| 7.2.1 | Conduction Angles | 206 |
| 7.2.2 | Efficiency of Class A, AB, B, and C Amplifiers | 210 |
| 7.2.3 | $I_0$ and Load Line Selection | 214 |
| 7.2.4 | Class A to Class B Modes in Actual RF Amplifiers | 215 |
| 7.3 | Operation in Overdriven Conditions | 219 |
| 7.3.1 | $I_0 = I_{max}$ | 219 |
| 7.3.2 | $I_0$ by the Load Line Selection | 221 |
| 7.4 | Harmonic Flattening and Class F and Inverse Class F Amplifiers | 221 |
| 7.4.1 | Second- and Third-Harmonic Flattening | 221 |

| 7.4.2 | Class F and Inverse Class F Amplifiers | 225 |
| 7.5 | Class E and Doherty Amplifiers | 227 |
| 7.5.1 | Class E Amplifier | 228 |
| 7.5.2 | Doherty Amplifier | 230 |
| | References | 232 |

## 8 RF Designs in Wireless Communications Systems — 233

| 8.1 | Introduction | 233 |
| 8.2 | Baseband, Modulation, and Passband | 235 |
| 8.2.1 | Baseband and M-ary Scheme | 235 |
| 8.2.2 | Passband of Modulated RF Carriers | 240 |
| 8.3 | Cascaded Noise Figure and Linearity Specifications | 242 |
| 8.3.1 | Cascaded Noise Figure | 242 |
| 8.3.2 | Cascaded $P_{1dB}$ and $IP_3$ | 244 |
| 8.4 | Specifications for RF Transmitters | 246 |
| 8.4.1 | Peak-to-Average Power Ratio | 246 |
| 8.4.2 | Phase Noise | 248 |
| 8.4.3 | EVM | 249 |
| 8.4.4 | Spectral Regrowth | 250 |
| 8.5 | Specifications for RF Receivers | 252 |
| 8.5.1 | Receiver Sensitivity and Noise Figure | 253 |
| 8.5.2 | Interference in Receivers | 255 |
| 8.6 | Grounding in RF Circuits | 258 |
| | References | 260 |

## 9 Passive Components in RF Circuits — 261

| 9.1 | Capacitors, Inductors, and Resistors | 261 |
| 9.1.1 | Capacitors | 263 |
| 9.1.2 | Inductors | 269 |

| | | |
|---|---|---|
| 9.1.3 | Resistors | 275 |
| 9.2 | Devices for RF Power Wave Manipulations | 276 |
| 9.2.1 | Directional Couplers | 276 |
| 9.2.2 | Circulators | 279 |
| 9.2.3 | Power Dividers and Combiners | 280 |
| 9.2.4 | 90-Degree Hybrid Couplers | 283 |
| 9.2.5 | Low–Passive-Intermodulation Components | 285 |
| | References | 285 |
| | | |
| **10** | **RF Measurements** | **287** |
| 10.1 | Measurement Techniques of S-Parameters | 287 |
| 10.1.1 | The VNA and Calibration Techniques | 287 |
| 10.1.2 | TRL Calibration | 291 |
| 10.1.3 | Port Extension Technique | 296 |
| 10.2 | Spectrum Measurements | 301 |
| 10.2.1 | Resolution Band Width (RBW) | 303 |
| 10.2.2 | Video Band Width | 306 |
| 10.3 | Phase Noise and Frequency Stability | 306 |
| 10.3.1 | Types of Modulation | 307 |
| 10.3.2 | Phase Noise | 309 |
| 10.3.3 | Frequency Stability | 312 |
| 10.4 | Noise Figure Measurement | 312 |
| 10.4.1 | Y-Factor Method | 312 |
| 10.4.2 | Uncertainty and Temperature Correction | 316 |
| 10.4.3 | Cold-Source Method | 317 |
| 10.5 | Power and Nonlinearity Measurements | 318 |
| 10.5.1 | Power Measurements | 318 |
| 10.5.2 | Nonlinearity Measurements | 319 |
| | References | 321 |

| 11 | **RF Switches** | **323** |
|---|---|---|
| 11.1 | Realization of Switching Functions Using Two-Impedance Devices | 324 |
| 11.2 | PIN Diode RF Switches | 326 |
| 11.3 | FET-Based RF Switches | 328 |
| 11.3.1 | RF Performance of FET Switches | 329 |
| 11.3.2 | DC Bias Networks for FET Switches | 334 |
| 11.4 | RF Switch Selection in Applications | 337 |
| | References | 339 |

| **About the Author** | **341** |
|---|---|

| **Index** | **343** |
|---|---|

# Preface

While most readers do not closely study a preface—for arguably good reasons—this book's readers would benefit from perusing at least the next couple of paragraphs before moving on to the chapters. This should help readers, particularly those relatively new to the field, to follow the discussion and flow of this book more easily.

Many technical terms and acronyms are used throughout the book. They are explained when they appear the first time but may be used again without a definition. Those reading the book out of sequence can refer to the extensive index to locate the place where terms are first discussed and defined. In fact, the index should be utilized as an integral part of the book.

This book is about the theory of RF circuits and systems and the practice of designing them. It is primarily intended for practicing RF engineers who are involved in PCB-based circuit designs and system integrations, particularly two groups of readers, described as follows:

- Engineers who are equipped with sufficient educational experience and knowledge but who are relatively new to the field. They are not always sure how to use textbook materials in a real-world engineering practice.
- Experienced engineers with a great deal of practical knowledge who may, nevertheless, need to brush up on the theory behind the practice.

The material selection of the book reflects this intention.

While the theory of the S and noise parameters and their applications in circuit designs (Chapters 1–4) and the basic operating principles of RF power amplifiers (Chapters 6–7) are equally applicable to other RF circuits, all the practical design techniques covered in this book are about PCB-based circuits. The discussion is centered on the components of RF amplifier circuits. The book does not cover other critical components for a complete RF system, such as mixers, oscillators, and power detectors, because they are usually integrated in the RF transceiver in a modern communication system. Another important but omitted subject is RF component filters.

The subject matter in this book was selected and presented with a primary motive: to be relevant to practitioners. Many of the circuit examples throughout this book and the comments on how analysis results can be used in everyday engineering practice are from my own experience. I hope some of this background information will be useful to my fellow engineers. Throughout the book, there are only a few instances where treatments are more highly mathematical. Examples are Section 2.2.4 (on the stability criteria of RF amplifiers) and Section 3.4.2 (on the relationship between Q factor and bandwidth). The concept of the invariance of mismatch factors of lossless networks, described in Section 2.4, is important in practice, but its proof is not necessary for many readers. These more analytic topics are included, because they are usually not covered in other textbooks, and because some of our more analytically minded readers may find them useful.

This book can also be used as a supplementary text for a college-level introductory course in RF and microwave engineering. Some treatments in the book, provided from a practical perspective, may help students better understand the subject matter in the context of actual applications.

This book does not touch on the topic of nonlinear models, a subject often inquired about by college students and sometimes by engineers. In practice, PCB-based RF circuit designs are still largely done on the bench, particularly for high-power and high-frequency applications, and often with only the aid of S-parameter simulation results. A lack of nonlinear models from device manufacturers contributes, partially, to this situation. For RF IC designs, the semiconductor fabs usually do provide some circuit models for nonlinear modeling simulations.

The book uses numerous practical examples to clarify the concepts and facilitate the discussion. Most of them are based on test data of components (performed by the author) from a now obsolete product line at California Eastern Laboratories (CEL). Since these data are generally not published and none of the parts are still available, I do not cite specific part numbers or data sources in these examples.

I appreciate the help of many individuals throughout my development of this book: First, I would like to thank Aileen Storry (formerly of Artech House). I would not have started the book without her encouragement. I would also like to thank Artech House's Rachel Gibson and Soraya Nair for their assistance during the writing of this book. I am particularly indebted to Artech House for allowing me extra time to complete the book.

I am sincerely grateful to CEL's management, particularly Paul Minton, for providing a pleasant and well-balanced working environment. Special thanks are due to Marc Sheade, my manager at CEL for more than 15 years, for his encouragement and friendship. My friend and colleague, Steve Morris reviewed most of the book and provided many valuable suggestions. I'd also like to thank Song Ling, who reviewed a chapter, and Paige Dong, who proofread the manuscript, for their help in the manuscript's preparation. I also had a number of fruitful discussions with Dr. Weishu Zhou on the subject of RF power amplifiers. His comments are greatly appreciated.

My deepest gratitude goes to my family, my wife Jin and daughter Paige, for their patience during the long hours I spent on this project and for their love and support in my life.

Finally, I devote this book to the memory of my late professor Arnold Honig who first introduced me to the field of microwave technology in his electron spin resonance (ESR) lab.

# 1

# Essentials of Radio Frequency (RF) Circuits and S-Parameters

## 1.1 Introduction

This book is about RF circuits and their applications. Traditionally, RF is thought of as radio waves in the air intended for radio listeners. In the modern days, RF is more often associated with various wireless communications devices, most notably, mobile phone systems. In these applications, the word "wireless" indeed implies that the communications between devices are carried by radio waves over the air. Here, however, we are discussing bringing RF signals onto circuits. In this sense, RF circuits are similar to the analog circuits with which the book's readers are assumed to be familiar (likely through a college course on electric circuits). In fact, at the fundamental physics level, both RF and analog circuits deal with electromagnetic (EM) fields changing in space and time (or EM waves) that are confined in a circuit instead of propagating in free space. The question, then, is how an RF circuit is different from an analog circuit. The major difference is that RF changes faster than analog, or that the signal frequency of RF is higher than that of analog. From an engineering perspective, the terms of RF and analog are beyond simple labeling for different frequency bands. They are actually two subcategories of the discipline of electric circuits. RF and analog circuits use different (albeit with some overlap) circuit parameters and have different sets of rules for circuit design and practical implementation.

We start with an example that explains why the analog circuit theory and techniques are inadequate when the signal frequency is sufficiently high; from this, we can estimate at what frequency point the RF circuit theory and practice should be employed. Consider a typical prototype analog circuit on a breadboard for the gain measurement of a voltage amplifier IC as shown in Figure 1.1. On the breadboard there are two sets of connectors for input and output connections to a signal generator and a resistor load respectively. An oscilloscope is used to measure the waveforms at the input and output connectors. The gain is simply the ratio of the amplitudes of the output and input waveforms. This simple setup works perfectly fine at low frequencies but becomes unreliable when the frequency of the analog signal reaches a certain point due to the EM nature of the analog signal. To quantitatively analyze this case, we state two facts from the theory of electromagnetic fields and waves, without derivations.

First, the voltage generated in the signal generator is in the form of a propagating wave from the generator to the amplifier and is described by:

$$v^+(t,x) = V_0 \sin\left[\omega\left(t - \frac{x}{s}\right)\right] \qquad (1.1)$$

where $\omega$ is the angular frequency, $s$ is the propagation velocity, and $x$ is the distance from a point along the wire to the generator; $x = 0$ and $x = L$ at the signal generator and the amplifier input, respectively. The negative sign of the

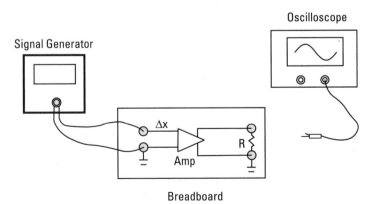

**Figure 1.1** Setup for gain measurement of analog voltage amplifiers. An adapter of BNC to banana plugs is used at the signal generator output.

$x$ term indicates that the propagation is to the direction that $x$ increases. $V_0$ is the amplitude of the wave. Modern oscilloscopes have two settings for input impedance, 50Ω and high (usually > 10 MΩ). When the input impedance setting is at 50Ω, $V_0$ is the measured value on the scope, assuming that the signal generator has a 50-Ω source impedance. At high input impedance the measured value is $2V_0$.

The second fact is that a propagating wave is completely reflected, in phase, at an open termination. Since voltage amplifiers usually have high input impedance (to avoid a loading effect to the source, among other reasons) that can be approximately considered as open in this case, the reflected wave is given by:

$$v^-(t,x) = V_0 \sin\left[\omega\left(t + \frac{x-L}{s} - \frac{L}{s}\right)\right] = V_0 \sin\left[\omega\left(t + \frac{x-2L}{s}\right)\right] \quad (1.2)$$

In constructing $v^-(t, x)$, we took into account that the mathematical expressions for $v^-$ and $v^+$ must be consistent since $v^-$ is a reflected wave from $v^+$. Specifically, there are three requirements for $v^-$: (1) that the $x$ term is positive because the propagation is to the opposite direction; (2) that $v^-$ starts at $x = L$, and thus $x$ in (1.1) is replaced by $x - L$ in (1.2); and (3) that an additional term $-L/s$ is added to account for the delay time for $v^-$ with respect to $v^+$.

For the simplicity of analysis we assume that the reflected traveling wave is completely absorbed at the generator and hence, there is no further reflection. (Section 1.4 analyzes a more general case where the reflected wave is partially absorbed at the generator.) In our calculation, it is also assumed that the propagation velocity, $s$, is the speed in free space (also known as the speed of light in a vacuum), $c = 3 \times 10^8$ m/s. Then the voltage pattern along the wire ($x = 0$ to $x = L$) is a superposition of the two traveling waves:

$$v(t,x) = v^+(t,x) + v^-(t,x) = V_0 \sin\left[\omega\left(t - \frac{x}{c}\right)\right] + V_0 \sin\left[\omega\left(t + \frac{x-2L}{c}\right)\right]$$

Using a trigonometric identity, we have:

$$v(t,x) = 2V_0 \sin\left[\omega\left(t - \frac{L}{c}\right)\right] \cos\left[\omega\left(\frac{x-L}{c}\right)\right] \quad (1.3)$$

Equation (1.3) is a standard standing wave equation, which indicates that the amplitude of the voltage waveform varies along the wire connecting the signal generator and the amplifier. At the amplifier input, $x = L$, and the amplitude is $2V_0$, which is the expected value in an oscilloscope measurement with the high impedance setting. As the location moves away from the amplifier, the amplitude decreases according to $\cos[\omega(x-L)/c] = \cos(2\pi\Delta x/\lambda)$, where $\Delta x = L - x$ is the distance to the amplifier and $\lambda = 2\pi c/\omega$ is the wavelength of the voltage wave. Figure 1.2 plots the function $\cos(2\pi\Delta x/\lambda)$ from $\Delta x/\lambda = 0$ to 10% in log scale. It can be seen that the function essentially remains constant up to $\Delta x/\lambda = 1\%$; after that point, it starts to drop off. At $\Delta x/\lambda = 10\%$, it is only 80% of the original value. An extreme case occurs at $\Delta x = \lambda/4$ where the amplitude is zero.

In our experiment, $\Delta x$ is the distance between the amplifier input and the connectors where the voltage is probed. The plot in Figure 1.2 implies that when the wavelength $\lambda$ is sufficiently short, the voltage measured by the scope is no longer a true measure of that at the amplifier input. For this simple measurement, the problem can be avoided by directly probing the pins of the IC. For a more complex network, however, this problem is clearly unmanageable in circuit analysis. Recall that in Kirchhoff's circuit laws, which are the foundation for analog circuit analysis, voltages need to be specified only at each node in a circuit network, and they are implicitly assumed to be constant between the circuit elements. The above analysis reveals that when a circuit dimension is about 1% of the wavelength, the EM nature of the signal starts to be observable. This point is approximately where the boundary between the analog and RF circuitries lies. Using the 1% criterion, for a circuit dimension of 10 cm, the corresponding frequency is 30 MHz. It is no coincidence that the starting frequencies of most standard vector network analyzers (VNAs) are in tens of megahertz. Chapter 10 details the applications of VNAs in RF device and circuit characterization.

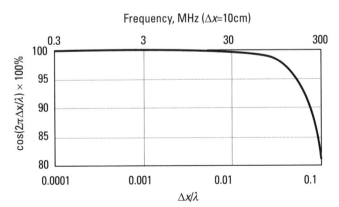

**Figure 1.2** $\cos(2\pi\Delta x/\lambda)$ versus $\Delta x/\lambda$.

In summary, while both analog and RF are fundamentally EM waves confined in a circuit, the wavelength with respect to the circuit dimension determines whether the spatial changes of a circuit variable can be neglected (as in analog circuits) or need to be considered (as in RF circuits).

From the above example, we make three observations. First, the reflection due to a high impedance load should be avoided. The solution to this problem in RF circuits is impedance control, which is introduced in Section 1.3. Section 1.3 also introduces a new measurable circuit quantity for RF circuits: the power wave. Our second observation is that voltage measurement is no longer reliable once a standing wave is present, as illustrated in the example. In fact, with the voltage measurement in a high-frequency circuit, the problem can be worse than being unreliable. Section 1.3.3 explains that the conventional voltage measurement using a scope probe may disturb a circuit drastically at certain frequencies. The third issue with the measurement setup is less obvious from our analysis. It is about transmission lines, which are discussed in Section 1.2.

The behavior of EM fields and waves is governed by a set of Maxwell's equations. In principle, any EM problem can be solved by Maxwell's equations with appropriate boundary conditions. However, like in any engineering field, it is much easier to deal with practical EM problems using a set of simplified parameters and rules, instead of working from the first principle of physics. This chapter introduces some basic circuit quantities and rules that are the foundation of the theory of RF circuits. All these parameters and rules are related to some fundamental physical quantities and principles; in other words, they can be derived. However, the derivations require a certain level of understanding of physics and a significant amount of manipulation of differential equations. Neither is essential or efficient for practicing RF engineers. For this reason, we will skip the derivations and, instead, present a few basic concepts and rules as the starting points for the discussion in this book. Interested readers can review derivations with various degrees of rigor on the subject in a number of textbooks, including [1–4]. Once Sections 1.2 and 1.3 establish the basic concepts and rules, the treatments in the remainder of this book will be relatively rigorous with a few exceptions. It is clearly stated when these exceptions occur.

Finally, please note the following regarding terminology: Some of the literature differentiates "RF" and "microwave," designating RF for a frequency range up to about 1 GHz and microwave for higher frequencies. In most practical cases, these two terms are used interchangeably without any substantive difference. This book uses the term RF for the entire spectrum of RF/microwave, which covers roughly 30 kHz to 300 GHz. Whenever the subject matter is frequency-range-specific, the book states it explicitly.

## 1.2 Transmission Lines

The wires connecting the signal generator and the breadboard in Figure 1.1 are intentionally drawn in a way that implies that the wires can be arbitrarily arranged. For low-frequency applications, the function of a wire is simply to electrically connect one circuit component to another. As such, with a few exceptions in high-voltage and high-current applications, the physical properties of a wire are generally not critical in circuit designs. The situation is remarkably different when the frequency is high enough that the effect of EM waves becomes a factor. Under this condition, the wire's physical properties, both mechanical and electrical, have a major impact on the wave behavior and, therefore, need to be taken into account in designs. In RF circuits, conductors that connect circuit elements are commonly referred to as transmission lines. Unlike a pair of loose wires, a transmission line must have a rigid mechanical structure so that its properties are stable. Other than waveguides, most transmission lines consist of at least two conductors. Two common types are coaxial cables and printed circuit board (PCB)-based transmission lines. A coaxial cable has a center conductor and an outer conducting layer with dielectric material in between as illustrated in Figure 1.3(a). A coaxial cable can be flexible, but the cable structure has to be maintained while being bent. PCB transmission lines have many variations, one of which, the microstrip line, is presented in Figure 1.3(b). It is important to note that at high frequencies the EM fields in these structures exist in the dielectric media and the surrounding air rather than inside the conductors (due to the so-called skin effect [2, 4]). As a result, along with the physical structure and dimensions, the dielectric material used in a transmission line is also a critical factor affecting the characteristics of a transmission line.

In addition to the function of connecting circuit elements, PCB transmission lines are also widely used as matching components in RF circuits, particularly at high frequencies, which is covered in Chapter 3. Sections 1.2.1–1.2.3 summarize three general transmission-line properties that are most relevant to RF circuit performance: characteristic impedance, electrical length, and attenuation. All three properties are transmission line–specific; that is, they are dependent on the physical structure and dimensions of the transmission line as well as its material properties.

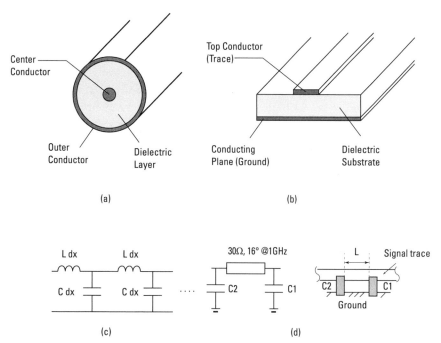

**Figure 1.3** Transmission lines: (a) coaxial cable, (b) microstrip line, (c) L-C equivalent circuit, and (d) electrical length and impedance representation of a transmission line.

### 1.2.1 Characteristic Impedance

This parameter was originally introduced in EM theory as a measure of the magnitude ratio of the electric field to the magnetic field (the E and H fields in EM theory) for a propagating wave [1, 2]. It has the unit of electrical resistance but is not related to any physical resistance. Its numerical value is affected by the medium in which the wave propagates, even though a medium is not necessary for the propagation of EM waves. (For this reason, it is also called wave impedance in some treatments of EM waves.) For example, the impedance for the free-space EM propagation is $377\Omega$. In the case of the transmission line, when a transmission line is modeled as a distributed L-C network [1, 4], the characteristic impedance of the transmission line (or simply transmission line impedance) $Z_0$ does have a circuit interpretation as $Z_0 = \sqrt{L/C}$ where $L$ and $C$ are unit length inductance and capacitance respectively, as illustrated in Figure1.3(c).

The most significant effect of the characteristic transmission line impedance on an RF circuit is that it affects the signal propagation at the interface between the transmission line and circuit components. As explained in Section 1.3, when the transmission line impedance does not match the component impedance, a signal reflection occurs. To avoid such a problem, standards for RF cable impedance are established in industries. Most industries use 50Ω as the standard. The only exception is the 75-Ω standard adopted by the CATV industry. As high-speed digital signal devices become ubiquitous, 100-Ω cables (usually in the twisted-pair form) for differential signals are also a major cable type in the marketplace. For PCB-based RF circuits, other than those connecting the circuit to the input and output connectors (or antennas), transmission lines used between circuit elements do not necessarily need to have a standardized impedance. In fact, different transmission line impedances are designed for various circuit functions, as demonstrated in a number of cases throughout the book. In principle, designers can choose any impedance for a PCB transmission line. However, there are practical limitations on the value of transmission line impedance, due to the constraints of substrate materials and physical dimensions. For example, in the case of a microstrip transmission line, a combination of a thick substrate and a relatively small dielectric constant can make the required trace width for a 50-Ω impedance too wide for a reasonable component layout. In general, the transmission line impedance should be part of the design considerations in PCB substrate selection.

### 1.2.2 Electrical Length

Strictly speaking, electrical length is an equivalent length that replaces the physical length of a transmission line in the phase-shift calculation. In practice, however, there are several variations in presentation of this parameter. The physical reason for the replacement of the physical length with an equivalent length is that EM waves propagate at a slower speed in a medium than in vacuum. The electrical length is essentially a measure of this effect for a specific transmission line.

In the cable industry, the velocity factor, $VF$, is usually used to specify the EM velocity for cables. It is defined as $VF = s/c$ where $s$ and $c$ are defined in (1.1) and (1.3). The $VF$ values for different cables are normally in the range of 50% to almost 100% [5, 6]. For a coaxial cable, if the space between the two conductors is uniformly filled with dielectric material, $VF$ is given by $VF = 1/\sqrt{\varepsilon_r}$, where $\varepsilon_r$ is the dielectric constant of the material. Cables used in RF circuits are mostly for connection purposes, and the value of $VF$ is not always practically important except when a specification of propagation delay is required. In that case, the delay of a cable assembly, $\Delta T_d$, can be calculated from VF as follows:

$$\Delta T_d = \frac{L}{s} = \frac{L}{VF \cdot c}$$

where $L$ is the cable length.

For PCB transmission lines, the space between conductors is only partially filled with dielectric material in many configurations. Microstrip is an example. In such cases, an effective dielectric constant denoted as $\varepsilon'_r$, or $\varepsilon_{eff}$ is used to replace $\varepsilon_r$. Evidently, $\varepsilon'_r \leq \varepsilon_r$, for the space is partially filled with the material of $\varepsilon_r$. The EM velocity in the PCB transmission line is then given by:

$$s = \frac{c}{\sqrt{\varepsilon'_r}} \quad (1.4)$$

When a section of PCB transmission line is employed in an RF circuit for matching purposes, the concept of electrical length becomes critical. As detailed in Section 1.3, the effect of a transmission line on an RF signal is a phase shift, which, according to the wave propagation equation (1.1), is given by:

$$\Delta\theta = \frac{2\pi f L}{s} \quad (1.5)$$

Here, $f$ is the frequency, and $L$ is the transmission-line length. Equation (1.5) is often written in a simpler form using the propagation constant (also called the wave number) $\beta = 2\pi f/s$ as

$$\Delta\theta = \beta L \quad (1.6)$$

If we define electrical length (or effective length), $L' = L\sqrt{\varepsilon'_r}$, by using (1.4) we obtain:

$$\Delta\theta = \frac{2\pi f L \sqrt{\varepsilon'_r}}{c} = \frac{2\pi f L'}{c} \quad (1.7)$$

Thus the idea of electrical length is that in calculating $\Delta\theta$, one can utilize $L'$ to replace $L$ while keeping the propagation velocity constant (the value of $c$). In practice, however, the term of electrical length is often used loosely. Sometimes, $\beta L$ in (1.6) is called electrical length. More confusing is that this parameter is customarily provided in a format of degrees/frequency (e.g., 45° at 915 MHz). In this form of electrical length, the specific frequency is usually chosen to be the operation frequency, or a convenient point for calculation, such as 1 GHz. It is also seen in practice that the electrical length is provided by $\varepsilon'_r$ along with the physical length. Despite various forms of electrical length, (1.5) or (1.7) is always the basis for calculating the phase shift.

It is worth noting that there are practically countless variations in PCB transmission line configuration; each has a different electrical length for a given physical length. It is the electrical length, rather than the physical length, that determines, along with the characteristic impedance, the effects of the transmission line on circuit performance. For this reason, transmission lines on an RF circuit schematic should be labeled with electrical length and characteristic impedance. Figure 1.3(d) illustrates an example. To implement a transmission line given in a schematic on a real PCB, one needs to know the substrate material and thickness first and then synthesize the physical dimensions using the given electrical length and impedance. This approach is not perfect in practice because some of the subtle differences in PCB layout that may affect circuit performance cannot be completely captured in a schematic, particularly when the frequency is sufficiently high. Furthermore, the determination of the physical length of a transmission line between two shunt components [labeled $L$ in Figure 1.3(d)] is not precisely defined. One obvious factor for this uncertainty is the component size. As a result, some additional tuning is almost always required when transferring an RF circuit from one PCB substrate to another. The practice of reference design becomes popular, for it allows a direct copy of PCB design and eliminates any need of circuit transfer through schematic. However, the reference design approach only works for a complete functional unit with a fixed-form factor, such as modules.

### 1.2.3 Attenuation

Unavoidably, a propagating EM wave in a transmission line attenuates with traveling distance. Similar to the characteristic impedance and effective dielectric constant, the attenuation characteristic of a transmission line is dependent on the physical properties of the transmission line. In addition, it is also strongly dependent on the frequency, as discussed here. In practice, for a typical PCB size and operation frequency, power loss due to transmission-line attenuation is usually small and often can be ignored. There are perhaps two exceptions: (1) the low-noise amplifier (LNA) (see Section 8.3.1) and (2) ultra high-power amplifiers, for which the PCB substrate selection is critical. For example, for a moderate 0.2-dB loss at 100W, the power dissipation inside the transmission line is about 5W, which may cause a serious thermal problem, not to mention the loss in output power. Here we outline some basic formulas on the subject with an emphasis on how to estimate a transmission-line loss in practice. A more detailed discussion can be found in [1–3].

The attenuation of an EM wave in a transmission line over a distance $l$ is described by an exponential law as

$$E(l) = E_0 e^{-\alpha l} \tag{1.8}$$

where $E$ represents any linear quantity associated with the EM wave, such as an electric or magnetic field or voltage, if the traveling voltage wave is used. The square of $E$ is proportional to power. $E_0$ is the initial value. $\alpha$, the attenuation constant, is generally transmission-line specific and dependent on frequency.

As implied in (1.8), the attenuation constant $\alpha$ is the only parameter needed to characterize the attenuation of a transmission line. For a typical frequency range of wireless applications (below 10 GHz), $\alpha$ has two major components; that is,

$$\alpha = \alpha_c + \alpha_d$$

where $\alpha_c$ is the attenuation constant due to power loss in the conductor and $\alpha_d$ the loss in the dielectric substrate.

The power loss in conductors, which is due to their finite conductivity, can be estimated using the formula for the skin effect [1, 2]. The result shows that $\alpha_c$ is proportional to the square root of frequency $\sqrt{f}$ and inversely proportional to the square root of the conductivity of the conductor $\sqrt{\sigma}$. That is,

$$\alpha_c = F_c \frac{\sqrt{f}}{\sqrt{\sigma}}$$

Here, $F_c$ includes all other factors and is generally dependent on the transmission line structure and dimensions as well as the dielectric constant of the substrate. Both the dielectric constant and conductivity have a limited range in terms of material selection; as a result, the dominant factor for $\alpha_c$ is usually the frequency in practical design considerations.

The power loss in dielectric material is characterized by a parameter called loss tangent, usually denoted as $\tan\delta$. The loss tangent is related to $\alpha_d$ by [1, 4]

$$\alpha_d = F_d \cdot \tan\delta \cdot f$$

Similarly, $F_d$ contains all other factors and is dependent on the transmission line structure and substrate material. Unlike the conductivity, the loss tangent of practically used substrate materials can vary by a factor of a hundred. Besides the factor $f$ in $\alpha_d$, $\tan\delta$ also slightly rises with frequency; thus $\alpha_d$ increases much faster than $\alpha_c$ with frequency.

For applications where the power loss in a transmission line needs to be controlled, one first needs to estimate $\alpha_c$ and $\alpha_d$ at the operation frequency. If $\alpha_c$ is the dominant factor, there are limited options available to the designer in terms of material selection other than shortening the transmission line. On the other hand, if $\alpha_d$ is the larger term in the attenuation constant, it may be possible to select a better material, if the one chosen is not yet the best. The downside is usually higher cost and a limitation on the choice of the dielectric constant. In terms of frequency range, $\alpha_c$ and $\alpha_d$ are usually comparable for a substrate material with a moderate $\tan\delta$ value in the range of a few gigahertz, while $\alpha_d$ becomes more significant at higher frequencies.

When using $\alpha$ to estimate a transmission-line loss, the unit for $\alpha$ has to be dealt with correctly. It is clear from (1.8) that the unit of $\alpha$ is [length]$^{-1}$. In the literature, $\alpha$ is often provided either in nepers (Np) per length or in decibels per length. Both nepers and decibels are a logarithm form of a number (usually a ratio), and they are defined as the following:

$$\mathrm{Np} = \ln(\ )$$

$$\mathrm{dB} = 20\log(\ )$$

It is important to note that the factor 20 in the decibel definition is correct for the ratio of a linear quantity, but a factor of 10 should be used if the quantity is power or proportional to power. It is then from (1.8) that

$$\alpha = \ln\left(\frac{E_0}{E(l)}\right) \cdot \frac{1}{l} = \mathrm{Np}\left(\frac{E_0}{E(l)}\right) \cdot \frac{1}{l} \tag{1.9}$$

and

$$\alpha = \frac{20\log\left(\frac{E_0}{E(l)}\right)}{20\log(e)} \cdot \frac{1}{l} = \frac{\mathrm{dB}\left(\frac{E_0}{E(l)}\right)}{8.686} \cdot \frac{1}{l} \tag{1.10}$$

Equations (1.9) and (1.10) state that $\alpha$ is the same numerically as the attenuation spec measured in nepers while it is the spec in decibels divided by 8.686. Practically, the attenuation spec of a transmission line is most often provided in decibels per unit length. For example, if the attenuation of a transmission line is specified as 2 dB/m, then the attenuation constant is $\alpha = 0.23$/m.

In more sophisticated treatments using the EM theory [1, 2], the wave propagation is often characterized by a complex propagation constant with the attenuation constant being the real part. Any quantity derived from this representation naturally includes the attenuation effect. On the other hand, this approach evidently complicates the formula manipulation in an analysis. For most functions of transmission lines that covered in this book, the transmission line loss is usually small enough to be ignored for practical purposes. Therefore, we assume a lossless transmission line in this discussion. When it is necessary to consider the transmission line loss, we deal with it separately.

A comprehensive list of formulas for the above three parameters, namely, characteristic impedance, effective dielectric constant, and attenuation constant, can be found for a large variety of transmission-line configurations in Wadell's handbook [7]. In addition, various calculation tools are available, either as free online versions or as part of a commercial electronic design automation (EDA) package [8]. These tools usually allow users to calculate the electrical parameters from the physical parameters and vice versa.

## 1.3 Power Waves and the Reflection Coefficient

### 1.3.1 Power Waves

The power wave is another critical concept in RF circuit theory. It was first employed, in a systematic approach, by Kurokawa in 1965 [9] in circuit analysis using the scattering matrix, which is discussed in depth in Section 1.5. The essence of Kurokawa's treatment was that EM waves (or RF signals) were first represented by the voltage and current waves on a transmission line and that the power waves then were defined in terms of the voltage and current waves. Most modern textbooks on the subject have followed the same or a similar approach. However, power waves, rather than voltage or current waves, are real measurable quantities in RF circuits. In Section 10.1.1, we will explain how the power waves are measured using a VNA. In addition, on the theory side, the concept of voltage and current waves is not necessary in our analytical treatments other than to establish the relationship between the impedance and power waves described in (1.13). For these reasons, this book directly introduces the power wave as a fact by summarizing some of its key characteristics instead of deriving them from other circuit quantities. Mathematically, the book's presentation uses the technique of phasor representation of traveling waves [3, 4].

We start with a general description of a power wave denoted as $a$ traveling along a lossless transmission line:

$$a = \rho e^{-j\beta x} \quad (1.11)$$

Here $\beta$ is defined in (1.6), and $\rho$ is the amplitude of the wave. The physical meaning of $\rho$ is that $|\rho|^2$ is the average (rms value) power flowing across any cross-section of the transmission line. For those familiar with the Poynting vector in EM theory, $\rho$ has the same physical meaning as the magnitude of the Poynting vector (after integration over the cross-section area). $x$ is the coordinate of the physical distance from a reference point. For a transmission line of length $\ell$, $\beta\ell$ is the electrical length defined in Section 1.2. The sign of the $\beta x$ term in the exponent is a matter of convention and can be either positive or negative. However, an opposite sign for the $\omega t$ term must be chosen when converting the phasor to the *sine* (or *cosine*) form. To be consistent with the sign convention in (1.1), (1.11) chooses a negative sign. Also, power waves traveling in opposite directions must have opposite signs for the $\beta x$ term.

## 1.3.2 Reflection Coefficient

Equation (1.11) describes a wave that propagates to $x = \infty$ when the transmission line has an infinite length. When the transmission line is terminated with a load, (1.11), is, obviously, only valid up to the termination point where the incident power wave is generally partially reflected and partially absorbed by the load. The situation is illustrated in Figure 1.4(a), where the incident and reflected waves are denoted as $a$ and $b$, respectively. The power wave reflection is characterized by reflection coefficient $\Gamma$ defined as

$$\Gamma = \left.\frac{b}{a}\right|_{\text{termination}} \quad (1.12)$$

Analogous to impedance, which is defined as the ratio of voltage to current at a load, $\Gamma$ is defined as the ratio of the reflected wave to the incident wave at the load. $\Gamma$ is generally a complex number, accounting for changes both in the magnitude and phase of the reflected wave with respect to the incident one. It should be noted that the power ratio of the reflected and incident waves is given by $|\Gamma|^2$. The arrow next to $\Gamma$ in Figure 1.4 points to the component being considered. It is obvious in this case since there is only one

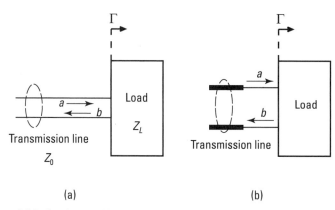

**Figure 1.4** (a) Reflection coefficient defined in power waves and (b) circuit representation.

component. When two components are connected, however, the distinction between them becomes necessary.

The incident and reflected waves actually travel in the same transmission line but in opposite directions as depicted in Figure 1.4 (a). For the convenience of illustration, however, it is customary to separate wave $a$ and wave $b$ in a circuit representation, which is shown in Figure 1.4(b). We follow this circuit convention in our discussions. Nevertheless, this representation should not be interpreted as indicating that the reflected wave is on a separate return path as in the case of a current loop in low frequency or DC circuits. Section 9.2.1 explains how two traveling waves existing in the same physical space are separated in measurements. Also, when a transmission line is pertinent to the discussion, it is represented explicitly as illustrated in Figure 1.4(b). The lines between the transmission line and the load are for illustration purposes only.

The physical origin for the power wave reflection at a termination point is the impedance mismatch between the two sides of the termination. A quantitative account of the relationship between the power wave reflection coefficient and the impedance mismatch requires at least a description of EM waves in terms of voltage and current waves [1, 3]. Again, we leave out the proof here and only state the result as follows.

To be precise, the term "load" used in this discussion means a one-port passive network [10] that is characterized by its impedance $Z_L$. In reference to Figure 1.4(a), the reflection coefficient at the termination point is given by

$$\Gamma = \frac{Z_L - Z_0}{Z_L + Z_0} \qquad (1.13)$$

From the standpoint of power transfer, power reflected is power not delivered to the load. In many cases, a strong reflection can also cause severe signal interference. Therefore, power reflection is generally undesirable in RF circuits. It follows from (1.13) that if $Z_0 = Z_L$ then $\Gamma = 0$. That is, when the load and transmission line impedances are matched there is no power reflection. In fact, (1.13) explains the reason for the impedance standardization in the RF industry: to eliminate power reflection between components. Equation (1.13) also explains two other cases that are of interest in practice: when $Z_L = 0$ and $\infty$, corresponding to short and open terminations, respectively, $\Gamma = -1$ and $\Gamma = 1$. In both cases, the reflection is 100%. However, for the short circuit the reflected wave is out-of-phase (180°) with the incident wave; for the open circuit, it is in-phase.

To get a numerical sense of the power loss due to reflection, consider two cases:

1. $Z_L$ is within 10% of $Z_0$. Then, $|\Gamma|^2 \approx 0.2\%$, meaning that only 0.2% of power is reflected. Therefore, a 10% mismatch between the load and line is a very good specification. On the other hand, it is also a very challenging design task.
2. $Z_L$ is half of $Z_0$. Then, $|\Gamma|^2 = 11\%$, which may still be tolerable in some practical cases.

In the general electric circuit theory, a one-port passive component is fully characterized by its impedance in terms of its circuit behavior. The same is true in RF circuits. However, the reflection coefficient $\Gamma$, defined in (1.12), proves to be more convenient in most cases. $\Gamma$ actually has a one-to-one relationship with the impedance $Z_L$ for a given $Z_0$, which means one can pick either parameter in circuit analysis and expect an identical result. The inverse function of (1.13) is

$$Z_L = Z_0 \frac{1+\Gamma}{1-\Gamma} \tag{1.14}$$

Then we can define the normalized impedance as:

$$z' = \frac{Z_L}{Z_0} = \frac{1+\Gamma}{1-\Gamma} \tag{1.15}$$

This rearrangement is nontrivial in that (1.15) indicates that in the calculation of $Z_L$ from $\Gamma$, $Z_0$ is simply a scaling factor. By utilizing the normalized impedance $z'$, $\Gamma$ can be treated as an independent component parameter in circuit analysis without referencing any specific $Z_0$. In fact, this scheme is the basis for the Smith chart introduced in Chapter 2. Also, under this scheme, $Z_0$ in (1.13) is no longer necessarily the actual impedance of the transmission line that is connected to the load; instead, it is a parameter used in relating $Z_L$ to $\Gamma$. For this reason, $Z_0$ is referred to as the reference impedance or system impedance. This concept is also applied to the S-parameters, the focus of our discussion in Section 1.5. When analyzing a circuit that consists of multiple components, each component must be specified with the same reference impedance. Phrases such as "50-$\Omega$ environment" and "75-$\Omega$ environment" in the technical literature refer to the choice of $Z_0$ for the system. This book generally specifies $Z_0$ only when necessary. In numerical works, however, the value of $Z_0$ has to be taken into account correctly.

### 1.3.3 Impedance Transformation

In (1.12), it is emphasized that $\Gamma$ is defined at the termination point. Once the reflection occurs at the termination, it can be visualized that in the steady state there are two opposite traveling waves at any point along the transmission line. Thus, the concept of the reflection coefficient can be extended to any point away from the termination. Consider a case shown in Figure 1.5(a) where the reflection coefficient $\Gamma'$ at a distance $\ell$ from the reflection point is given by

$$\Gamma' = \frac{b'}{a'}\bigg|_{\ell} \qquad (1.16)$$

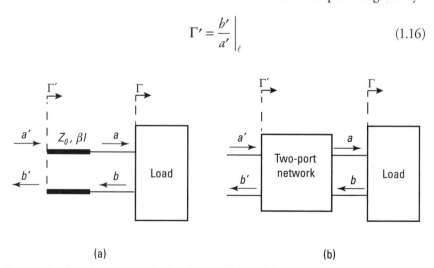

**Figure 1.5** Transformations of reflection coefficient: (a) through a transmission line and (b) through a generic two-port network.

If $x = 0$ is chosen at the termination point, then $x = -\ell$ at the point for $\Gamma'$. According to (1.11), $a'$ and $b'$ are related to $a$ and $b$ by

$$a' = ae^{j\beta\ell}; b' = be^{-j\beta\ell} \tag{1.17}$$

Note that the exponents of $a'$ and $b'$ have opposite signs, indicating the opposite traveling directions of the two waves. Then, $\Gamma'$ is related to $\Gamma$ as

$$\Gamma' = \frac{b}{a} e^{-j2\beta\ell} = \Gamma e^{-j2\beta\ell} \tag{1.18}$$

indicating that the magnitude of the reflection coefficient remains constant along the transmission line, which is expected given the assumption of a lossless transmission line. The additional phase term $-2\beta\ell$ is the phase shift associated with a traveling distance of $2\ell$, which is also expected.

We have discussed how a reflection coefficient $\Gamma$ is transformed to $\Gamma'$ through a transmission line. Section 1.5 shows that the transmission line is a special case of two-port networks [10]. In fact, a reflection coefficient (or impedance) can be transformed through any two-port network. This process, depicted in Figure 1.5(b), is usually referred to as impedance transformation, although the quantity to be transformed can be the reflection coefficient as well. The concept of impedance transformation has a profound importance in the theory and practice of RF circuits, as is seen repeatedly throughout this book. In this section we will continue to study the case of impedance transformation through a transmission line.

It is clear from (1.18) that the effect of a transmission line on the reflection coefficient is a phase shift, which seems relatively trivial. If the same transformation is expressed in terms of impedance, however, some of its significant effects in circuit applications will be revealed. To see this, we derive an expression for the impedance transformation from $Z_L$ to $Z'_L$ through a transmission line characterized by impedance $Z_c$ and electrical length $\beta\ell$ as illustrated in Figure 1.6.

We can choose $Z_c$ as the reference impedance without a loss of generality, and the corresponding reflection coefficient is denoted as $\Gamma_c$. Then using (1.14) and (1.18), the impedance at $\ell$ is obtained as

$$Z'_L = Z_c \frac{1+\Gamma'_c}{1-\Gamma'_c} = Z_c \frac{1+\Gamma_c e^{-j2\beta\ell}}{1-\Gamma_c e^{-j2\beta\ell}} = Z_c \frac{e^{j\beta\ell}+\Gamma_c e^{-j\beta\ell}}{e^{j\beta\ell}-\Gamma_c e^{-j\beta\ell}} \tag{1.19}$$

where $\Gamma'_c$ is the reflection coefficient at $\ell$. Substituting $\Gamma_c$, which, according to (1.13), is

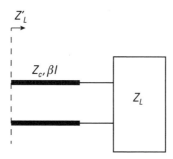

**Figure 1.6** Impedance transformation through a transmission line.

$$\Gamma_c = \frac{Z_L - Z_c}{Z_L + Z_c}$$

into (1.19) gives

$$\begin{aligned}Z'_L &= Z_c \frac{(Z_L + Z_c)e^{j\beta\ell} + (Z_L - Z_c)e^{-j\beta\ell}}{(Z_L + Z_c)e^{j\beta\ell} - (Z_L - Z_c)e^{-j\beta\ell}} \\ &= Z_c \frac{Z_L + jZ_c \tan\beta\ell}{Z_c + jZ_L \tan\beta\ell}\end{aligned} \quad (1.20)$$

since $e^{jx} = \cos x + j\sin x$ (Euler's formula). Apparently, for the same transformation through a transmission line, the transformation function in terms of impedance is much more complex than that in the reflection coefficient [see (1.18)], which is another reason that the reflection coefficient is often the preferred parameter in RF circuitry. In practice, if the transformation of $Z_L \to Z'_L$ is to be performed, the Smith chart (Chapter 2) is normally utilized, rather than directly using (1.20) to do the calculation. However, (1.20) does provide a convenient analytical explanation for a special case of $\beta\ell = \pi/2$, which is of great significance in RF circuits. This condition is often referred to as the quarter-wavelength transmission line, since $\beta\ell = \pi/2 \Rightarrow \ell = \lambda/4$. In fact, it is not limited to the quarter wavelength; for any electrical length that satisfies the conditions, $\beta\ell = \pi(1/2 + n)$, with $n = 0, 1, \cdots$, we have $\tan\beta\ell \to \infty$. Consider three cases under this condition:

- For $Z_L \neq Z_c$, $Z'_L = Z_c^2/Z_L$. That is, $Z_c$ is the geometric mean of $Z_L$ and $Z'_L$.
- For $Z_L = 0$, $Z'_L = jZ_c \tan\beta\ell \to \infty$. That is, a short load becomes an open after the transformation.
- For $Z_L = \infty$, $Z'_L = -jZ_c/\tan\beta\ell \to 0$. That is, an open load becomes short after the transformation.

The third case explains why a traditional oscilloscope cannot be used to probe voltages on an RF circuit: the high impedance at the scope input is transformed through the probe wires to a short at certain frequencies, thus totally disrupting the circuit. Most modern oscilloscopes are equipped with a special probe that eliminates the impedance transformation problem associated with the probe wires. However, the spatial variation due to the standing wave pattern, which was discussed in Section 1.1, is still a problem for the high frequency voltage measurement using a probe. Generally, it is critically important to realize that in RF circuits an open load and a short load are often interchangeable, because all it takes to convert one to the other is a section of transmission line. On the other hand, all three cases are utilized in a controlled manner to achieve certain functionalities in RF circuit designs, as will be demonstrated throughout the book.

## 1.4 RF Source

In circuit theory, a signal source, either voltage or current, can be represented by two independent circuit parameters: a signal generator and an internal impedance either in series (for voltage) or parallel (for current) with the generator as illustrated in Figure 1.7(a, b). Both quantities are measurable. For example, the source voltage $V_s$ is simply the voltage measured at the terminals $S-S'$ with an open load, and the source impedance is what is measured at the same terminals with $V_s$ set to zero. Similarly, in RF circuits a source can also be specified by two independent and measurable quantities as shown in Figure 1.7(c): The power generator characterized by available power, $P_{av}$, and the source reflection coefficient $\Gamma_S$ ($\Gamma_S$ can be considered an independent parameter even though it depends on the choice of the reference impedance. (See the discussion in Section 1.3.2.) The measurement of $\Gamma_S$ is simply that for the reflection coefficient at the output of the source with $P_{av} = 0$. In a real-world application, a question may arise as to whether $\Gamma_S$ remains the same with the generator turned on and off. This text is not concerned with this level of complexity. For $P_{av}$, the measurement is less straightforward. Next, we will show that for a given $\Gamma_S$ when the reflection coefficient of a load $\Gamma_L$ is the complex conjugate of $\Gamma_S$, i.e., $\Gamma_L = \Gamma^*_S$, the power delivered to the load reaches a maximum value. This maximum possible power that can be delivered to a load is defined as $P_{av}$. In comparison with a voltage source, the measurement condition for an RF generator is $\Gamma_L = \Gamma^*_S$ rather than $Z_L = \infty$ (open), but the concept is the same.

With $P_{av}$ and $\Gamma_S$ defined for a source, we now consider how much power a source delivers to a load, in the general case shown in Figure 1.8. We use

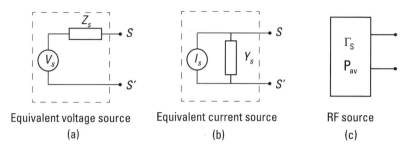

Equivalent voltage source  
(a)

Equivalent current source  
(b)

RF source  
(c)

**Figure 1.7** Representation of signal sources: (a) voltage source, (b) current source, and (c) RF source.

the concept of power waves in our analysis. We first introduce source power wave $b_s$. The physical meaning of $b_s$ is that when a source is connected to a transmission line of $Z_0$ (the same as the reference impedance for $\Gamma_S$), the source sends a power wave of amplitude $|b_s|$ down the transmission line. If the transmission line is terminated by a matched load (that is, $Z_L = Z_0$), then $|b_s|^2$ is the power dissipated in the load. Thus, $b_s$ is a measurable quantity and is dependent on the choice of $Z_0$ as well. When $\Gamma_S \neq 0$, $|b_s|^2$ is not equal to $P_{av}$, due to the internal reflection. In this case, not all available power comes out of the source. This section determines that $|b_s|^2 \propto P_{av}$ with the proportional factor being dependent of $\Gamma_S$. Also, in a general condition ($\Gamma_S \neq 0$ and $\Gamma_L \neq 0$), reflection occurs at both the source and load. If $b_s$ is considered a traveling wave, the wave front will be bounced back and forth between the source and load indefinitely with a reduced magnitude after each reflection. Here we only consider a steady-state solution in which two waves in opposite directions, labeled as $a$ and $b$ in Figure 1.8, are assumed. Ultimately, this section shows that when the steady-state solution is calculated as a superposition of a series of reflected waves, the same result is reached. Note that the

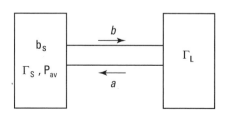

**Figure 1.8** Source power wave and power waves between source and load.

steady-state solution only works for a continuous signal. If a signal consists of a series of pulses such as those in high-speed digital systems, multiple reflections do potentially cause intersymbol interference and in practice definitely should be avoided.

Clearly, waves $a$ and $b$ must satisfy the boundary conditions described by the reflection coefficients at the source and load. Now, however, the terms of incident and reflected waves become arbitrary depending on which component is referenced. Hence, the notations $a$ and $b$ no longer have the original meaning defined in (1.12). In reference to Figure 1.8, wave $b$ has two components: $b_s$ and the reflected wave from wave $a$ at the source while wave $a$ is simply the reflected wave from $b$ at the load. Thus, we have an equation system for variables $a$ and $b$:

$$b = b_s + \Gamma_S a \qquad (1.21)$$

$$a = \Gamma_L b \qquad (1.22)$$

In writing (1.21) and (1.22), the transmission line connecting the source and load is ignored. The example at the end of this section shows that this omission has no effect on the result. Solving the equation system yields:

$$b = \frac{b_s}{1 - \Gamma_S \Gamma_L} \qquad (1.23)$$

$$a = \frac{b_s \Gamma_L}{1 - \Gamma_S \Gamma_L} \qquad (1.24)$$

When $\Gamma_L = 0$ (no reflection from the load), $b = b_s$ and $a = 0$, as expected. For $\Gamma_S = 0$, a condition for which most practical sources are designed, $b = b_s$ and $a = \Gamma_L b_s$.

Now we can evaluate the power dissipated in the load $P_L$. Obviously $P_L$ is the difference between the incident power to the load $|b|^2$ and the reflected power from the load $|a|^2$; that is,

$$P_L = |b|^2 - |a|^2 \qquad (1.25)$$

Substitution of (1.23) and (1.24) into (1.25) gives

$$P_L = \frac{|b_s|^2 \left(1 - |\Gamma_L|^2\right)}{|1 - \Gamma_S \Gamma_L|^2} \qquad (1.26)$$

When $\Gamma_L = 0$, (1.26) leads to $P_L = |b_s|^2$, which is consistent with the definition of $b_s$. To study a more general case when $\Gamma_L \neq 0$, we rewrite (1.26) as

$$P_L = |b_s|^2 M_g$$

where $M_g$ is called matching factor and is defined by

$$M_g = \frac{1-|\Gamma_L|^2}{|1-\Gamma_S \Gamma_L|^2} \qquad (1.27)$$

The functional form of $M_g$ appears repeatedly in our discussions on power transfer between two circuit components. The question here is: For a given $\Gamma_S$, what is the value of $\Gamma_L$ for which $M_g$ (and $P_L$) is maximized? The answer is

$$\Gamma_L = \Gamma_S^* \qquad (1.28)$$

Under this condition, the load is said to be complex conjugate matched to the source, and $P_L$ reaches the maximum value, equal to the available power of the source $P_{av}$. A proof is briefly outlined as follows.

We first consider the phase requirement. For the numerator, since the term is $|\Gamma_L|^2$, the phase has no effect on its value. The term $|1-\Gamma_S\Gamma_L|$ in the denominator reaches a minimum when $\Gamma_S\Gamma_L$ is a real number, meaning that the phases of $\Gamma_S$ and $\Gamma_L$ must be equal with opposite signs. Then the task is reduced to finding the value of $|\Gamma_L|$ such that the function

$$\frac{1-|\Gamma_L|^2}{\left(1-|\Gamma_S||\Gamma_L|\right)^2}$$

is maximized. The solution of the equation

$$\frac{\partial}{\partial |\Gamma_L|}\left\{\frac{1-|\Gamma_L|^2}{\left(1-|\Gamma_S||\Gamma_L|\right)^2}\right\} = 0$$

is $|\Gamma_L| = |\Gamma_S|$. Thus, we proved that (1.28) is the condition for the maximum value of $M_g$, which is given by

$$M_{g\_max} = \frac{1}{\left(1-|\Gamma_S|^2\right)} \qquad (1.29)$$

Since $\Gamma_S < 1$, we have $M_{g\_max} > 1$. Considering $M_g(\Gamma_L = 0) = 1$, the physical meaning of $M_{g\_max}$ is the extra power gained from matching (i.e., making $\Gamma_L = \Gamma^*_S$) in comparison with the case of no matching ($\Gamma_L = 0$). For this reason, sometimes $M_{g\_max}$ is referred to as matching gain. Furthermore, since $P_{av}$ is defined as the maximum deliverable power to a load, it follows from (1.26) that

$$P_{av} = P_L(\Gamma_L = \Gamma^*_S) = \frac{|b_s|^2}{\left(1 - |\Gamma_S|^2\right)} \tag{1.30}$$

Equation (1.30) shows that $|b_s|^2$ is indeed proportional to $P_{av}$ with the proportional factor being $1 - |\Gamma_S|^2$. Now we can relate the power delivered to a load $P_L$ to the available power of a source $P_{ab}$ by combining (1.30) and (1.26):

$$P_L = P_{av} \frac{\left(1 - |\Gamma_S|^2\right)\left(1 - |\Gamma_L|^2\right)}{|1 - \Gamma_S \Gamma_L|^2} = P_{av} M \tag{1.31}$$

Here,

$$M = \frac{\left(1 - |\Gamma_S|^2\right)\left(1 - |\Gamma_L|^2\right)}{|1 - \Gamma_S \Gamma_L|^2} \tag{1.32}$$

is called the mismatch factor (also called mismatch loss), which quantifies the effect of a mismatch (between a source and a load) on the power transfer from the source to the load.

From (1.27) and (1.32), $M$ and $M_g$ are related by

$$M = M_g \left(1 - |\Gamma_S|^2\right)$$

From this relationship it can be seen that $M \leq 1$. Under the condition of complex conjugate matching, $M = 1$, corresponding to the maximum power transfer. This condition is analogous to that of a voltage source where the maximum power transfer occurs when the source and load resistances are equal. The terminology for these two parameters may appear confusing: $M_{g\_max}$ is called matching gain, while $M$ is the mismatch factor. The answer lies in their respective reference points. $M_{g\_max}$ measures how much power gain can be achieved with matching in reference with the case of no matching.

In other words, the gain is from the condition of $\Gamma_L = 0$ to that of $\Gamma_L = \Gamma^*_S$. In comparison, $M$ is a measure of how much power is lost when $\Gamma_L$ (or $\Gamma_S$) deviates from the optimal condition $\Gamma_L = \Gamma^*_S$. Both parameters are important concepts in RF circuit design, as demonstrated in Chapters 2 and 3.

The following summarizes the preceding discussion:

- An RF source is fully characterized by the available power $P_{av}$ and the source reflection coefficient $\Gamma_S$ (after the reference impedance $Z_0$ is chosen).
- Source power wave $b_s$ is introduced to facilitate analysis using power waves. $|b_s|^2$ is related to $P_{av}$ by (1.30). When $\Gamma_S = 0$, $P_{av} = |b_s|^2$, which is almost always the condition for a standard RF source in practice.
- The power delivered to a load $P_L$ is related to $P_{av}$ by (1.31) for a general condition $\Gamma_L \neq 0$ and $\Gamma_S \neq 0$. The mismatch factor $M$ characterizes the efficiency in power transfer.
- If the source and load are complex conjugate matched, i.e. $\Gamma_L = \Gamma^*_S$, $M = 1$ and $P_L = P_{av}$, corresponding to the condition for maximum power transfer.

For the reader who still wonders how a sequence of an infinite number of reflections between the source and load can be represented by two steady-state waves, we work out the derivation as follows. The discussion also includes a transmission line with finite length as a general case.

Figure 1.9 illustrates the process we are interested in where a power wave $b_s$ starts at the source and continuously propagates between the source and load.

The wave described in a phasor form follows these rules:

1. Each time it travels over a distance $L$ in the transmission line, it has a factor $e^{-j\beta L}$ to account for the phase delay.
2. For each reflection at the source or load, it has an additional factor of the reflection coefficient $\Gamma_S$ or $\Gamma_L$.

Then a series of incident waves at the load can be constructed as shown in Figure 1.9. The steady-state solution is a superposition of these waves as:

$$b = b_s e^{-j\beta L} \sum_0^\infty \left( \Gamma_S \Gamma_L e^{-j2\beta L} \right)^n$$

$$= \frac{b_s e^{-j\beta L}}{1 - \Gamma_L e^{-j\beta L} \cdot \Gamma_S e^{-j\beta L}}$$

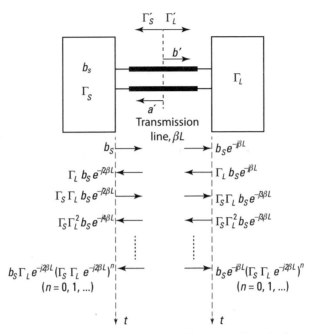

**Figure 1.9** Steady-state solution as a superposition of a series of reflected waves.

In the last step the formula for the sum of an infinite geometric series is used. Now if we move the reference point to the middle of the transmission line and use (1.17) and (1.18), the power wave at that point toward the load is given by

$$b' = b \cdot e^{j\beta L/2} = \frac{b_s e^{-j\beta L/2}}{1 - \Gamma_L e^{-j\beta L} \cdot \Gamma_S e^{-j\beta L}}$$

$$= \frac{b'_s}{1 - \Gamma'_S \Gamma'_L}$$

where $b'_s$, $\Gamma'_L$, and $\Gamma'_S$ are the values at the midpoint. Similarly, it can be shown that at the midpoint,

$$a' = \frac{b'_s \Gamma'_L}{1 - \Gamma'_S \Gamma'_L}$$

Thus we have shown the steady-state solution of a propagating wave experiencing an infinite number of reflections is equivalent to the solution in (1.23) and (1.24) where two steady-state waves propagating in opposite directions are assumed. In fact, this is the essence of the theory of multiple reflections. It is also worth noting that in the above analysis, we start with a propagating wave $b_s$, then work out the reflected waves for each subsequent reflection up to $t \to \infty$. However in reality the steady-state solution is a sum of the current wave and all reflected waves started earlier all the way back to $t \to -\infty$. There is no mathematical difference between the two cases but this subtle difference should not be lost in the analysis.

## 1.5 Scattering Parameters (S-Parameters)

### 1.5.1 Two-Port S-Parameters

So far, we have shown how incident and reflected power waves are used as terminal variables to characterize a one-port network. The same technique can be extended to the general case of an N-port network. This section is mainly concerned with the two-port network, since it represents the majority of RF circuits and components that are commonly used. In the general circuit theory [10], among four terminal variables $V_1$, $V_2$, $I_1$ and $I_2$, two are chosen to be independent (or excitation) variables and the other two dependent (or response) variables. Then a two-port network can be characterized by one of six possible sets of network parameters depending on the choice of independent variables. Some parameters are more useful than the others in practice. The most common one is perhaps the $Z$ parameter for which $I_1$ and $I_2$ are the independent variables. In RF circuits, we also have four terminal variables, namely, incident power waves, $a_1$ and $a_2$ and reflected power waves, $b_1$ and $b_2$ as shown in Figure 1.10. For practical purposes, only one configuration is meaningful: $a_1$ and $a_2$ as independent variables and $b_1$ and $b_2$ as dependent ones. The corresponding network parameters that relate $(b_1, b_2)$ to $(a_1, a_2)$, as described by the following equations, are called S-parameters.

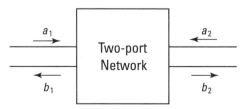

**Figure 1.10** Two-port network defined by power waves.

$$\begin{cases} b_1 = S_{11}a_1 + S_{12}a_2 \\ b_2 = S_{21}a_1 + S_{22}a_2 \end{cases} \quad (1.33)$$

They can also be written in a matrix form:

$$\begin{pmatrix} b_1 \\ b_2 \end{pmatrix} = \begin{pmatrix} S_{11}, S_{12} \\ S_{21}, S_{22} \end{pmatrix} \begin{pmatrix} a_1 \\ a_2 \end{pmatrix} \quad (1.34)$$

For this reason, the term "scattering matrix" is sometimes used in the literature. However, the term S-parameters is the most widely used in the RF industry by far. As in the case of the reflection coefficient, S-parameters are also dependent on the choice of reference impedance. More on this in a moment.

It is clear that $S_{11}$ and $S_{22}$ are a measure of reflection, since

$$S_{11} = \left.\frac{b_1}{a_1}\right|_{a_2=0} ; S_{22} = \left.\frac{b_2}{a_2}\right|_{a_1=0}$$

More precisely, $S_{11}$ and $S_{22}$ are the reflection coefficients at one port of the network, while the other port is perfectly terminated. $S_{21}$ and $S_{12}$ are measures of power transfer (gain or loss) through the network from port 1 to port 2 and from 2 to 1 respectively as

$$S_{21} = \left.\frac{b_2}{a_1}\right|_{a_2=0} ; S_{12} = \left.\frac{b_1}{a_2}\right|_{a_1=0}$$

The condition, $a_i = 0$ ($i = 1$ and $2$), in the above equations, appears similar to those used in two-port networks defined by voltages $V_i$ and currents $I_i$ [e.g., for Z-parameters, $Z_{11} = V_1/I_1$ (at $I_2 = 0$)]. However, it is this condition that reveals the significance of the power wave (versus voltage and current) as a circuit variable in RF circuits. This condition requires the port in question to be perfectly terminated to avoid any power reflection, as illustrated in Figure 1.11 for the $S_{11}$ and $S_{21}$ measurements under the condition $a_2 = 0$. In contrast, the condition, $V_i = 0$ or $I_i = 0$ requires either an open or a short termination at one of the ports. Either condition causes a complete reflection at the termination. Part or all of the reflected power dissipates inside the RF device, which can severely affect the device operation condition and sometimes even cause oscillation.

The termination conditions ($a_i = 0$) also implicitly define the reference impedance for the S-parameters since the transmission line impedance and load impedance must be matched to achieve $a_2 = 0$ in the measurement setup

# Essentials of Radio Frequency (RF) Circuits and S-Parameters

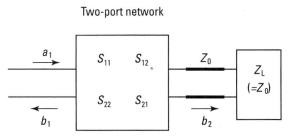

**Figure 1.11** Configuration for $S_{11}$ and $S_{21}$ measurements of a two-port network.

shown in Figure 1.11. This impedance is the reference impedance for the S-parameters measured in such a system. In practice, measured S-parameters are almost always in a 50-$\Omega$ system.

In practice, S-parameters are generally assumed to be linear parameters (i.e., independent on power), and circuit analysis using the S-parameters is often referred to as linear or small-signal analysis. In power applications, sometimes large-signal S-parameters—the parameters specified at a certain power level—are also used. This book consistently uses S-parameters as linear parameters. For calculation, an S-parameter is usually expressed in the linear (also known as polar) form of magnitude and phase, e.g., $S_{21} = 5.6\angle 128°$, and $S_{11} = 0.82\angle 18°$. Sometimes, the real-imaginary format is also used. However, in specification and measurement, magnitude is usually the only quantity of interest. In that case, an S-parameter $S_{ij}$ is more often given in a logarithm form defined by

$$S_{ij}(\text{dB}) = 20\log\left(\left|S_{ij}\right|\right) \tag{1.35}$$

S-parameters are at the core of the design and characterization of linear RF circuits, especially RF amplifiers. Chapter 2 discusses various applications of the S-parameters in circuit design and analysis. Sections 1.5.1.1–1.5.1.3 briefly explain how the S-parameters are used in three types of circuit elements.

### 1.5.1.1 S-Parameters for a Transmission Line

Consider a section of transmission line specified with $Z_c$ and $\beta\ell$ in place of the two-port network in Figure 1.10. For simplicity of analysis, $Z_c$ is assumed to be the same as the reference impedance. Then, in this case, an incident wave at one port goes through the transmission line without any reflection, with a delay of $e^{-j\beta\ell}$. Thus, we have

$$S_{21} = S_{12} = e^{-j\beta\ell}$$

and

$$S_{11} = S_{22} = 0$$

or in matrix form,

$$S = \begin{pmatrix} 0, & e^{-j\beta\ell} \\ e^{-j\beta\ell}, & 0 \end{pmatrix} \quad (1.36)$$

When $Z_c \neq Z_0$, the matrix still has the symmetry: $S_{21} = S_{12}$ and $S_{11} = S_{22}$, but the expressions for both terms are quite complex. Since actual calculations can be easily done with an EDA tool, these expressions are omitted here.

### 1.5.1.2 S-Parameters for Lossy Two-Port Networks

The term "lossy networks" in RF applications usually (but not always) refers to devices/circuits whose main function is to selectively let RF signal pass through. Examples are the "on" path of an RF switch (see Chapter 11) and the in-band of a bandpass filter. For such a device, the main specification for circuit applications is called insertion loss, which is a measure of power loss associated with the insertion of the device in an RF path. There are two factors contributing to this loss: power reflection and power dissipated inside the device. The behavior of a lossy device in a circuit can be fully characterized by the S-parameters in the measurement setup shown in Figure 1.12.

In Figure 1.12, a common practice for port convention is assumed; that is, port 1 is assigned as input and port 2 as output. Now consider an incident power $|a|^2$ at port 1. From the conservation of energy, it should equal the sum of three terms: power reflected $|b|^2$, power dissipated inside the device $D$, and power coming out of the device $|c|^2$; namely,

$$|a|^2 = |b|^2 + |c|^2 + D \quad (1.37)$$

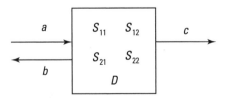

**Figure 1.12** Insertion loss measurement. (*D* represents power dissipation inside the device.)

Dividing both sides of (1.37) with $|a|^2$ and using the definition of the S-parameters, we have

$$|S_{21}|^2 = 1 - |S_{11}|^2 - D' \qquad (1.38)$$

Here, $D' = D/|a|^2$ is the normalized power dissipation inside the device. $|S_{21}|^2$ in (1.38) is the insertion loss (IL) expressed in an S-parameter. In practice, IL is almost always given in a decibel term defined as:

$$\text{Insertion Loss} = -20\log(|S_{21}|) \qquad (1.39)$$

Strictly speaking, IL should be $1 - |S_{21}|^2$, but the definition for IL in (1.39) turns out to be convenient to remember since a 0-dB IL means no loss (not to imply that the number 0 is always associated with nonexistence). Similarly, return loss (RL) is defined as:

$$\text{Return Loss} = -20\log(|S_{11}|) \qquad (1.40)$$

In this case $|S_{11}|^2$ is a true measure of power loss due to reflection. However, the RL in decibels is a little counterintuitive numerically in that a 0-dB return loss means total reflection, while no reflection ($S_{11} = 0$) corresponds to an infinite value of RL in decibels.

In both definitions of (1.39) and (1.40), a negative sign is used so that the IL and RL are positive numbers in decibels. (Note that $|S_{21}| < 1$ and $|S_{11}| < 1$.) On the other hand, it is not uncommon in reality that they are provided as negative numbers. This is simply because the measurement results in the decibel form of $S_{21}$ and $S_{11}$ are negative. For most practical purposes, whether an IL (or RL) is written as a positive or a negative figure is inconsequential. However, when doing a numerical calculation, the sign has to be taken care of correctly. We now return to the general expression (1.38) for the IL. For the two terms on the right-hand side, the term $|S_{11}|^2$, which is related to the RL, can be improved, in principle, by an external matching network, whereas the power dissipation $D'$ is fixed for a chosen device. For a set of measured data, if a better IL is desirable, then the question is whether improving the RL makes this possible. To answer the question, one has to convert IL and RL data from decibel form to linear form and then use (1.38) to calculate $D'$. Two cases are shown in Table 1.1.

**Table 1.1**
Calculation of Power Dissipation from IL and RL

|        | IL (decibels) | RL (decibels) | $|S_{21}|^2$ | $|S_{11}|^2$ | $D'$  |
|--------|---------------|---------------|--------------|--------------|-------|
| Case 1 | 1             | 15            | 0.794        | 0.032        | 0.174 |
| Case 2 | 0.5           | 10            | 0.891        | 0.1          | 0.009 |

In case 1, the power dissipation $D'$ is almost six times larger than the reflection power $|S_{11}|^2$, implying that a better RL will result in little improvement in insertion loss. In contrast, the IL can be reduced by improving the RL in case 2, since $D'$ is only about 10% of $|S_{11}|^2$.

Besides RL, the voltage standing wave ratio (VSWR) is also used in some occasions as a specification for the reflection of a device. VSWR is related to $S_{11}$ by [1, 3]:

$$\text{VSWR} = \frac{1+|S_{11}|}{1-|S_{11}|}.$$

It is customary to write a VSWR spec in the form of VSWR:1, although VSWR itself is sufficient.

We conclude our discussion on RL by noting that unintended power reflection occurs whenever the RF power wave experiences a discontinuity. Typical examples of the discontinuity for a PCB-based circuit are coaxial to PCB connectors and a transmission line trace to discrete components. In general, as the operation frequency increases the effects associated with these discontinuities become more prominent. Numerically, a 20-dB RL, which corresponds to 1% power reflection, is a very good specification for a PCB circuit even at a moderate frequency (low gigahertz). This is because the imperfection of the circuit components alone can easily cause more than 1% reflection. For this reason, it is necessary to put forth a significant effort if the design is aimed at an RL better than 20 dB in the gigahertz range.

### 1.5.1.3 S-Parameters for Amplifiers

The S-parameters for an amplifier (or a transistor in a configuration for amplifier applications, discussed further in Chapter 5) are highly asymmetric with port 1 and port 2. By our port convention, $S_{21}$ is a measure of the linear gain of the device in the reference impedance condition (50-Ω, practically) without any matching. For most practical amplifier designs, a device with $|S_{21}|^2 > 1$

is desirable but not absolutely necessary, as explained in Chapter 2. In comparison, $S_{12}$ is called a reverse gain (also known as isolation or feedback). It is a measure of coupling from the output to the input. When used in a buffer amplifier, a device with smaller value of $S_{12}$ is more desirable since it provides higher isolation. As to $S_{11}$ and $S_{22}$, they are the input and output reflection coefficients of the device directly measured in a 50-Ω system. When evaluating a transistor for amplifier applications, smaller values of $S_{11}$ and $S_{22}$ imply more convenient and consistent matching networks in terms of circuit realization. On the other hand, a large value of $S_{11}/S_{22}$ (that is still less than 1) allows more matching gain. Chapter 2 presents a quantitative analysis using the S-parameters for amplifier designs.

### 1.5.2   N-Port Networks and the SnP File Format

In a general case of an N-port network, each port can still have terminal variables, $a_i$ and $b_i$ ($i = 1$ to $n$) and correspondingly (1.33) can be generalized as

$$b_i = \sum_{j=1}^{n} S_{ij} a_j$$

and $S_{ij}$ is given by

$$S_{ij} = \left. \frac{b_i}{a_j} \right|_{\text{All } a_k = 0 \, (k \neq j)}$$

In terms of the measurement capability of N-port S-parameters, most advanced VNAs offer a four-port option. Beyond that, a special switching matrix system is usually required to perform the S-parameters measurement. The basic idea of the measurement is still simple (i.e., sending an incident wave at port $j$ and measuring the outgoing wave at port $i$ while terminating all other ports).

All commercial VNAs allow the measured S-parameter data to be exported in a special file format called SnP, where "n" is the number of ports. An SnP file can be directly imported to any EDA system. Most manufacturers provide S-parameters for their products in this format too. Figure 1.13 shows an example of an S2P file. As in many programming languages, an exclamation mark "!" indicates a comment line. The option line contains useful information for data configurations: "Hz" indicates the frequency unit (other options are kHz, MHz and GHz); "S MA" indicates the S-parameters with

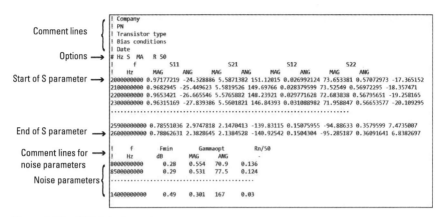

**Figure 1.13**  S2P data format.

the linear-Mag/Phase format (with the other two options for S-parameters log-Mag/Phase and Real/Imaginary); and "R 50" indicates the system impedance. If noise parameters, which will be discussed in Chapter 4, are available, they are listed after the S-parameters as shown in Figure 1.13. Most of the time in practice, engineers only need to load an SnP file into a simulator without being concerned with the data format. However, if one wants to use the S-parameters to make a quick evaluation of the device performance, understanding the data format of SnP files becomes necessary.

# References

[1]  Collin, R. E., *Foundations for Microwave Engineering*, Second Edition, McGraw-Hill, 1992.

[2]  Jackson, J. D., *Classical Electrodynamics*, Third Edition, John Wiley and Sons, 1999.

[3]  Pozar, D. M., *Microwave Engineering*, Second Edition, John Wiley and Sons, 1998.

[4]  Ramo, S., J. R. Whinnery, and T. Van Duzer, *Fields and Waves in Communication Electronics*, Second Edition), John Wiley and Sons, 1993.

[5]  Kaiser, K., *Transmission Lines, Matching, and Crosstalk*, CRC Press, 2006.

[6]  Gottlieb, I. M., *Practical RF Power Design Techniques*, TAB Books, 1993.

[7]  Wadell, B. C., *Transmission Line Design Handbook*, Norwood, MA: Artech House, 1991.

[8] Yeom, K., *Microwave Circuit Design A Practical Approach Using ADS*, Prentice Hall, 2015.

[9] Kurokawa, K., "Power Waves and the Scattering Matrix," *IEEE Trans. on Microwave Theory and Techniques*, Vol. 13, No. 2, 1965.

[10] Alexander, C. K., and M. N. O. Sadiku, *Fundamentals of Electric Circuits*, McGraw-Hill, 2000.

# 2
# Circuit Analysis Using S-Parameters

Following up on Chapter 1's introduction to S-parameters, this chapter discusses how they can be used in RF circuit design. In principle, the materials presented in this chapter are applicable to any device characterized by the S-parameters. However, the discussions on power gains and stability are most relevant to amplifier design. The invariant property of the mismatch factor covered in Section 2.4 is an important concept in design of lossless networks.

## 2.1 Power Gains

The concept of impedance matching is introduced in Section 1.4 in the context of power transfer from a source to a load. We have shown that complex conjugate matching yields a maximum power transfer between the source and the load. The same concept can be applied to an RF circuit that consists of more components. Figure 2.1 depicts a simple case in which a device characterized by the S-parameter matrix $S$ is connected to a source and a load specified by $(P_{av}, \Gamma_S)$ and $\Gamma_L$, respectively. This section analyzes the power transfer from the source to the load for this case. Specifically, we wish to derive a formula for the ratio of the power delivered to the load $P_L$ to the available power $P_{av}$ of the source. If the device is an amplifier, it is expected that $P_L > P_{av}$; that is, the system has gain. For this reason, this ratio is referred to as transducer power gain (or simply transducer gain) in the literature, although it is perfectly applicable to the case of no gain or loss. The term transducer normally implies an energy or signal conversion from one form to another. No such conversion occurs in the systems to be considered here. Nevertheless, this terminology is standard in the field of RF engineering. In the process of deriving the transducer gain, which is obviously a function of $\Gamma_S$, $\Gamma_L$, and $S$, we also derive formulas for other two gains, namely, available gain and power gain. These two gains account for the effects of $\Gamma_S$ and $\Gamma_L$ on the transducer gain separately. The input and output mismatch factors, denoted as $M_S$ and $M_L$ in Figure 2.1, are of significant importance in practical design and are also included in our analysis.

We start our analysis by introducing two new sets of source and load parameters at the device output (port 2) and input (port 1) respectively as shown in Figure 2.1. The first set consists of equivalent source parameters at the device output, $P_{av\_out}$ and $\Gamma_{out}$. The concept here is the same as that in the Thevenin and Norton equivalent sources. Similarly, at the device input are the equivalent load parameters $\Gamma_{in}$ and $P_{L\_in}$. Strictly speaking, unlike $P_{av}$ of the source, the power dissipation $P_L$ is not a component parameter of the load. However, for the convenience of analysis, these two power terms can be treated the same way analytically because of the symmetry of the system with respect to $\Gamma_S$ and $\Gamma_L$.

## 2.1.1 $\Gamma_{out}$ and $\Gamma_{in}$

We first consider the equivalent reflection coefficients. With reference to the convention of power waves in Figure 2.1, $\Gamma_S$ and $\Gamma_L$ are now written as

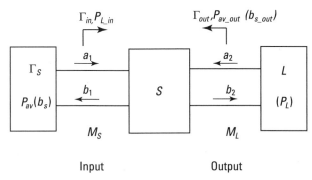

**Figure 2.1** Block diagram for circuit analysis using S-parameters.

## Circuit Analysis Using S-Parameters

$$\Gamma_S = \frac{a_1}{b_1} \tag{2.1}$$

and

$$\Gamma_L = \frac{a_2}{b_2} \tag{2.2}$$

[Note that the notations in (2.1) and (2.2) are different from that in (1.12).]

Equation (1.33) for the S-parameters remains unchanged. We consider $\Gamma_{out}$ first. Substituting (2.1) into (1.33) gives

$$b_1 = S_{11} \Gamma_S b_1 + S_{12} a_2 \tag{2.3}$$

$$b_2 = S_{21} \Gamma_S b_1 + S_{22} a_2 \tag{2.4}$$

Then, the equivalent reflection coefficient at the output of the network terminated with $\Gamma_S$ at the input is given by

$$\Gamma_{out} = \frac{b_2}{a_2} = S_{22} + \frac{S_{21} S_{12} \Gamma_S}{1 - S_{11} \Gamma_S} \tag{2.5}$$

Notice that the system is symmetric with respect to the interchange of port 1 and 2 as well as the source and the load. The equivalent reflection coefficient at the input is obtained as:

$$\Gamma_{in} = \frac{b_1}{a_1} = S_{11} + \frac{S_{21} S_{12} \Gamma_L}{1 - S_{22} \Gamma_L} \tag{2.6}$$

### 2.1.2  $P_{av\_out}$ and $P_{L\_in}$

We need the equivalent source power wave at the output $b_{s\_out}$ in derivations of these equivalent powers. To this end, we note, from the discussion in Section 1.4, that

$$b_{s\_out} = b_2 \big|_{\Gamma_L = 0}$$

With the new notations defined in Figure 2.1, (1.21) can be written as

$$a_1 = b_s + \Gamma_S b_1 \tag{2.7}$$

Substituting (2.7) into (1.33) and letting $a_2 = 0$ (since $\Gamma_L = 0$) we have

$$b_1 = S_{11}(b_s + \Gamma_s b_1) \tag{2.8}$$

$$b_2 = S_{21}(b_s + \Gamma_s b_1) \tag{2.9}$$

Solving (2.8) and (2.9) for $b_2$, $b_{s\_out}$ is found to be

$$b_{s\_out} = b_2\big|_{\Gamma_L=0} = S_{21}b_s + S_{21}\Gamma_s \frac{S_{11}b_s}{1-S_{11}\Gamma_s} = \frac{S_{21}b_s}{1-S_{11}\Gamma_s} \tag{2.10}$$

$P_{av\_out}$ and $b_{s\_out}$ are related by the same function as that in (1.30):

$$P_{av\_out} = \frac{|b_{s\_out}|^2}{1-|\Gamma_{out}|^2}$$

$$= \frac{|S_{21}|^2}{(1-|\Gamma_{out}|^2)|1-S_{11}\Gamma_s|^2}|b_s|^2. \tag{2.11}$$

Then, by replacing $b_s$ and $\Gamma_S$ with $b_{s\_out}$ and $\Gamma_{out}$ in (1.26), the power dissipation at the load is given by

$$P_L = \frac{|b_{s\_out}|^2(1-|\Gamma_L|^2)}{|1-\Gamma_{out}\Gamma_L|^2} \tag{2.12}$$

Similarly, using (1.26) again at the input with $\Gamma_L$ replaced with $\Gamma_{in}$ leads to

$$P_{L\_in} = \frac{|b_s|^2(1-|\Gamma_{in}|^2)}{|1-\Gamma_{in}\Gamma_s|^2} \tag{2.13}$$

With $P_{av\_out}$, $\Gamma_{out}$ and $P_L$ at the output and $P_{L\_in}$ and $\Gamma_{in}$ at the input derived, we are ready to introduce the available gain $G_A$ and the power gain $G_P$, which are defined at the device input and output, respectively. First, $G_A$ is defined as

$$G_A = \frac{P_{av\_out}}{P_{av}}.$$

$G_A$ is a measure of how much the source available power is increased (or decreased, if $G_A < 1$) by the device. Making use of (1.30) and (2.11), we find that

$$G_A = \left\{ \frac{1-|\Gamma_S|^2}{|1-S_{11}\Gamma_S|^2\left(1-|\Gamma_{out}|^2\right)} \right\} \cdot |S_{21}|^2 \qquad (2.14)$$

Here, (2.14) is expressed as the product of two factors: $|S_{21}|^2$ is the unmatched power gain of the device, and the other factor is a function of $\Gamma_S$ and $S$ and is independent on $\Gamma_L$. As a result, $G_A$ is independent of $\Gamma_L$ too, as expected by its definition.

Next, we define the power gain $G_P$:

$$G_P = \frac{P_L}{P_{L\_in}}$$

The power gain is a ratio of the power dissipated at the load to the power delivered by the source at the device input. Unlike $P_{av}$ and $P_{av\_out}$, which are independent of $\Gamma_L$, $P_L$ and $P_{L\_in}$ are dependent on $\Gamma_S$ as indicated by (2.12) and (2.13). However, their ratio $G_P$ is not. Consequently, we can pick a special case of $\Gamma_S = 0$ to simplify the derivation as follows:

$$P_L(\Gamma_S = 0) = \frac{|S_{21}|^2|b_s|^2\left(1-|\Gamma_L|^2\right)}{|1-S_{22}\Gamma_L|^2}$$

$$P_{L\_in}(\Gamma_S = 0) = |b_s|^2\left(1-|\Gamma_{in}|^2\right)$$

Then

$$G_P = \left\{ \frac{1-|\Gamma_L|^2}{|1-S_{22}\Gamma_L|^2\left(1-|\Gamma_{in}|^2\right)} \right\} \cdot |S_{21}|^2 \qquad (2.15)$$

Equation (2.15) can be proved in general by using the expressions for $P_L$ and $P_{L\_in}$ in (2.12) and (2.13) without the assumption of $\Gamma_S = 0$. The proof is straightforward but lengthy. Clearly, $G_P$ has the same functional form as $G_A$ with interchanges of $\Gamma_{in} \leftrightarrow \Gamma_{out}$, $S_{11} \leftrightarrow S_{22}$ and $\Gamma_S \leftrightarrow \Gamma_L$. For a given device (the S-parameters are given), analysis of functions $G_A(\Gamma_S)$ and $G_P(\Gamma_L)$ allows the designer to evaluate the effect of $\Gamma_S$ and $\Gamma_L$ on the network gain separately, which is illustrated in Section 2.3.

Finally, we derive the transducer gain $G_T$, which is defined as

$$G_T = \frac{P_L}{P_{av}}.$$

$G_T$ can be expressed in two different ways:

$$G_T = \frac{P_L}{P_{av\_out}} \cdot \frac{P_{av\_out}}{P_{av}} = \frac{P_L}{P_{av\_out}} \cdot G_A \qquad (2.16)$$

or

$$G_T = \frac{P_L}{P_{L\_in}} \cdot \frac{P_{L\_in}}{P_{av}} = \frac{P_{L\_in}}{P_{av}} \cdot G_P \qquad (2.17)$$

From (1.31), the ratio $P_L/P_{av\_out}$ in (2.16) is simply the load mismatch factor at the output and is given by

$$M_L = \frac{P_L}{P_{av\_out}} = \frac{\left(1-|\Gamma_{out}|^2\right)\left(1-|\Gamma_L|^2\right)}{|1-\Gamma_{out}\Gamma_L|^2} \qquad (2.18)$$

Thus, $G_T$ is

$$G_T = \frac{\left(1-|\Gamma_S|^2\right)}{|1-S_{11}\Gamma_S|^2} |S_{21}|^2 \frac{\left(1-|\Gamma_L|^2\right)}{|1-\Gamma_{out}\Gamma_L|^2} \qquad (2.19)$$

Similarly, the ratio $P_{L\_in}/P_{av}$ in (2.17) is the source mismatch factor at the input:

$$M_S = \frac{P_{L\_in}}{P_{av}} = \frac{\left(1-|\Gamma_{in}|^2\right)\left(1-|\Gamma_S|^2\right)}{|1-\Gamma_{in}\Gamma_S|^2} \qquad (2.20)$$

$G_T$ can then be expressed in yet another form:

$$G_T = \frac{\left(1-|\Gamma_S|^2\right)}{|1-\Gamma_{in}\Gamma_S|^2} |S_{21}|^2 \frac{\left(1-|\Gamma_L|^2\right)}{|1-S_{22}\Gamma_L|^2} \qquad (2.21)$$

The arrangement of (2.19) and (2.21) follows Gonzalez [1], which distinctly shows three contributions to $G_T$: $|S_{21}|^2$, the unmatched power gain of the device, and the two other factors related to the parameters at the input and the output in the form of matching factors, $M_g$ defined in (1.27). (Note that only at one port the equivalent reflection coefficient, $\Gamma_{in}$ or $\Gamma_{out}$, is used, and at the other port $S_{22}$ or $S_{11}$ is used.) From this expression it is clear that we can achieve a practically useful transducer gain through the matching gains even if $|S_{21}|^2 < 1$.

It is also interesting to note that (2.16) to (2.20) show that

$$G_T = G_A M_L = G_P M_S \tag{2.22}$$

Equation (2.22) imposes a constraint on the selection of circuit parameters of $G_T$, $M_S$, and $M_L$: We have the freedom to choose only two out of three parameters by selecting a pair of $\Gamma_S$ and $\Gamma_L$, but not all three. For example, if the design of an amplifier requires a gain of $G_{T0}$ and a perfect output matching, we can first select $\Gamma_S$ such that $G_A = G_{T0}$ (assuming that $G_{T0}$ is in the possible range of $G_A$) and then select $\Gamma_L$ so that $M_L = 1$. This process leads to the desired $G_T$ and $M_L$ but leaves $M_S$ unattended. In practical designs, some trade-off is often made between $M_S$ and $M_L$. Section 2.3 shows how to determine the actual values of $\Gamma_S$ and $\Gamma_L$ for the above specified $G_A$ and $M_L$, as well as how to systematically evaluate the balance between $M_S$ and $M_L$.

In the literature, another gain, the unilateral power gain, denoted as $G_{TU}$ (or simply $U$), is often seen. It is the transducer gain in (2.21) with the condition $S_{12} = 0$; that is,

$$G_{TU} = \frac{\left(1-|\Gamma_S|^2\right)}{|1-S_{11}\Gamma_S|^2} |S_{21}|^2 \frac{\left(1-|\Gamma_L|^2\right)}{|1-S_{22}\Gamma_L|^2}$$

Under this condition, the transducer gain can be maximized by simply matching the source and the load to $S_{11}$ and $S_{22}$, respectively:

$$G_{TU\_max} = G_{TU}\left(\Gamma_S = S_{11}^*, \Gamma_L = S_{22}^*\right) = \frac{1}{1-|S_{11}|^2} |S_{21}|^2 \frac{1}{1-|S_{22}|^2} \tag{2.23}$$

Thus the assumption of $S_{12} = 0$ can significantly simplify analysis. Furthermore, $G_{TU\_max}$, as expressed in (2.23), is determined by the S-parameters only and therefore is used in some cases as a figure of merit of the device. However, for practical transistors, $S_{12}$ is only negligible at very low frequencies.

At the frequency range suitable for RF amplifiers the magnitude of $S_{12}$ of a given transistor is typically too large to be ignored. A figure of merit that is more practically relevant is the maximum available gain (Section 2.3).

We conclude this section with a comment on techniques for network analysis using S-parameters. For power gains we use power waves as circuit variables along with the original definitions of the S-parameters and reflection coefficients to analyze the circuit. The derivations are somewhat tedious even for this relatively simple system. For a more complex circuit that consists of multiple devices, an analysis using this approach quickly becomes unmanageable. One solution to this problem is a technique called signal flow graphs, which can significantly ease the task. However, since this kind of formula manipulation is rarely needed in engineering practice, this technique has limited value for practicing engineers. Accordingly, this book does not cover signal flow graphs. Interested readers should refer to [1, 2].

## 2.2 Stability of Amplifiers

In the introduction to the reflection coefficient for a one-port network in Section 1.3, it is specifically assumed that the network is passive. When the network is passive, the reflection coefficient is restricted to be less than 1, since the reflected power cannot be higher than the incident power. When an active device is involved, however, this restriction is no longer valid. A reflection coefficient with a magnitude larger than unity implies a potentially unstable system as illustrated in Figure 2.2. In this system, an active one-port network with $|\Gamma_a| > 1$ is terminated with a load of $\Gamma_L$. The incident wave $a_n$ in Figure 2.2 can be considered as being randomly generated from the noise power in the system. Using the same analysis leading to (1.24), we obtain

$$b_n = \frac{a_n \Gamma_a}{1 - \Gamma_a \Gamma_L} \quad (2.24)$$

Here the source power wave $b_s$ in (1.24) is replaced with the noise power wave $a_n$. If $|\Gamma_a| > 1$, then there exists a passive load $\Gamma_{La}$ such that $\Gamma_a \Gamma_{La} = 1$. For this particular load, $b_n \to \infty$, meaning that the system is unstable. In practice, this instability is often manifested by an oscillation. In contrast, if $|\Gamma_a| < 1$ no such $\Gamma_{La}$ exists, and the system is stable.

The reader may question how (1.24), which is valid only for a steady state as mentioned in Section 1.4, can be applied to a case where the incident power wave is randomly generated. The answer, from a physical perspective,

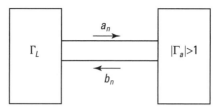

**Figure 2.2** One-port network with $|\Gamma_a| > 1$

is that a noise power wave can still be considered in a steady state during a finite, albeit very short, period of time (i.e., the so-called coherent time). Analytically, if the noise were modeled as an impulse input, the impulse response of a system shown in Figure 2.2 would indicate an instability (usually in the form of exponential growth) when $|\Gamma_a| > 1$. This way, the steady-state requirement for the noise power is avoided altogether. A more detailed account of this technique [3] is outside the scope of this book. Regardless of justification, for the discussion in this section, we state that $|\Gamma_a| < 1$ is a necessary and sufficient condition for a one-port active network to be definitely stable. Chapter 5 discusses whether a system actually oscillates or not when $|\Gamma_a| > 1$.

The stability problem is almost always associated with amplifiers whose gains are excessively high at certain frequency ranges. Some equations derived in Section 2.1 for amplifier gains turn out to be handy in stability analysis as well. Accordingly, Section 2.2.1 introduces a technique of graphic representation and then discusses how to use the analytic results in combination with the graphic representation to characterize the stability of a network.

## 2.2.1 Analysis in the $\Gamma_S$ and $\Gamma_L$ Planes

Section 2.1 derives equations for the gains and mismatch factors of a network characterized by the S-parameters and terminated at an input and output with $\Gamma_S$ and $\Gamma_L$, respectively. The results indicate that these circuit parameters are functions of $\Gamma_S$ and $\Gamma_L$ in addition to $S$. In fact, in practical RF designs, the first major task for the designer can often be reduced to the determination of the values of $\Gamma_S$ and $\Gamma_L$ for the chosen device so that the target specifications are satisfied. Here we show that the equations that characterize the stabilities are also a function of $\Gamma_S$ or $\Gamma_L$. As the reflection coefficient is a complex number, analysis of these functions is frequently said to be performed in the $\Gamma_S$ or $\Gamma_L$ planes. In our discussion, the source and the load are assumed to be passive, which means that the values of $\Gamma_S$ and $\Gamma_L$ are within their respective unit circles, $|\Gamma_S| = 1$ and $|\Gamma_L| = 1$. As proved in Section 2.3, a useful property of these functions is that their constant contours are circles in the $\Gamma_S$ or $\Gamma_L$ planes. First, let's outline two general properties of circles in a complex plane.

For a complex variable $\Gamma$, by definition, a circle in the $\Gamma$ plane with the center at $\Gamma_0$ and the radius of $R$ takes the form

$$\left|\Gamma - \Gamma_0\right|^2 - R^2 = 0 \tag{2.25}$$

Expansion of (2.25) leads to

$$|\Gamma|^2 - \Gamma\Gamma_0^* - \Gamma_0\Gamma^* + \left|\Gamma_0\right|^2 - R^2 = 0 \tag{2.26}$$

This is the standard form that will be used to determine the center and the radius of a circle in a complex plane in the following discussion.

Another category of complex function we need in our discussion is the so-called bilinear transformation, which has a standard form:

$$W = \frac{A\Gamma + B}{C\Gamma + D} \tag{2.27}$$

where $W$ and $\Gamma$ are complex variables and $A$, $B$, $C$, and $D$ are complex constants. It can be verified by solving $\Gamma$ in (2.27) that the function of $\Gamma(W)$ has the same functional form as (2.27) with a different set of constants. This function has the following property: Circles in the $W$ plane map into circles in the $\Gamma$ plane, and vice versa. The proof is easier for the $W \to \Gamma$ mapping. Let $W_0$ and $\rho$ be the center and the radius of a circle in the $W$ plane (i.e., $|W - W_0|^2 = \rho^2$). Then from (2.27), we have

$$\left|W - W_0\right|^2 = \left|\frac{A\Gamma + B}{C\Gamma + D} - W_0\right|^2 = \rho^2 \tag{2.28}$$

Introducing two new constants as $A_1 = A - CW_0$ and $B_1 = B - DW_0$, and after some manipulation, (2.28) becomes

$$\left|A_1\Gamma + B_1\right|^2 = \rho^2 |C\Gamma + D|^2 \tag{2.29}$$

By expanding (2.29), we can see that $\Gamma$ is in the standard circle equation of (2.26), and the center $\Gamma_0$ and the radius $R$ are given by

$$\Gamma_0 = \frac{\rho^2 C^* D - B_1 A_1^*}{\left|A_1\right|^2 - \rho^2 |C|^2} \tag{2.30}$$

$$R = \frac{\rho|A_1 D - B_1 C|}{\left||A_1|^2 - \rho^2 |C|^2\right|} \tag{2.31}$$

## 2.2.2 Stability Circles

Let's return to the stability issue. As explained in Section 2.2, the reflection coefficient of a one-port network must be less than 1 for the network to be stable. Applying this principle to the amplifier circuit represented in Figure 2.1, we see that for a stable circuit, the reflection coefficient $\Gamma_{in}$ at the input of the network that is terminated with $\Gamma_L$ at the output must satisfy the following inequality:

$$|\Gamma_{in}(\Gamma_L)| < 1 \tag{2.32}$$

By the same token, the following must also be true:

$$|\Gamma_{out}(\Gamma_S)| < 1 \tag{2.33}$$

We first consider (2.32) for $\Gamma_{in}(\Gamma_L)$. An area in the $\Gamma_L$ plane within which (2.32) is true is called the stable region in the $\Gamma_L$ plane. Outside the stable region is the unstable region. Between these two regions is the boundary represented by a locus that satisfies the condition:

$$|\Gamma_{in}| = \left|S_{11} + \frac{S_{21} S_{12} \Gamma_L}{1 - S_{22} \Gamma_L}\right| = 1 \tag{2.34}$$

Following the convention in the literature, we define a new parameter $\Delta$ as

$$\Delta = S_{11} S_{22} - S_{21} S_{12} \tag{2.35}$$

which is actually the determinant of the $S$ matrix. Then (2.34) can be written as

$$|S_{11} - \Delta \Gamma_L| = |1 - S_{22} \Gamma_L| \tag{2.36}$$

Taking the square of both sides of (2.36) and rearranging terms, we obtain

$$\left(|\Delta|^2 - |S_{22}|^2\right)|\Gamma_L|^2 + \left(S_{22} - S_{11}^* \Delta\right)\Gamma_L + \left(S_{22}^* - S_{11} \Delta^*\right)\Gamma_L^* + |S_{11}|^2 - 1 = 0 \tag{2.37}$$

A comparison with the standard circle equation (2.26) shows that the points in the $\Gamma_L$ plane that satisfy (2.34) are on a circle with the center and the radius given by

$$\Gamma_{0L} = \frac{S_{11}\Delta^* - S_{22}^*}{|\Delta|^2 - |S_{22}|^2} \quad (2.38)$$

$$R_L = \left| |\Gamma_0|^2 - \frac{|S_{11}|^2 - 1}{|\Delta|^2 - |S_{22}|^2} \right|^{1/2} = \frac{|S_{21}S_{12}|}{||\Delta|^2 - |S_{22}|^2|} \quad (2.39)$$

The circle described by (2.38) and (2.39) is called the stability circle, as it is the boundary between the stable and unstable regions. While this analysis leads to the equation for the boundary, it does not specify which side of the circle these regions are. In fact, either side can be the stable region (with the other side unstable), depending on the values of the S-parameters. To resolve this ambiguity, we note that the center of the $\Gamma_L$ plane where $\Gamma_L = 0$ is a special point in that $|\Gamma_{in}(\Gamma_L = 0)| = |S_{11}|$. Then from the magnitude of $S_{11}$, we can determine whether or not the center is in the stable region. Generally, manufacturers do not make a device that has $|S_{11}| > 1$. Therefore, for all practical purposes, the point $\Gamma_L = 0$ is always in the stable region. Once this rule is established, the stability region can be easily determined by the location of the center of the $\Gamma_L$ plane with respect to the stability circle. There are four cases that are usually seen in practice. They are shown in Figure 2.3. In Figure 2.3(a, b) the entire unit circle, $|\Gamma_L| = 1$, is in the stable region, meaning that the network is stable with any passive load. The device is said to be unconditionally stable under this condition. In Figure 2.3(c, d), the unit circle is only partially in the stable region. This condition is termed as conditionally stable or potentially unstable. We use both terms in the book depending on the context of the discussion. In the case of conditional stability, we use stability margin in this book to describe the distance of a given point in the stable region to the nearest boundary of the stability circle. In general, the larger the stability margin the less likely it is that the system will oscillate. There is, however, no quantitative relationship between the stability margin and the likelihood of oscillation. In practical designs, it is ultimately an engineering judgment call. An example in Section 2.3.3 illustrates this point.

From a simple geometric argument, we can see that the unconditional stability can be verified by the following two inequalities:

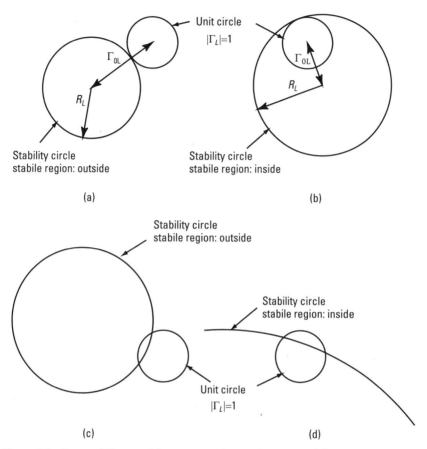

**Figure 2.3** Four stability conditions commonly seen in practice: (a) The unit circle is completely outside the stability circle. (b) The unit circle is completely inside the stability circle. (c) The center of the unit circle is outside the stability circle. (d) The center of the unit circle is inside the stability circle.

- The unit circle is outside the stability circle [case (a)]: $|\Gamma_{0L}| - R_L > 1$
- The unit circle is enclosed inside the stability circle [case (b)]: $R_L - |\Gamma_{0L}| > 1$

Combining the two inequalities, we can write the condition for unconditional stability as

$$\left||\Gamma_{0L}| - R_L\right| > 1 \qquad (2.40)$$

Equation (2.40) is equivalent to the statement that the unit circle of $\Gamma_L$ is entirely in the stable region. Therefore, it can be used as a necessary and sufficient criterion for unconditional stability.

The stability condition at the output side of the network, (2.33), can be analyzed the same way. Again, from the symmetry between the input and output, the stability circle in the $\Gamma_S$ plane is given by

$$\Gamma_{0S} = \frac{S_{22}\Delta^* - S_{11}^*}{|\Delta|^2 - |S_{11}|^2} \qquad (2.41)$$

$$R_S = \frac{|S_{21}S_{12}|}{\left||\Delta|^2 - |S_{11}|^2\right|} \qquad (2.42)$$

If $|\Gamma_{out}(\Gamma_S)| < 1$ holds for all values of $\Gamma_S$ inside the unit circle, the device is unconditionally stable. The necessary and sufficient criterion for this condition is given by

$$\left||\Gamma_{0S}| - R_S\right| > 1 \qquad (2.43)$$

We will argue later that the inequalities in (2.40) and (2.43) are equivalent. That is, only one is needed in determining the unconditional stability.

### 2.2.3 Stability Criteria of *K* and *μ* Parameters

Thus far, we have determined that for an amplifier shown in Figure 2.1 with a passive termination at both ports, the conditions in (2.40) and (2.43) guarantee the unconditional stability of the amplifier. It should be noted at this point that only the S-parameters are present in (2.40) and (2.43), which implies that once **S** is given we can determine whether an unconditionally stable amplifier is achievable or not. However, as evidenced in (2.38) to (2.43), the functions involved are rather complex and inconvenient for practical usage. In practice, a simpler criterion, the *K* factor, is used instead. It is defined as

$$K = \frac{1 - |S_{11}|^2 - |S_{22}|^2 + |\Delta|^2}{2|S_{21}S_{12}|} \qquad (2.44)$$

It can be shown that the condition $K > 1$ plus one of the auxiliary conditions [1, 4] is the necessary and sufficient criterion for the system to be unconditionally stable. (Section 2.2.4 provides a more detailed discussion on this topic.) The $K$ factor is also known as the Rollett factor, as Rollett first proposed it in 1962 [5]—although in his original paper the $K$ factor was presented in immittance parameters rather than the S-parameters. In a later publication in 1966, Ku [6] explicitly expressed the $K$ factor in the form of (2.44). Ever since, the $K$ factor has been adopted as a standard criterion for determination of the stability characteristics of a device. In theory, the $K$ factor itself alone is not sufficient for the determination of unconditional stability. Several additional conditions in various forms can be found in the literature as an auxiliary condition. They are all equivalent when $K > 1$ in the sense that one can be derived from the other. The most common one is perhaps $\Delta < 1$. In addition, the $K$ factor only offers a conclusion as to whether the system is unconditionally stable or not. Its value itself is not related to any physically meaningful quantity. In 1992, a different criterion using $\mu$ or $\mu'$ parameters was proposed [4] to overcome these two perceived shortcomings of the $K$ factor: Either $\mu > 1$ or $\mu' > 1$ is a single criterion for necessary and sufficient conditions for unconditional stability. In addition, both $\mu$ and $\mu'$ have a well-defined geometric meaning in terms of the stability circles in the $\Gamma_L$ plane (for $\mu$) and $\Gamma_S$ (for $\mu'$) as defined below.

$$\mu = \left\| \Gamma_{0L} \right| - R_L \right| = \frac{1 - |S_{11}|^2}{\left| S_{22} - S_{11}^* \Delta \right| + \left| S_{21} S_{12} \right|} \tag{2.45}$$

$$\mu' = \left\| \Gamma_{0S} \right| - R_S \right| = \frac{1 - |S_{22}|^2}{\left| S_{11} - S_{22}^* \Delta \right| + \left| S_{21} S_{12} \right|} \tag{2.46}$$

Equations (2.45) and (2.46) state that $\mu(\mu')$ is the minimum distance from the center of the plane to the unstable region. The proof can be found in [4]. The parameters defined in (2.45) and (2.46) are positive, which are are special cases when the centers are in the stable regions (see [4], for the general case). In principle, the $\mu$ (or $\mu'$) parameter is a better criterion for the reasons mentioned above. However, $K$ is still widely used in practice. In addition to the historical reasoning, another possible reason is that for practical devices the auxiliary condition is almost always true if $K > 1$. So for all practical purposes the condition $K > 1$ is as good as $\mu > 1$. For the case when $K < 1$ or $\mu < 1$, the numerical value of $\mu$ provides a quantitative comparison between two circuits in terms of likelihood of oscillations, but there is no established method to determine whether the circuit oscillates or not by using the $\mu$ value.

We still have to make an engineering judgment based on several other factors. Chapter 5 discusses some of these in detail. In summary, the advantage of $\mu$ over $K$ is limited in practice.

The importance of stability in RF circuit designs cannot be overstated. While a circuit with deficiency in specifications may still function perhaps 90% of the time, a circuit with deficiency in stability can result in a complete system failure if it oscillates. Section 5.2 is devoted to practical approaches in dealing with (in)stability. The next subsection (Section 2.2.4) is for readers interested in the analytic treatment in stability using the $K$ and $\mu$ parameters.

## 2.2.4 Additional Comments

This section outlines some of the basic concepts and analytic approaches used in the derivations of the $K$ and $\mu(\mu')$ parameters. The step-by-step manipulations are left to the reader. More detailed derivations can also be found in many textbooks such as [1, 2, 7].

For reference, some auxiliary conditions commonly seen in the literature are listed as follows [1, 4]:

$$B_1 = 1 + |S_{11}|^2 - |S_{22}|^2 - |\Delta|^2 > 0$$

$$B_2 = 1 - |S_{11}|^2 + |S_{22}|^2 - |\Delta|^2 > 0$$

$$|\Delta| < 1$$

$$1 - |S_{11}|^2 > |S_{12} S_{21}|$$

$$1 - |S_{22}|^2 > |S_{12} S_{21}|$$

We start with the unconditional stability condition $|\Gamma_{in}(\Gamma_L)| < 1$. (Again we only work on the input side of the network.) In this scheme, if $\Gamma_{in}$ and $\Gamma_L$ are considered two variables, then a function that relates them can be either in the original form,

$$\Gamma_{in} = f(\Gamma_L, S) = \frac{S_{11} - \Delta \Gamma_L}{1 - S_{22} \Gamma_L} \tag{2.47}$$

or in its inverse form

$$\Gamma_L = f^{-1}(\Gamma_{in}, S) = \frac{S_{11} - \Gamma_{in}}{\Delta - S_{22} \Gamma_{in}} \tag{2.48}$$

Equation (2.47) is simply a rearrangement of (2.6).

In writing the functions in (2.47) and (2.48), the S-parameters are explicitly included to emphasize that the characteristics of the functions are determined by the values of the four S-parameters. It is also clear that both functions are in the standard bilinear form of (2.27). Therefore a circle in one plane is mapped into a circle in the other plane. In the context of the stability circle, we can think of the unit circle, $|\Gamma_{in}| = 1$, as the stability circle in the $\Gamma_{in}$ plane, since this circle happens to be the boundary for stability condition, $|\Gamma_{in}| < 1$. If this stability circle is mapped to the $\Gamma_L$ plane, the mapped circle is what we call the stability circle in Section 2.2.3. The relative location of this circle with respect to the unit circle $|\Gamma_L| = 1$ determines the stability condition, as discussed in Section 2.2.3.

Alternatively, we can tackle the same problem by mapping the unit circle $|\Gamma_L| = 1$ into the $\Gamma_{in}$ plane. Since all valid $\Gamma_L$ points are inside the unit circle, if the mapped circle in the $\Gamma_{in}$ plane is completely inside the unit circle $|\Gamma_{in}| = 1$, then the inequality $|\Gamma_{in}(\Gamma_L)| < 1$ holds for all values of $\Gamma_L$ with $|\Gamma_L| < 1$. That is, the device is unconditionally stable. Otherwise the device is potentially unstable. Figure 2.4 illustrates both mapping scenarios. From Figure 2.4(a), the criterion for the mapped circle being inside the unit circle in the $\Gamma_{in}$ plane is

$$|C_{in}| + R_{in} < 1 \tag{2.49}$$

where $C_{in}$ and $R_{in}$ represent, respectively, the coordinate of the center and the radius of the circle. They are given by

$$C_{in} = \frac{S_{11} - S_{22}^* \Delta}{1 - |S_{22}|^2} \tag{2.50}$$

$$R_{in} = \frac{|S_{21} S_{12}|}{1 - |S_{22}|^2} \tag{2.51}$$

The proof is straightforward if we start with (2.48) and use the results from (2.28) through (2.31). (Note that $W_0 = 0$ and $\rho = 1$ in this case.) Thus, there can be two criteria for unconditional stability: (2.40) from the stability analysis in the $\Gamma_L$ plane and (2.49) from that in the $\Gamma_{in}$ plane. They must be equivalent based on our above argument, but they evidently have different functional forms. Furthermore, by using the same technique on the stability circles in the $\Gamma_S$ and $\Gamma_{out}$ planes, we can reach two additional criteria. The reader can work out the specific inequalities for these criteria by using the network symmetry in (2.40) and (2.49).

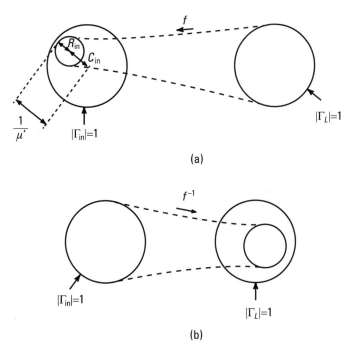

**Figure 2.4** Unit circle mappings between $\Gamma_L$ and $\Gamma_{in}$ planes: (a) $|\Gamma_L| = 1$ mapped to $\Gamma_{in}$ plane and (b) $|\Gamma_{in}| = 1$ mapped to $\Gamma_L$ planes.

Using (2.50) and (2.51), we can see that the unconditional stability condition expressed in (2.49) is the same as $\mu' > 1$. Furthermore, it is clear from Figure 2.4(a) that $(1 - 1/\mu')$ is the minimum distance between the stable and unstable regions in the $\Gamma$ in plane when $\mu' > 1$. The same conclusion can be made for $\mu$. In addition to the geometric meanings, there is another distinction between the $K$ factor and the $\mu$ (or $\mu'$) parameter in terms of the effect of a lossless matching network to the DUT: Such a network has no effect on the $K$ factor. Using Rollett's phrase in his original paper [5], the $K$ factor is invariant. In contrast, the $\mu$ parameter will generally be affected by an output, but not input, lossless network and $\mu'$ by an input, but not output, lossless network. We will outline the proof at the end of Section 2.4.

In principle, any of the four criteria can be used to determine whether a device is unconditionally stable or not. However, these criteria are in different functional forms and are generally too complex for practical use. After the $K$ factor was proposed the industry quickly adopted it as a standard parameter for stability considerations. Mathematically, the derivation for $K > 1$ being a stability criteria can start from any of the four inequalities discussed. The difference in the selection of the starting point is the main reason that the derivations for the $K$ factor often seem to be considerably different among textbooks. The fact that the inequality $K > 1$, plus an auxiliary condition, is equivalent to all other four inequalities indirectly establishes that all other criteria are equivalent too.

The need for an auxiliary condition for $K > 1$ to be the necessary and sufficient condition for unconditional stability can be understood from the following observation: All terms in $K$ are squared terms, whereas there is a linear term in the original inequalities [e.g., the $S_{22}$ term in (2.40)]. If we follow a $K$ factor derivation closely it is clear that at a certain point a squaring operation on an inequality takes place. It is this operation that prevents $K > 1$ from being the sufficient condition alone for unconditional stability, since squaring an inequality always yields an extra solution that is not valid for the original inequality. An auxiliary condition simply excludes that extra solution. Obviously, the squaring operation can happen at different steps in the derivation, resulting in different forms of the auxiliary condition. Also, it can be shown that all auxiliary conditions are equivalent (i.e., one can be derived from any other) provided that $K > 1$ holds. The proof for the equivalence between specific conditions can be found in the literature, including [1, 4]. We can also argue that if the condition $K > 1$ plus any one of the auxiliary conditions is necessary and sufficient for unconditional stability, then all the auxiliary conditions must be equivalent.

## 2.3 Conjugate Matching and Constant Circles of Gains and Mismatch Factors

Section 2.2 covers the device stability that is analyzed in a network configuration (Figure 2.1). We now turn our attention to the following question: For the same network, what are the values of $\Gamma_S$ and $\Gamma_L$ that yield an optimal system performance (i.e., those that minimize the mismatch factors and thus maximize the gain). This section explains that the answer to this question is closely tied to the stability condition of the device. The derivations for the results used in this section are generally tedious and can be found in many textbooks on the subject (e.g., [1, 2, 8]). Therefore, they are omitted here.

### 2.3.1 Simultaneous Conjugate Matching

According to the conclusion regarding conjugate matching in Section 1.4, if a pair of $\Gamma_S$ and $\Gamma_L$ exists such that conjugate matching is simultaneously achieved at the input and output of the device, then the device can be perfectly matched at both ports, corresponding to a maximum transducer gain. Mathematically, this condition can be expressed as

$$\Gamma_{in}^{*}(\Gamma_L) = \Gamma_S \qquad (2.52)$$

$$\Gamma_{out}^{*}(\Gamma_S) = \Gamma_L \qquad (2.53)$$

It is customary to write the solutions, denoted as $\Gamma_{SC}$ and $\Gamma_{LC}$, for the equation system of (2.52) and (2.53) in four parameters defined as:

$$A_1 = 1 + |S_{11}|^2 - |S_{22}|^2 - |\Delta|^2$$

$$A_2 = 1 + |S_{22}|^2 - |S_{11}|^2 - |\Delta|^2$$

$$B_1 = S_{11} - \Delta S_{22}^{*}$$

$$B_2 = S_{22} - \Delta S_{11}^{*}$$

Then $\Gamma_{SC}$ and $\Gamma_{LC}$ are given by

$$\Gamma_{SC} = \frac{A_1 \pm \sqrt{A_1^2 - 4|B_1|^2}}{2B_1} \qquad (2.54)$$

$$\Gamma_{LC} = \frac{A_2 \pm \sqrt{A_2^2 - 4|B_2|^2}}{2B_2} \qquad (2.55)$$

Valid solutions for $\Gamma_{SC}$ and $\Gamma_{LC}$ do not always exist. One obvious constraint is that both $\Gamma_{SC}$ and $\Gamma_{LC}$ have to be within the unit circle in their respective $\Gamma_S$ and $\Gamma_L$ planes to be a realizable solution (passive source and load). There are several other constraints on the values of $A_1$, $A_2$, $B_1$, and $B_2$ for $\Gamma_{SC}$ and $\Gamma_{LC}$ to be physically meaningful. For example, consider the case, $A_1^2 < 4|B_1|^2$. Under this condition $\Gamma_{SC}$ can be written as $\left(A_1 \pm j\sqrt{4|B_1|^2 - A_1^2}\right)/2B_1$. Since $A_1$ is a real number, it follows immediately that $|\Gamma_{SC}| = 1$. Therefore, there is no valid solution when $A_1^2 < 4|B_1|^2$. For practical purposes, the situation is actually much simpler: If the device is unconditionally stable, then valid solutions exist. Otherwise they do not. Therefore, we can use the stability factor $K$ or $\mu$ to determine whether simultaneous conjugate matching can be achieved or not. The mathematical proof is convoluted because of the various constraints described here. However, it can be justified from a simple argument as follows: If the system is unconditionally stable, then for any point in the $\Gamma_S$ plane, the conjugate matching, $\Gamma_L = \Gamma_{out}^{*}$, can always be achieved at the device output. Let $\Gamma_S$ run over the entire unit circle; there must be a point that yields a maximum transducer gain. On the other hand, if the system is potentially unstable, we will have $\Gamma_{out} = 1$ at the boundary of stability, which corresponds to $G_A \to \infty$, according to (2.14). As a result, the maximum transducer gain cannot be determined in this case.

From (2.21), the maximum transducer gain under the simultaneous conjugate matching is

$$G_{T\max} = \frac{\left(1-|\Gamma_{SC}|^2\right)}{|1-\Gamma_{inC}\Gamma_{SC}|^2}|S_{21}|^2\frac{\left(1-|\Gamma_{LC}|^2\right)}{|1-S_{22}\Gamma_{LC}|^2} \qquad (2.56)$$

Here $\Gamma_{inC} = \Gamma_{in}(\Gamma_{LC})$. Substituting (2.54) and (2.55) into (2.56) gives the expression for $G_{T\max}$ in terms of S-parameters only:

$$G_{T\max} = \frac{|S_{21}|}{|S_{12}|}\left(K - \sqrt{K^2-1}\right) \qquad (2.57)$$

In practice, $G_{T\max}$ is usually referred to as maximum available gain (MAG), although some variations, such as maximum power gain, also appear in manufacturers' datasheets. MAG is a key specification in device selection. Generally, for practical devices, $K$ increases while the ratio $|S_{21}|/|S_{12}|$ decreases with frequency. Since the factor $K - \sqrt{K^2-1}$ monotonically increases as $K \to 1$ (from the $K > 1$ side), both factors in (2.57) increase as the frequency decreases in the region where $K > 1$. Therefore, at $K = 1$, $G_{T\max}$ reaches the maximum stable gain, denoted as MSG. It is defined as:

$$MSG = G_{T\max}(K=1) = \frac{|S_{21}|}{|S_{12}|} \qquad (2.58)$$

If $K < 1$, (2.57) is no longer valid. MSG defined in (2.58) is yet kept as a gain specification for the region of $K < 1$, despite of its lack of practical meaning regarding stability. Figure 2.5 shows a typical datasheet plot of different gains versus frequency for a Si RF transistor.

The gain curves versus frequency in Figure 2.5 allow a designer to quickly determine the suitability of a transistor for the target circuit specifications. If at the operation frequency, the device is unconditionally stable and MAG meets the gain specifications, then the designer simply needs to calculate the values for $\Gamma_{SC}$ and $\Gamma_{LC}$, which can usually be done with an EDA tool or a computer program.

If a device is only conditionally stable, the design decision is more complex. There are two main practical approaches in this situation. One is to

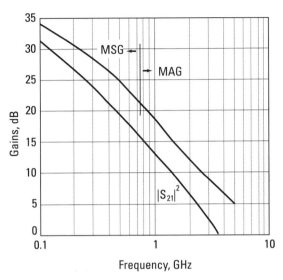

**Figure 2.5** Gains versus frequency of a Si RF transistor.

stabilize the device first by adding a resistance at input or output and then to employ the simultaneous conjugate matching. This approach is relatively easy to implement and ensures that the circuit is unconditionally stable. The drawback of it is that the added resistive element generally degrades the circuit performance such as noise figure in LNAs or output power and efficiency in power amplifiers. The second technique is to limit the source and load impedances to the stable regions. This approach permits a full utilization of the device capabilities, but the circuit is only conditional stable, which is acceptable in certain practical conditions. Sections 2.3.2 and 2.3.3 are mainly concerned with the selection of $\Gamma_S$ and $\Gamma_L$ based on a set of performance requirements under the condition of conditional stability. Chapter 5 offers a more detailed discussion on dealing with the stability issue in practical amplifier design.

### 2.3.2 Constant Circles of Gains and Mismatch Factors

Recall that under the condition of conditional stability, maximizing the gain cannot be a design target, since $G_A$ (or $G_P$) approaches infinity when $\Gamma_S$ (or $\Gamma_L$) moves toward the stability boundary. Also, mismatch is unavoidable under this condition. Consequently, the first step of the design is to consider the balance among the specifications of stability margin, gain, and mismatch factors, all of which are functions of $\Gamma_S$ and $\Gamma_L$. An analytic approach for this purpose is exceedingly difficult because of the interactive nature of the circuit with the input and output matching conditions affecting each other simultaneously as evidenced in the expressions for $\Gamma_{in}$ and $\Gamma_{out}$ [(2.5) and (2.6)]. In the so-called unilateral case, where $S_{12} = 0$ is assumed, the input and output can be designed independently since $\Gamma_{in} = S_{11}$ and $\Gamma_{out} = S_{22}$. However, in most practical cases, the magnitude of $S_{12}$ cannot be ignored. Accordingly, let's now consider a graphic technique that proves to be effective in design and evaluation of the circuit specifications when the device is conditional stable. The technique is based on the constant-circle property of the available gain and the power gain as well as the mismatch factors.

The concept of this property is straightforward. The derivations of these circle properties are lengthy and are not needed for practical engineers. Accordingly, this section simply shows the proof for one case and provides the results for the rest. Interested readers can work out the derivations themselves or consult [1, 2].

Recall that $G_A$ and $G_P$ are respective functions of $\Gamma_S$ and $\Gamma_L$ as indicated in (2.14) and (2.15) and so are $M_S$ and $M_L$ once $\Gamma_{in}$ (through $\Gamma_L$) and $\Gamma_{out}$ (through $\Gamma_S$) are fixed. [See (2.20) and (2.18).] Then with the same technique for the stability circles, we can show that the constant contours of $G_A$ and $M_S$ are circles in the $\Gamma_S$ plane and that those of $G_P$ and $M_L$ are circles on the $\Gamma_L$ plane. In the following we derive the circle equation for $M_S$ as an example to further illustrate the concept of this property.

The expression for $M_S$ (2.20) is repeated here:

$$M_S = \frac{\left(1-|\Gamma_{in}|^2\right)\left(1-|\Gamma_S|^2\right)}{|1-\Gamma_S\Gamma_{in}|^2} \tag{2.59}$$

Let $\Gamma_{in}$ be a constant $\Gamma_{in0}$. (That is, $\Gamma_L$ is fixed.) Then $M_S$ can be considered a function of $\Gamma_S$ only. Section 2.2.2 shows that the solution for $|\Gamma_{in}(\Gamma_L)| = 1$ forms a circle in the $\Gamma_L$ plane. Similarly, we determine here that for a constant value of $M_S$, denoted as $M_{S0}$, the solution of the equation

$$M_S(\Gamma_S, \Gamma_{in0}) = M_{S0} \tag{2.60}$$

forms a circle in the $\Gamma_S$ plane.

It follows from (2.59) and (2.60) that

$$M_{S0}(1-\Gamma_{in0}\Gamma_S)(1-\Gamma_{in0}^*\Gamma_S^*) = 1-|\Gamma_{in0}|^2 -|\Gamma_S|^2 +|\Gamma_{in0}|^2|\Gamma_S|^2 \tag{2.61}$$

Expanding (2.61) gives

$$|\Gamma_S|^2\left[|\Gamma_{in0}|^2(1-M_{S0})-1\right]+M_{S0}\Gamma_{in0}\Gamma_S+M_{S0}\Gamma_{in0}^*\Gamma_S^*+1-|\Gamma_{in0}|^2-M_{S0}=0 \qquad (2.62)$$

Dividing (2.62) by $(|\Gamma_{in0}|^2(1-M_{S0})-1)$, we find that

$$|\Gamma_S|^2+\frac{M_{S0}\Gamma_{in0}\Gamma_S+M_{S0}\Gamma_{in0}^*\Gamma_S^*}{|\Gamma_{in0}|^2(1-M_{S0})-1}+\frac{1-|\Gamma_{in0}|^2-M_{S0}}{|\Gamma_{in0}|^2(1-M_{S0})-1}=0 \qquad (2.63)$$

Equation (2.63) is in the standard circle form of (2.26). The center of the circle in the $\Gamma_S$ plane is

$$\Gamma_{MS0}=\frac{M_{S0}\Gamma_{in0}^*}{1-|\Gamma_{in0}|^2(1-M_{S0})} \qquad (2.64)$$

and the radius of the circle, $R_{MS0}$, is determined by the equation

$$\frac{1-|\Gamma_{in0}|^2-M_{S0}}{1-|\Gamma_{in0}|^2(1-M_{S0})}=R_{MS0}^2-|\Gamma_{MS0}|^2 \qquad (2.65)$$

Substituting (2.64) into (2.65) we obtain the final expression for $R_{MS0}^2$:

$$R_{MS0}^2=\left|\frac{M_{S0}\Gamma_{in0}^*}{1-|\Gamma_{in0}|^2(1-M_{S0})}\right|^2+\frac{1-|\Gamma_{in0}|^2-M_{S0}}{1-|\Gamma_{in0}|^2(1-M_{S0})}$$

$$=\frac{(1-M_{S0})\left(1-|\Gamma_{in0}|^2\right)^2}{\left[1-|\Gamma_{in0}|^2(1-M_{S0})\right]^2}$$

Thus,

$$R_{MS0}=\frac{\sqrt{1-M_{S0}}\left(1-|\Gamma_{in0}|^2\right)}{1-|\Gamma_{in0}|^2(1-M_{S0})} \qquad (2.66)$$

The same property holds true for $M_L$; that is, the constant contour of $M_{L0}$ is a circle on the $\Gamma_L$ plane for a given $\Gamma_{out0}$. The center and radius of the circle are given by

$$\Gamma_{ML0} = \frac{M_{L0}\Gamma_{out0}^*}{1-|\Gamma_{out0}|^2(1-M_{L0})} \qquad (2.67)$$

and

$$R_{ML0} = \frac{\sqrt{1-M_{L0}}\left(1-|\Gamma_{out0}|^2\right)}{1-|\Gamma_{out0}|^2(1-M_{L0})} \qquad (2.68)$$

Similarly, it can be proved that the constant contours of $G_A$ and $G_P$ are circles in the $\Gamma_S$ and $\Gamma_L$ planes, respectively. The centers and radii of these circles are as follows:

- Constant circle for $G_A = G_{A0}$:

Center: $$\Gamma_{Ga0} = \frac{g_a C_a^*}{1+g_a\left(|S_{11}|^2-|\Delta|^2\right)} \qquad (2.69)$$

Radius: $$R_{Ga0} = \frac{\sqrt{1-2Kg_a|S_{12}S_{21}|+g_a^2|S_{12}S_{21}|^2}}{\left|1+g_a\left(|S_{11}|^2-|\Delta|^2\right)\right|} \qquad (2.70)$$

- Constant circle for $G_P = G_{P0}$:

Center: $$\Gamma_{Gp0} = \frac{g_p C_p^*}{1+g_p\left(|S_{22}|^2-|\Delta|^2\right)} \qquad (2.71)$$

Radius: $$R_{Gp0} = \frac{\sqrt{1-2Kg_p|S_{12}S_{21}|+g_p^2|S_{12}S_{21}|^2}}{\left|1+g_p\left(|S_{22}|^2-|\Delta|^2\right)\right|} \qquad (2.72)$$

The constants in (2.69) to (2.72) are defined as: $g_a = G_{A0}/|S_{21}|^2$; $g_p = G_{P0}/|S_{21}|^2$; $C_a = S_{11} - \Delta S_{22}^*$; $C_p = S_{22} - \Delta S_{11}^*$; $K$ and $\Delta$ are defined in (2.35) and (2.44).

Finally, as indicated in (2.6), $\Gamma_{in}$ and $\Gamma_L$ are related through a bilinear transformation. As such, a constant $G_P$ circle described by (2.71) and (2.72) in the $\Gamma_L$ plane is mapped into a circle in the $\Gamma_{in}$ plane, and the center and radius are:

$$C_{in0}(G_{P0}) = \frac{(1 - S_{22}\Gamma_{Gp0})^*(S_{11} - \Delta\Gamma_{Gp0}) - R_{Gp0}^2 S_{22}^* \Delta}{|1 - S_{22}\Gamma_{Gp0}|^2 - R_{Gp0}^2|S_{22}|^2} \quad (2.73)$$

$$R_{in0}(G_{P0}) = \frac{R_{Gp0}|S_{21}S_{12}|}{\left||1 - S_{22}\Gamma_{Gp0}|^2 - R_{Gp0}^2|S_{22}|^2\right|} \quad (2.74)$$

Here the notation for the center is changed to $C_{in0}$ to avoid any confusion with that in (2.60). $C_{in0}$ and $R_{in0}$ are explicitly expressed as functions of $G_{P0}$ to emphasize that the circle corresponds to a constant $G_P$ value. To derive (2.73) and (2.74), we can first write $\Gamma_L$ as a function of $\Gamma_{in}$ from (2.48) and then use the formulas for the center and radius in (2.30) and (2.31) to obtain the results. Note that the points on the circle described by (2.73) and (2.74) are $\Gamma$ in values. According to (1.28), $\Gamma_S = \Gamma_{in}^*$ is the condition for perfect matching. Then by plotting the conjugate of the $\Gamma_{in}$ circle (the $\Gamma_{in}^*$ circle) in the $\Gamma_S$ plane we can easily visualize where $\Gamma_S$ should be selected to minimize the mismatch. Since $|\Gamma_{in}^* - C_{in0}^*| = |\Gamma_{in} - C_{in0}|$, the $\Gamma_{in}^*$ circle is simply the one with the center at $C_{in0}^*$ and the same radius given in (2.74). This process is illustrated in the examples in Section 2.3.3.

The same analysis and conclusion can be applied to the mapping of constant $G_A$ circles from the $\Gamma_S$ to $\Gamma_{out}$ planes. The results are as follows:

$$C_{out0}(G_{A0}) = \frac{(1 - S_{11}\Gamma_{Ga0})^*(S_{22} - \Delta\Gamma_{Ga0}) - R_{Ga0}^2 S_{11}^* \Delta}{|1 - S_{11}\Gamma_{Ga0}|^2 - R_{Ga0}^2|S_{11}|^2} \quad (2.75)$$

$$R_{out0}(G_{A0}) = \frac{R_{Ga0}|S_{21}S_{12}|}{\left||1 - S_{11}\Gamma_{Ga0}|^2 - R_{Ga0}^2|S_{11}|^2\right|} \quad (2.76)$$

For the same reason as in the case of $\Gamma_{in}^*$ in the $\Gamma_S$ plane, the circle of $\Gamma_{out}^*$ is usually plotted in the $\Gamma_L$ plane with the center given by $C_{out0}^*$.

The constant circle properties we outline here demonstrate mathematical elegance but are not entirely necessary from the circuit design perspective. In practice, the designer rarely needs to work numerically with these circle properties. However, plots based on these circle properties provide designers with a systematic approach in the determination of the desired values of $\Gamma_S$ and $\Gamma_L$ through a trade-off process. We refer to this design method of using constant circles as the constant-circle method [9]. All major brands of EDA software provide some capabilities for plotting these constant circles. It is the easy availability of the plotting tools that makes the constant-circle method practical.

### 2.3.3 Linear Circuit Design Using the Constant Circle Method

The actual implementation of the constant circle plotting tools varies from one software program to another. Generally, these tools enable a user to plot constant circles for the functions defined in Section 2.3.2 based on the S-parameters and a set of constant values provided by the user. Usually, users can also perform a circle mapping using these tools such as the mapping of a constant $G_P$ (or $G_A$) circle in the $\Gamma_L$ (or $\Gamma_S$) plane to a $\Gamma^*_{in}$ (or $\Gamma^*_{out}$) circle in the $\Gamma_S$ (or $\Gamma_L$) plane according to (2.73) to (2.76). The plotting capability of the mismatch factors $M_S$ and $M_L$ is less universal. If the software does not provide this plotting function, $M_S$ (or $M_L$) can be calculated numerically using the program's built-in functions according to (2.20) [or (2.18)].

As is customary in practice, we use the decibel form for numbers in dealing with gains and mismatch factors.

Table 2.1, which lists a set of S-parameters at 1.6 GHz of a GaAs FET, is used to illustrate the design process.

Readers with some familiarity with the constant circle plotting capability of specific EDA software will follow this discussion more easily. Figures 2.6 and 2.7 show constant circle plots for several $G_A$ and $G_P$ values along with their corresponding $\Gamma^*_{out}$ and $\Gamma^*_{in}$ curves in the $\Gamma_S$ and $\Gamma_L$ planes. Also shown are the stability boundaries. We may immediately notice that all these curves intercept the respective unit circle at two identical points. It can be

**Table 2.1**
*S* Parameters Used in Constant-Circle Plots

| $S_{11}$ | $S_{12}$ | $S_{21}$ | $S_{22}$ | *K* | MSG |
|---|---|---|---|---|---|
| 0.539∠−84.0° | 0.068∠56.9° | 7.909∠104.1° | 0.474∠−38.8° | 0.72 | 20.66 |

proved [7] that this feature is generally true for any $G_A$ and $G_P$ values. We also observe that a $G_A$ (or $G_P$) curve with higher gain and its corresponding $\Gamma^*_{out}$ (or $\Gamma^*_{in}$) curve are closer to the stability boundaries. This is important in design consideration regarding the trade-off between the gain and the stability margin. This kind of plots also provides an easy visual aid in the evaluation of mismatch. For example, if a point on the $G_A = 18.5$ dB curve is chosen for $\Gamma_S$ and a point on the $G_P = 20.5$ dB curve for $\Gamma_L$, from the distance between $\Gamma_S$ and $\Gamma^*_{in}$ and that between $\Gamma_L$ and $\Gamma^*_{out}$, one can quickly see that this set of $\Gamma_S$ and $\Gamma_L$ yields a good mismatch factor at the output but a very poor $M_S$ at the input. In fact, the mismatch factors can be quantitatively estimated with plots of constant mismatch circles. As an example, Figure 2.6 plots a constant mismatch circle of $M_S = -0.5$ dB for point $A$ in the $\Gamma_L$ plane (see Figure 2.7). It can be seen that the $G_A = 19.5$ dB curve barely intercepts with this circle, indicating that once point $A$ is chosen for $\Gamma_L$, for a $M_S$ better than $-0.5$ dB, $G_A$ needs to be greater than 19.5 dB.

We now consider the case discussed in Section 2.1 after (2.22), setting our target for $G_{T0}$ at 18.5 dB. Then the task is to select a $\Gamma_S$ corresponding to

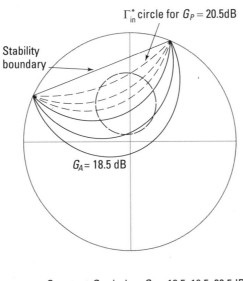

——————  Constant $G_A$ circles, $G_A = 18.5, 19.5, 20.5$dB
– – – –  $\Gamma^*_{in}$ circles for $G_P = 18.5, 19.5, 20.5$ dB circles
– – – –  $M_S = -0.5$dB circle for $\Gamma_L$ at point A (on Fig. 2.7)

**Figure 2.6** Constant circles on the $\Gamma_S$ plane.

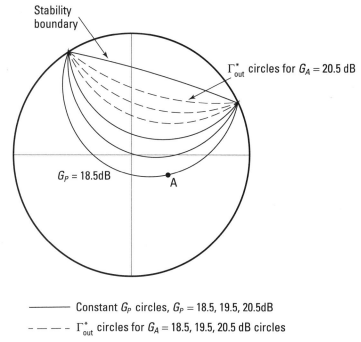

- ———— Constant $G_P$ circles, $G_P$ = 18.5, 19.5, 20.5dB
- – – – – $\Gamma^*_{out}$ circles for $G_A$ = 18.5, 19.5, 20.5 dB circles

**Figure 2.7** Constant circles on the $\Gamma_L$ plane.

$G_A$ = 18.5 dB, followed by a selection of $\Gamma_L$ that yields $M_L$ = 1. In the final step, we estimate the input mismatch factor $M_S$.

Figure 2.8 demonstrates the process. The $\Gamma_S$ and $\Gamma_L$ planes are on the same chart this time. To avoid overcrowding, only curves that are relevant to the process are plotted. Readers need to keep track of which plane a particular curve is on. The first step of selecting $\Gamma_S$ is straightforward since any point on the $G_A$ = 18.5 dB circle works. There is no rule as to where to pick the initial point on the selected constant circle. Usually it should be somewhere close to the center of the $\Gamma_S$ plane, because that is where the matching is easier for circuit realization (discussed further in Chapter 3). In this case, point $B$ is selected. The corresponding point on the $\Gamma^*_{out}$ circle is $B'$. This is where $\Gamma_L$ should be selected to have $M_L$ = 1. Next, we estimate the input mismatch factor $M_S$ for the selected $\Gamma_S$ and $\Gamma_L$. To this end, two constant $M_S$ circles are plotted for $M_S$ = −2 and −3 dB for the chosen $\Gamma_L$. It is clear from the plots that point $B$ is almost exactly on the $M_S$ = −3 dB circle. Therefore, we conclude that the input mismatch is about −3 dB. The plotting software usually provides the coordinates of the selected point on these curves. From the coordinates of points $B$

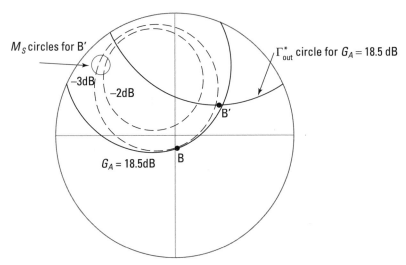

**Figure 2.8** Design using constant circle plots for $G_T = 18.5$ dB and $M_L = 1$.

and $B'$, we obtain the desired values for $\Gamma_S$ and $\Gamma_L$: $\Gamma_S = 0.094\angle-78.6°$ and $\Gamma_L = 0.451\angle 33.5°$. We can further verify our results by observing that the $G_P = 21.5$ dB curve (not shown in the plot) almost goes through point $B'$. Then according to (2.22), to get $G_T = 18.5$ dB for $G_P = 21.5$ dB, $M_S$ needs to be $-3$ dB, which is in agreement with the result from the constant circle technique. For real applications, a mismatch factor of $-3$ dB, which corresponds to a 3-dB return loss, is too high in many cases.

In practical designs using a device that is conditionally stable, the plan is often to have moderate mismatch factors at both ports, with the preference perhaps given to one port, while still maintaining the gain reasonably below the MSG value for the stability margin. For such a plan, the constant circle approach turns out to be particularly useful because it provides a visual aid in a trial-and-error process. This process is illustrated in the following example.

Using the same S-parameters, we now consider a case that requires that the input and output mismatch factors at least meet the following conditions: $M_S = -0.5$ dB and $M_L = -1$ dB. We need to determine the gain for achieving these requirements and the corresponding $\Gamma_S$ and $\Gamma_L$. The design process is shown in Figure 2.9. Again, we use one chart for both the $\Gamma_S$ and $\Gamma_L$ planes.

From the previous example, we know that a gain of 18.5 dB for either $G_A$ or $G_P$ is too low to achieve the required mismatch factors. We start with the $\Gamma_L$ plane by selecting point $C$ on the $G_P = 19$ dB constant circle. Since $M_S = -0.5$ dB is required, only the $M_S = -0.5$ dB curve is plotted on the $\Gamma_S$ plane.

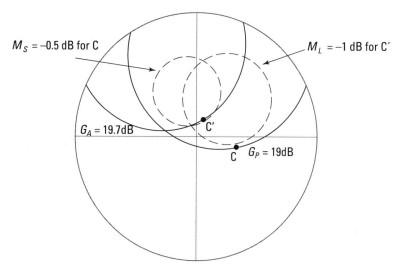

**Figure 2.9** Design using constant circle plots for $G_T = 19$ dB, $M_L = -1$ dB and $M_S = -0.5$ dB.

By trial and error we can quickly determine that the $G_A = 19.7$ dB circle is approximately the lowest $G_A$ value (and thus has the largest stability margin) that is still within the $M_S = -0.5$ dB circle, as shown in Figure 2.9. Point $C'$ on this circle is selected and the corresponding $M_L = -1$ dB circle is plotted in the $\Gamma_L$ plane. The selected $\Gamma_L$ (point $C$) is slightly outside this circle, which means that the target mismatch factors $M_S = -0.5$ dB and $M_L = -1$ dB cannot be met with $G_P = 19$ dB. We can either relax the mismatch requirement or increase the gain. Let us assume that the former is the choice. Again from the software we obtain the selected $\Gamma_S$ and $\Gamma_L$: $\Gamma_S$ (point $C'$) = 0.153∠68.2° and $\Gamma_L$ (point $C$) = 0.336∠−15.7°. It is left to the reader to verify that the $M_L = -1.2$ dB circle almost exactly goes through point $C$. Therefore, for the chosen $\Gamma_L$ and $\Gamma_S$, we have $M_S = -0.5$ dB and $M_L = -1.2$ dB and $G_T$ in decibels is given by $G_T$ (dB) = $G_A$ (dB) + $M_L$ (dB) = 19.7 − 1.2 = 18.5 dB or $G_T$ (dB) = $G_P$ (dB) + $M_S$ (dB) = 19 − 0.5 = 18.5 dB, exactly as (2.22) predicts. Also, we observe that the two different matching choices, (1) $M_S = -3$ dB, $M_L = 0$ dB in the previous case, and (2) $M_S = -0.5$ dB, $M_L = -1.2$ dB in this case, lead to the same transducer gain of $G_T = 18.5$ dB. This feature allows the designer to make some trade-offs between $M_S$ and $M_L$ while maintaining $G_T$ unchanged. It is left to the readers to determine the $G_T$ value for which the original mismatch targets $M_S = -0.5$ dB and $M_L = -1$ dB can be achieved.

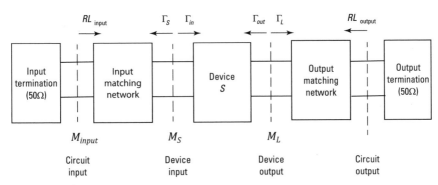

**Figure 2.10** Block diagram for a typical RF amplifier circuit.

## 2.4 Invariance of the Mismatch Factors of Lossless Networks

Sections 2.1–2.3 show how to select $\Gamma_S$ and $\Gamma_L$, based on the S-parameter analysis of the stability factor, gain, and mismatch factors, in a linear amplifier design. Once this is accomplished, the next step is circuit realization of the selected $\Gamma_S$ and $\Gamma_L$. This is important because in real applications, circuits are specified with a particular set of input and output termination conditions, with the vast majority of cases being 50Ω. It is the designer's responsibility to create the required matching networks that transform the specified input and output terminations to the target values of $\Gamma_S$ and $\Gamma_L$. Figure 2.10 depicts this notion. The matching networks for this purpose are usually made of nonresistive components such as inductance, capacitance, and transmission lines to avoid any unnecessary power loss. This section deals with a special property of lossless matching networks, namely, the invariance of the mismatch factor theorem.

We use the input side in Figure 2.10 for our consideration. Note the difference between the device input and circuit input in Figure 2.10. Our previous analysis of the mismatch factor is always at the device input, whereas the circuit input is where the circuit parameter is specified and measured. The invariant property of lossless two-port networks states that if the matching network is lossless, the mismatch factors at the two ports are the same; that is, $M_{\text{input}} = M_S$. This property is the basis on which we can analyze the circuit performance using the device characteristics alone without actual matching networks included. The lossless matching networks can be designed separately; this is the focus of Chapter 3.

The return loss at the circuit input, $S_{11}$, is always measured with $\Gamma_S = 0$. Thus, according to (2.20), we have $M_{\text{input}} = 1 - |S_{11}|^2$. This relationship, in

# Circuit Analysis Using S-Parameters

**Table 2.2**
Conversion of Return Loss RL to Mismatch Factor M

| RL (dB) | 0.1 | 5 | 10 | 20 |
|---|---|---|---|---|
| M (dB) | −16.4 | −1.65 | −0.45 | −0.04 |

combination with $M_{input} = M_S$, allows us to directly relate the measured input return loss to the mismatch factor at the device input. Numerically, the return loss RL is related to the mismatch factor $M$ (in the linear form) by

$$\text{RL} = -10\log(1-M)$$

In practice, the mismatch factor is usually in decibels. Table 2.2 is a conversion table for a few values.

For the convenience of our discussion, in Figure 2.11 we redraw the input matching network of the circuit in Figure 2.10. Note that the S-parameters in the following analysis are the ones for the matching network rather than the device. Also, the network in Figure 2.11 is actually identical to that in Figure 2.1 with the indices changed from $S$ to 1 and $L$ to 2. The following derivation directly uses some of the results obtained in Section 2.1.

Using the configuration in Figure 2.11, we can show how the invariant property is useful in matching network synthesis. So far, we have phrased the matching problem as transformation from a 50-Ω source ($\Gamma_1 = 0$) to the desired $\Gamma_S$. The same matching network can be constructed from a transformation from $\Gamma^*_S$ to $\Gamma^*_1(=0)$. The equivalence of these two transformations can be seen by terminating port 2 of the matching network in Figure 2.11 with a conjugate matching $\Gamma_2 = \Gamma^*_S$. Since $M_2 = 1$ in this case, if the network is lossless, $M_1 = 1$ is also true, according to the invariant property. This implies that $\Gamma_2 = \Gamma^*_S$ must be transformed to $\Gamma^*_1(=0)$. This property proves convenient in practice as it allows network synthesis to be started from either side.

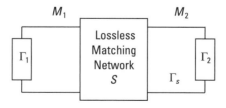

**Figure 2.11** A lossless matching network.

In the design and evaluation of multiple-section networks, the invariant property ensures that the mismatch factors at any two points within the network are the same, which provides flexibility in picking a convenient point to evaluate the matching characteristics of the network. (See, for example, the pigtail method in Chapter 10.)

As outlined here, the invariance theorem of the lossless network is an important concept in practice. The physical meaning of this theorem is obvious. Recall that the mismatch factor is defined as a ratio of the power dissipation to the available power (1.31). If the network is lossless, both powers should be the same at port 1 and port 2, and then, so are the mismatch factors $M_1$ and $M_2$. The mathematical proof is somewhat lengthy and is usually not covered in the literature. This section provides a proof as follows for interested readers.

We start with a general property of $S$ for a lossless network; it is a unitary matrix [2, 7]. For a two-port network, this property can be explicitly expressed as:

$$|S_{11}|^2 + |S_{21}|^2 = 1 \tag{2.77}$$

$$|S_{12}|^2 + |S_{22}|^2 = 1 \tag{2.78}$$

$$S_{11}^* S_{12} + S_{21}^* S_{22} = 0 \tag{2.79}$$

Equation (2.79) implies that the moduli of variables $S_{ij}$ ($i,j = 1$ or $2$) must satisfy the following condition. (We can easily see it by writing $S_{ij} = |S_{ij}| e^{\theta_{ij}}$.)

$$|S_{11}||S_{12}| = |S_{21}||S_{22}|$$

or

$$|S_{11}|^2 |S_{12}|^2 = |S_{21}|^2 |S_{22}|^2 \tag{2.80}$$

By eliminating $|S_{11}|^2$ and $|S_{22}|^2$ in (2.77), (2.78), and (2.80), we obtain

$$|S_{21}|^2 = |S_{12}|^2 \tag{2.81}$$

Similarly, if we eliminate $|S_{12}|^2$ and $|S_{21}|^2$, the result is

$$|S_{11}|^2 = |S_{22}|^2 \tag{2.82}$$

To simplify the derivation, in what follows, we assume that port 1 is terminated at 50Ω; that is $\Gamma_1 = 0$.

According to (1.32), $M_1$ and $M_2$, for the case of $\Gamma_1 = 0$, can be written as

$$M_1 = \frac{\left(1-|\Gamma_{in}|^2\right)\left(1-|\Gamma_1|^2\right)}{|1-\Gamma_{in}\Gamma_1|^2} = 1-|\Gamma_{in}|^2 = 1-\left|S_{11}+\frac{S_{21}S_{12}\Gamma_2}{1-S_{22}\Gamma_2}\right|^2 \quad (2.83)$$

$$M_2 = \frac{\left(1-|\Gamma_{out}|^2\right)\left(1-|\Gamma_2|^2\right)}{|1-\Gamma_{out}\Gamma_2|^2} = \frac{\left(1-|S_{22}|^2\right)\left(1-|\Gamma_2|^2\right)}{|1-S_{22}\Gamma_2|^2} \quad (2.84)$$

where $\Gamma_{out} = S_{22}$ is used since $\Gamma_1 = 0$. By slightly rearranging (2.83), $M_1$ becomes

$$M_1 = \frac{|1-S_{22}\Gamma_2|^2 - |S_{11}(1-S_{22}\Gamma_2)+S_{21}S_{12}\Gamma_2|^2}{|1-S_{22}\Gamma_2|^2} \quad (2.85)$$

The denominators in (2.84) and (2.85) are the same; we only need to prove that their numerators are equal. Let the numerator in (2.85) be $N$, and the expansion of it gives

$$N = 1-|S_{11}|^2 + \left(|S_{22}|^2 - |S_{11}|^2|S_{22}|^2 + S_{11}S_{22}S_{21}^*S_{12}^* + S_{21}S_{12}S_{11}^*S_{22}^* - |S_{21}|^2|S_{12}|^2\right)|\Gamma_2|^2$$
$$+ \left(|S_{11}|^2 S_{22} - S_{11}^*S_{21}S_{12} - S_{22}\right)\Gamma_2 + \left(|S_{11}|^2 S_{22}^* - S_{11}S_{21}^*S_{12}^* - S_{22}^*\right)\Gamma_2^*$$

Here the order is arranged in terms of $|\Gamma_2|^2$, $\Gamma_2$ and $\Gamma_2^*$.

Consider the $|\Gamma_L|^2$ term first. Using (2.77), the first two terms become $|S_{22}|^2|S_{21}|^2$; both the third and fourth terms are equal to $-|S_{22}|^2|S_{21}|^2$ when (2.79) and its complex conjugate are applied; then using (2.78), the $|\Gamma_2|^2$ term is $-|S_{21}|^2|\Gamma_2|^2$.

Next for the $\Gamma_2$ term, using (2.77) again, the first and third terms are reduced to $(|S_{11}|^2 - 1)S_{22} = -|S_{21}|^2 S_{22} = -S_{21}^*S_{21}S_{22} = S_{11}^*S_{21}S_{12}$. [(2.79) is used in the last step.] Thus, the first and third terms are canceled out with the second one, and the $\Gamma_2$ term is zero. By the same process it can be shown that the $\Gamma_2^*$ term is also zero. Then we have

$$N = 1-|S_{11}|^2 - |S_{21}|^2|\Gamma_2|^2 = \left(1-|S_{11}|^2\right)\left(1-|\Gamma_2|^2\right) = \left(1-|S_{22}|^2\right)\left(1-|\Gamma_2|^2\right)$$

Here (2.77) is used in the first step and (2.82) in the second. Thus, we have proved $M_1 = M_2$. Although this proof was done for a special case of $\Gamma_1 = 0$, it is only slightly more involved to prove that for a lossless passive network the mismatch factors at the input and output are equal for any values of $\Gamma_1$ and $\Gamma_2$ that are less than unity. Finally, we are ready to prove the statement made in Section 2.2.4, that is, the $\mu'(\mu)$ parameter is not affected by a lossless network at the output(input). We simply need to realize that $\mu'$ is defined by the mapping of the unit circle $|\Gamma_L| = 1$ (see Figure 2.4) and that a lossless network transforms a unit circle into another unit circle. The proof is left to the interested reader. Hint: using (2.78) and (2.79) in (2.50) and (2.51). We will further discuss this topic in Section 5.2 in the context of circuit stabilization.

## References

[1] Gonzalez, G., *Microwave Transistor Amplifiers Analysis and Design*, Second Edition, Prentice Hall, 1997.

[2] Pozar, D. M., *Microwave Engineering*, Second Edition, John Wiley and Sons, 1998.

[3] Dorf, R. C., and R. H. Bishop, *Modern Control Systems*, Seventh Edition, Addison-Wesley, 1995 (and later editions).

[4] Edwards, M. L., and J. H. Sinksy, "A New Criterion for Linear 2-Port Stability Using a Single Geometrically Derived Parameter," *IEEE Trans. on Microwave Theory Techniques*, Vol. 40, No. 12, 1992.

[5] Rollett, J. M., "Stability and Power Gain Invariants of Linear Twoports," *IRE Trans. Circuit Theory*, Vol. 9, No. 1, 1962.

[6] Ku, W. H., "Unilateral Gain and Stability Criterion of Active Two-Ports in Terms of Scattering Parameters," *Proc. IEEE (Letters)*, Vol. 54, No. 11, 1966.

[7] Collin, R. E., *Foundations for Microwave Engineering*, Second Edition, McGraw-Hill, 1992.

[8] Vendelin, G. D., A. M. Pavio, and U. L. Rohde, *Microwave Circuit Design Using Linear and Nonlinear Techniques*, John Wiley and Sons, 1990.

[9] Dong, M., and B. Urborg, "Constant-Circle Approach for Low Noise Microwave Amplifier Design," *Microwave Product Digest*, Sept. 2003.

# 3

# Matching Networks and the Smith Chart

Chapter 2 demonstrates that a linear RF amplifier design can be performed in two major steps: determination of $\Gamma_S$ and $\Gamma_L$ through an S-parameter analysis of the device based on the circuit requirements on stability, gain, and return losses (plus noise performance for LNAs, discussed in Chapter 4); and design of input and output matching networks. Chapter 2 details the first step; this chapter covers the second part of the design, focusing on the Smith chart, which is a graphic representation of impedances in the network/circuit analysis and data presentation. It was created by Phillip H. Smith in the 1930s [1] and remains a principle approach to the design of RF matching networks, even in the era of modern computers.

## 3.1 Circuit Realization of Matching Networks

This section outlines the process of the circuit realization of a matching network that is depicted in Figure 3.1(a). To this end, we have to work with circuit components. Here, we limit our discussion to the components known as the lumped elements, including resistance, inductance, and capacitance, and transmission lines. A major omission, at this point, is the transformer,

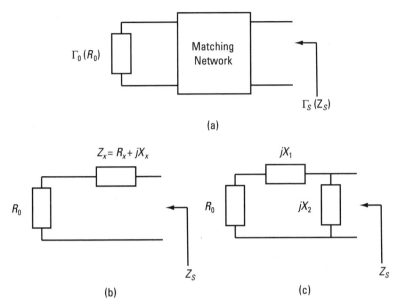

**Figure 3.1** (a) Impedance transformation through a matching network; (b) circuit realization of a matching network using an impedance component; and (c) circuit realization using two reactive components.

which is generally considered an analog circuit element despite the upper frequency limits of transformers being pushed well into the RF domain. We also assume (except in our discussion of the component $Q$ factors in Section 3.4.1) that all components are ideal; that is, no lossy and parasitic elements are included in them. Chapter 9 discusses the parasitic effects of lumped elements in more detail.

Figure 3.1(b, c) illustrates an obvious approach to devising the matching circuit. In this approach, $\Gamma_0$ and $\Gamma_S$ are first converted to their corresponding impedances using (1.13), and then the desired circuit elements are determined through solving circuit equations. However, this seemingly simple method is actually impractical. Sections 3.1.1 and 3.1.2 present two cases to demonstrate this point.

### 3.1.1  Case 1: Resistor Allowed for the Matching Circuit

In this case, the matching circuit can be a simple impedance $Z_x = R_x + jX_x$, either in series or parallel with $R_0$, depending on the value of $R_s$ (relative to $R_0$). The combined impedance should be equal to $Z_s$. For the series case shown in Figure 3.1(b), we have

$$R_0 + R_x + jX_x = R_s + jX_s$$

Thus, $R_x = R_s - R_0$ and $X_x = X_s$. Since $R_x$ has to be positive, the series case only works for $R_s > R_0$. When $R_s < R_0$, $Z_x$ should be in parallel with $R_0$. In either case, the synthesis is simple mathematically. The presence of $R_x$, however, makes the matching circuit lossy. As a result, this type of circuit has limited practical value.

### 3.1.2 Case 2: Lossless Matching Networks

Without resistive elements, at least two reactive elements, denoted as $jX_1$ and $jX_2$ in Figure 3.1(c), are generally required. The connection order can be either series-parallel [first series and then parallel, as shown in Figure 3.1(c)] or parallel-series. Both configurations yield a solution mathematically. For certain $Z_s$ values, however, only one configuration corresponds to a physically realizable circuit. Section 3.5 further explains the choice of matching-network configurations more in its discussion of the Smith chart. For the series-parallel case, a quick circuit analysis leads to

$$(R_0 + jX_1) \| jX_2 = \frac{(R_0 + jX_1)(jX_2)}{R_0 + jX_1 + jX_2} = R_s + jX_s \qquad (3.1)$$

By letting the real and imaginary parts of both sides be equal, we can obtain a quadratic equation system with two unknowns of $X_1$ and $X_2$. There are some tricks that can simplify the algebra for this circuit analysis. However, this particular problem will not be solved here. The main purpose of this exercise is to show that this approach, based on circuit theory, is generally too complex to serve as a practical technique for network design. There are a number of analytic techniques developed for the analysis and synthesis of linear networks [2–4]. However, this book does not delve into these techniques. Instead, this chapter focuses on the Smith chart, which proves to be a powerful and practical tool for the network design task considered in this section.

Sections 3.2–3.5 are concerned with basic principles related to the Smith chart and some conclusions through analytic derivations. Some of the detailed analytic derivations in Sections 3.2–3.5 are provided mainly for readers who wish to acquire an in-depth understanding of the subject; note, however, that such an understanding is not necessary for practitioners. In fact, most functions and features of the Smith chart introduced here may appear insignificant when considered individually in this modern computer age. It is the computerized Smith chart, a combination of computing capability and the graphic representation of the Smith chart, that makes the Smith chart technique especially effective. Section 3.5 explains the computerized Smith chart in the context of its applications in RF circuit designs.

## 3.2 Construction of the Smith Chart

Figure 3.2 shows a Smith chart in its original form. The concept of this chart is actually straightforward despite its complex appearance. The building blocks of the Smith chart are the so-called curvilinear coordinates. In the case of the Smith chart, there are two or three such coordinate systems plotted in the same Γ-plane. The term "curvilinear coordinates," which is more often seen in mathematics texts (e.g., [5]), implies a non-Cartesian coordinate system

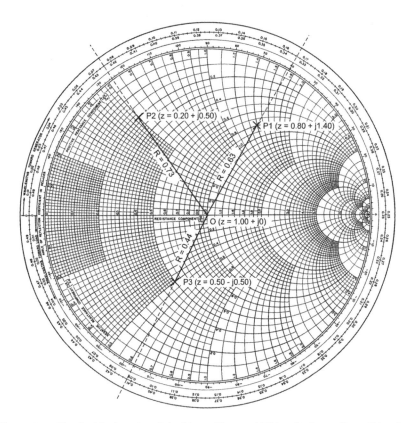

**Figure 3.2** The Smith chart in original form. (*Source:* Wikipedia: https://en.wikipedia.org/wiki/Smith_chart.)

whose variables are related to the Cartesian variables through a set of functions. We do not specifically discuss this topic; instead we illustrate the concept by constructing the Smith chart in this section.

We start with an expression of the reflection coefficient $\Gamma$ in the real and imaginary form

$$\Gamma = \Gamma_r + j\Gamma_i$$

Here $\Gamma_r$ and $\Gamma_i$ are two independent variables. In this form, any value of $\Gamma$ can be represented by a point in a $\Gamma_r$-$\Gamma_i$ Cartesian coordinate system, which is a standard technique in complex number analysis. For the discussion in this chapter, we only consider $\Gamma$ points inside the unit circle [i.e., $|\Gamma| \leq 1$ (Section 2.2)].

$\Gamma$ can also be expressed in a polar form:

$$\Gamma = \rho e^{\theta}$$

The two sets of the variables are related by

$$\rho = \sqrt{\Gamma_r^2 + \Gamma_i^2} \tag{3.2}$$

$$\theta = \tan^{-1}\frac{\Gamma_i}{\Gamma_r} \tag{3.3}$$

Equation (3.2) indicates that all points in a Cartesian coordinate system, $\Gamma_r$ and $\Gamma_i$, that correspond to a constant $\rho$ are on a circle with the center at the origin and the radius of $\rho$. Similarly, from (3.3), the curve for a constant $\theta$ is a straight line starting at the origin. Figure 3.3 shows curves for several values of $\rho$ and $\theta$. In fact, this is how a polar coordinate system is constructed: It consists of a group of circles whose radii are the values of $\rho$ and a group of straight lines whose angles from the horizontal line are the values of $\theta$. In this coordinate system, an intercept of a $\rho$ curve and a $\theta$ curve uniquely determines the value of $\Gamma$ at that point. For example, in Figure 3.3, point A corresponds to $\rho = 0.5$ and $\theta = 45°$.

The polar chart is perhaps the most common curvilinear coordinate. Now we consider a graphic chart of two overlaid coordinate systems in the $\Gamma$ plane as shown in Figure 3.4; for clarity of presentation, only a few curves (including straight lines) are plotted in this chart. If the curve spacing were much more condensed and all curves were labeled, then for any point in this plane, its value in either format, Cartesian or polar, could be estimated by reading the corresponding coordinates. In other words, this type of chart provides

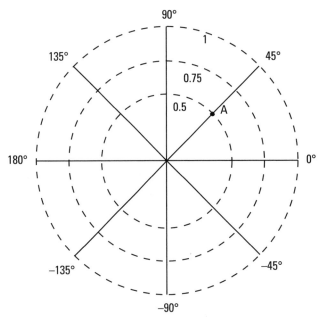

**Figure 3.3** A polar coordinate for complex numbers.

numerical conversion graphically, instead of by calculation, between the two sets of variables, $(\Gamma_r, \Gamma_i)$ and $(\rho, \theta)$. For instance, the Cartesian coordinates for the same point A, can be estimated as $\Gamma_r = 0.35$ and $\Gamma_i = 0.35$ from the plot, which leads to a numerical conversion: $(0.35, 0.35) \leftrightarrow (0.5, 45°)$.

This discussion is rather rudimentary, but it captures the essence of the Smith chart: It is an overlay of multiple coordinate systems whose variables are related by [repeat (1.13)]:

$$\Gamma = \frac{Z_L - Z_0}{Z_L + Z_0} \tag{3.4}$$

On the Smith chart, two additional coordinate systems, namely, impedance and admittance, are included in addition to the polar coordinates. For this purpose, we need to derive the equations for impedance and admittance as a function of $\Gamma$.

First, we revisit the concept of normalized impedance. Section 1.3 explains that the reflection coefficient $\Gamma$ implicitly depends on the choice of the reference impedance $Z_0$. Consequently, a variable that is a function of $\Gamma$ is generally dependent on $Z_0$ as well. To eliminate this $Z_0$ dependence, the

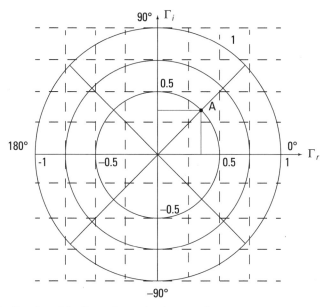

**Figure 3.4**  Overlay of the Cartesian and polar coordinates.

normalized impedance and admittance are used in the Smith chart. They are defined as

$$z' = \frac{Z}{Z_0} = r + jx \tag{3.5}$$

$$y' = \frac{1}{z'} = g + jb \tag{3.6}$$

Note that impedance and admittance are reciprocal of each other, but generally, $r \neq 1/g$ and $x \neq 1/b$. For brevity, we simply use the terms of impedance and admittance in our discussion in this section. They generally should be understood as normalized ones. The reference impedance is only required when performing a numerical conversion between the component values of the Smith chart and of an actual circuit.

We now consider the impedance coordinates, $r$ and $x$, defined in (3.5). Both $r$ and $x$ are a function of the Cartesian variables $\Gamma_r$ and $\Gamma_i$ [see (1.15)] as

$$z' = r + jx = \frac{1+\Gamma}{1-\Gamma} = \frac{1+\Gamma_r + j\Gamma_i}{1-\Gamma_r - j\Gamma_i} \tag{3.7}$$

The real and imaginary parts of both sides must be equal, thus we have

$$r = \frac{1 - \Gamma_r^2 - \Gamma_i^2}{(1 - \Gamma_r)^2 + \Gamma_i^2} \tag{3.8}$$

$$x = \frac{2\Gamma_i}{(1 - \Gamma_r)^2 + \Gamma_i^2} \tag{3.9}$$

Just as $\rho$ and $\theta$, the new set of variables, $r$ and $x$, expressed in (3.8) and (3.9), is also a curvilinear coordinate system. By letting $r$ and $x$ be a constant and through some manipulations, we can rearrange (3.8) and (3.9) as

$$\left(\Gamma_r - \frac{r}{r+1}\right)^2 + \Gamma_i^2 = \frac{1}{(r+1)^2} \tag{3.10}$$

$$(\Gamma_r - 1)^2 + \left(\Gamma_i - \frac{1}{x}\right)^2 = \left(\frac{1}{x}\right)^2 \tag{3.11}$$

Equation (3.10) is a circle equation in the $\Gamma$ plane with the center at $(r/(r + 1), 0)$ and the radius of $1/(r + 1)$. Figure 3.5 shows curves for a few specific $r$ values. Note how the constant $-r$ circle evolves as $r$ moves from 0 to ∞; the center moves from 0 to 1 on the real axis while the radius moves from 1 to 0. Similarly, (3.11) is also a circle equation for the $x$ coordinate, with the center at $(1, 1/x)$ and the radius of $|1/x|$. Since $x$ can be either positive or negative, there are two locations for the center, resulting in two groups of constant $-x$ circles as shown in Figure 3.5, with the group of $x > 0$ in the upper half and $x < 0$ in the lower half of the chart. As in the case of the polar coordinate, the $r$-$x$ coordinate of an intercept, $r = r_0$ and $x = x_0$, uniquely determines the impedance at that point; that is, $z = r_0 + jx_0$. $x = 0$ is a singular point mathematically in (3.11). It corresponds to the horizontal axis. All points of pure resistance must be on this axis. Three points on this axis are of special importance: (1) On the left end of the axis is short, where $r = 0$; (2), At the center, $r = 1$, which is where the resistance is the same as the reference impedance; (3), On the right end is an open, where $r \to \infty$. This is also the point where all constant $-r$ circles collapse when $x \to \infty$, which is expected from the circuit point of view.

Similar to the impedance coordinates, a group of circles of constant values of $g$ and $b$ [see (3.6)] forms the admittance coordinate system. By taking the inverse of (3.7), we have

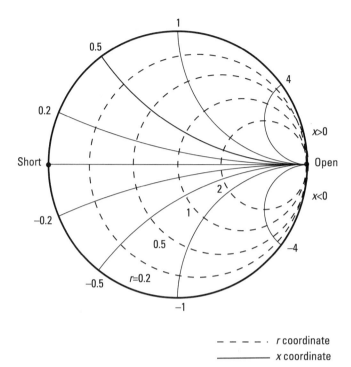

**Figure 3.5** The *r* and *x* coordinates for impedance.

$$y' = \frac{1-\Gamma}{1+\Gamma} \tag{3.12}$$

Comparing with (3.7), it is evident from (3.12) that $y'$ can be obtained from $z'$ by substituting $\Gamma$ with $-\Gamma$, which leads to

$$g(\Gamma) = r(-\Gamma); \text{ and } b(\Gamma) = x(-\Gamma) \tag{3.13}$$

The relationships in (3.13) imply that the coordinates of $g$ and $b$ at a point $\Gamma$ are the same as those of $r$ and $x$ at the point $-\Gamma$. Since $\Gamma \to -\Gamma$ is an operation of reflection in the origin or rotation of 180°, such an operation on the impedance coordinates creates the admittance coordinates as shown in Figure 3.6. As an example, the coordinate values of a reflection pair, points A and A', are also shown in Figure 3.6.

Generally, a Smith chart is an overlay of three coordinate systems: a polar plot of $\Gamma$ in the $\rho$-$\theta$ coordinates, an impedance plot of the $r$-$x$ coordinates, and an admittance plot of the $g$-$b$ coordinates. In the real world of engineering practice, however, the term "Smith chart" is used quite loosely. It is not

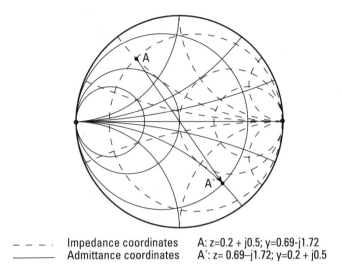

```
- - - -   Impedance coordinates    A: z=0.2 + j0.5; y=0.69-j1.72
_____   Admittance coordinates   A′: z= 0.69–j1.72; y=0.2 + j0.5
```

**Figure 3.6** Construction of admittance coordinates from impedance coordinates.

uncommon that a polar chart alone or a polar chart plus an impedance chart are referred to as a Smith chart. To fully utilize the Smith chart as a design tool, however, all three coordinate systems must be included in the chart. A traditional Smith chart such as the one shown in Figure 3.2 has especially fine coordinates that allow users to conveniently convert one parameter to another with reasonable accuracy. This fine coordinate scheme becomes unnecessary with the computerized Smith chart, since the coordinates can be easily calculated by the computer. (See Section 3.5 for details.)

## 3.3 The Smith Chart Representation of Networks

In reference to Figure 3.1, we consider a specific network topology in this section, the so-called ladder network [6, 7], for the impedance transformation (from $Z_0$ to $Z_s$) and its graphic representation on the Smith chart. As shown in Figure 3.7, a distinguishing feature of the ladder network is that any series branch, $z_i$, consists of only series components, while any parallel branch, $y_i$, consists of only parallel components. In addition, Figure 3.7 illustrates an important concept to be used in our discussion: the network impedance (or admittance) at a section denoted as $z_{Ni}$ or $y_{Ni}$. An "$N$" in the notations of the network parameters is used to emphasize the difference between the impedances for networks and components. For example, in Figure 3.7, $y_{N1}$ is the network admittance transformed from $y_0$ through a parallel admittance $y_1$, and the network impedance $z_{N2}$ is from $y_{N1}$ through a series impedance $z_2$. In addition, the term "section" is used here to avoid confusion with the term of node in Kirchhoff's current law (KCL). Consider $z_{N2}$ as an example. It is the network impedance at the second section of the network, while $y_{N3}$ is the admittance at the third section. In this case, they would be at the same node when applying KCL. Note that the choice of impedance or admittance as the parameter for a network point is purely for the convenience of discussion and has no physical significance. With this concept established, the impedance transformation problem illustrated in Figure 3.7 can be considered a series of transformations using a single component (either $z_i$ or $y_i$), each of which can be represented by an operation on the Smith chart, as described below.

Consider the simple circuit of two series impedances $z_A$ and $z_B$ in Figure 3.8(a). The combined impedance is simply $z_C = z_A + z_B$. The same series combination can be interpreted as an impedance transformation from $z_A$ to $z_C$ through a matching network consisting of a single impedance $z_B$, as depicted in Figure 3.8(b). Also highlighted in Figure 3.8(b) is that the series connection of $z_B$ is treated as a two-step operation of adding the real part (resistance) and the imaginary part (reactance) in sequence. Then, each step can be represented by a trace on a constant curve in an impedance coordinate system. The process is illustrated step-by-step in Figure 3.9.

The initial point is marked "$A$" in Figure 3.9, whose impedance coordinates are $(r_A, x_A)$. The addition of $z_B (= r_B + jx_B)$ splits into two steps: first $r_B$, then $jx_B$. In the first step, the reactance remains constant; the network moves from point $A$ along the constant $x_A$ curve in the direction that $r$ increases until reaching $r_C = r_A + r_B$. This is an intermediate point labeled as $M$ in Figure 3.9. In the second step, $r_C$ remains constant, and the network moves from point $M$ on the constant $r_C$ curve by amount of $x_B$. The direction of this move depends

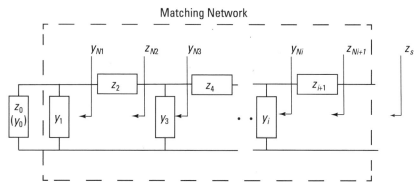

**Figure 3.7** Ladder network and the concept of network impedance.

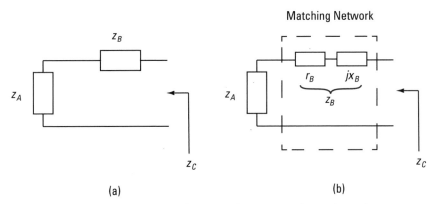

**Figure 3.8** Impedance combination interpreted as a network transformation.

on the sign of $x_B$. Assuming $x_B > 0$, then the end point, $x_C = x_A + x_B$, is on the increasing side of point $M$. The order of connection of $r_B$ and $jx_B$ can be altered as verified in the path $A \to M' \to C$ in Figure 3.9. The same technique can be applied to a matching network using a parallel component. In that case, the *g-b* coordinates of admittance should be used. Then an impedance transformation through the general matching network shown in Figure 3.7 can be represented on the Smith chart by an alternating sequence of motions along the constant impedance and admittance coordinates. Such a graphic representation is referred to as a Smith chart operation. Readers should take

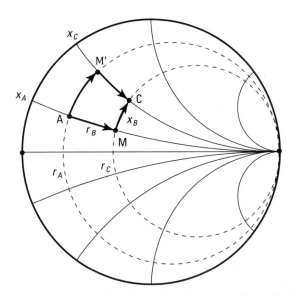

**Figure 3.9** The Smith chart operation for the impedance transformation in Figure 3.8.

care not to be confused by the difference in terminology between impedance transformation and transformation along the impedance coordinates. The former applies to any network transformation regardless of the parameter used for the network, while the latter implies a specific parameter [i.e., impedance (as opposed to admittance)].

In addition to impedance (in series branches) and admittance (in parallel branches), transmission lines can be used as circuit elements in matching networks. In fact, the Smith chart was initially created in an attempt to graphically represent the impedance transformation through a transmission line [1]. Recall that in terms of impedance, the transformation in question is governed by a rather complex function in (1.20), whereas in terms of reflection coefficient, the same transformation is expressed in a much simpler form in (1.18). According to (1.18), the transformation can be simply implemented with a clockwise rotation by an angle of $2\beta l$ on the Smith chart. Then instead of using (1.20), the impedance transformation problem, $Z_L \rightarrow Z'_L$, can be solved graphically on the Smith chart by a sequence of three simple steps: (1) locating $Z_L$ on the chart; (2) rotating by an angle of $2\beta l$; and (3) reading the impedance coordinates off the chart for $Z'_L$. This case demonstrates the significance of the invention of the Smith chart in the days before computers.

Figure 3.10 shows the Smith chart operations for the three special cases of the impedance transformation using the series quarter-wavelength transmission

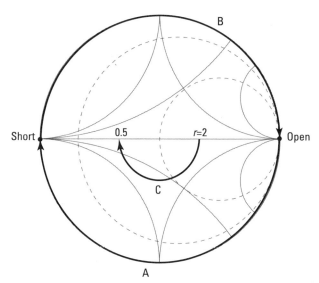

**Figure 3.10** Quarter wavelength transformations. A: open → short; B: short → open; C: $r=2 \rightarrow r=0.5$.

line discussed in Section 1.3 [after (1.20)]. The conclusions obtained analytically there become obvious on the Smith chart as seen in Figure 3.10. Section 3.5.1 considers more general cases.

Transmission lines can also be used in a parallel configuration in a matching network. Also, the transmission line impedance does not have to be the same as the reference impedance of the Smith chart. Section 3.5 covers these cases, along with a discussion of the lumped elements in applications of the Smith chart. First, Section 3.4 introduces another set of curves—the constant $Q$ lines—on the Smith chart.

## 3.4 Bandwidth of Matching Networks and $Q$ Lines on the Smith Chart

A lossless matching network can only consist of nonresistive components such as inductors, capacitors, and transmission lines. The characteristics of these components as a circuit element are obviously dependent on frequency. So far, in our discussions on matching networks, we have implicitly assumed a condition of single frequency. However, the bandwidth of frequency response is often part of the specification requirements for a matching network design. Analytic techniques in this regard are covered in the theory of broadband matching networks. For a comprehensive overview, see [6]. Generally, these techniques are too complex for practical RF engineers, with perhaps the exception of filter design specialists. This book does not treat this subject with any analytical rigor. Instead, Section 3.4.1 introduces the concept of the network $Q$ factor, which can be utilized in matching network designs as qualitative guidance for the bandwidth characteristics. More relevant, Section 3.5.3.3 shows that by incorporating the $Q$ factor in the Smith chart, the bandwidth consideration can be an integrated part of the matching network design using the chart.

### 3.4.1 Review of $Q$ Factors

The $Q$ factor (also known as the quality factor $Q$) is commonly used, as a specification, in three types of RF circuit applications: resonant circuits, $L$ and $C$ components, and matching networks. The definition of the $Q$ factor appears the same in each of these three cases, but, at the same time, in each case, the $Q$ factor has a somewhat different physical implication for practical purposes. Now, we briefly review the first two cases, and then provide a more detailed discussion on the $Q$ factor for matching networks.

### 3.4.1.1 Q Factor of Resonant Circuits

For most readers, the $Q$ factor was likely first introduced as a measure of relative bandwidth of a resistance-inductance-capacitance (RLC) resonant circuit, either in series or parallel configurations [7]. There are a variety of RF resonators in real-world applications. We are only concerned here with the *RLC* resonant circuits as they are most frequently used in practice. (Note, however, that the *RLC* resonator is also often used as an equivalent circuit in the analysis of other types of resonators). The $Q$ factor has two forms defined as follows:

- For series resonance:

$$Q_s = \frac{X}{R} \qquad (3.14)$$

- For parallel resonance:

$$Q_p = \frac{B}{G} \qquad (3.15)$$

where $R$ and $G$ are the resistance and conductance, and $X$ and $B$ are the reactance and susceptance in the respective resonators. It is important to note that the reactance $X$ and susceptance $B$ in (3.14) and (3.15) are the values of either inductance or capacitance since they are equal (but with opposite signs in the circuits) at the resonant frequency; that is, $X = \omega_0 L = 1/\omega_0 C$ and $B = \omega_0 C = 1/\omega_0 L$, where $\omega_0$ is the angular resonant frequency. When the frequency $\omega$ moves away from $\omega_0$, the response (the current for series and the voltage for parallel configurations) starts to roll off from the peak value. The $Q$ factor is a measure of the rate of this roll-off with frequency. A simple circuit analysis leads to the well-known formula

$$Q = \frac{\omega_0}{\Delta \omega} \qquad (3.16)$$

where $\Delta\omega$ is the 3dB bandwidth.

As indicated in (3.14) and (3.15), the essence of the $Q$ factor is the ratio of the imaginary part to the real part of an impedance (or admittance). This notion of the $Q$ factor can be applied to other cases that are not directly related to any resonant circuit. We consider two of them in Section 3.4.1.2.

### 3.4.1.2  $Q$ Factors of Inductors and Capacitors

Manufacturers of inductors and capacitors that are intended for RF applications usually provide $Q$ versus frequency data for their products. The $Q$ factor in this case is no longer associated with a resonant circuit; instead, it is mainly a measure of power loss due to the parasitic resistive element inside the components, known as equivalent series resistance (ESR). Chapter 9 discusses this topic in more detail. The $Q$ factors for the components are listed as follows:

- Inductor:

$$Q_L = \frac{X_L}{ESR} = \frac{\omega L}{ESR} \qquad (3.17)$$

- Capacitor:

$$Q_C = \frac{|X_C|}{ESR} = \frac{1}{\omega C \cdot ESR} \qquad (3.18)$$

It is customary in practice to define $Q$ as a positive parameter, which is the reason for the absolute value of $X_C$ in (3.18). A negative $Q$, which is also seen in the literature, simply means that the component is capacitive.

The component $Q$ factor can still be related to the $Q$ factor of a resonant circuit in the sense that it sets an upper limit for the $Q$ if the component in question is to be used in the resonator.

### 3.4.1.3  $Q$ Factor of Matching Networks

The network $Q$ factor is built upon the concept of network impedance (or admittance) introduced in Section 3.3. With reference to Figure 3.7, the network $Q$ factor at the $i$th section is associated with the network impedance, $Z_{Ni} = R_{Ni} + jX_{Ni}$ or network admittance, $Y_{Ni} = G_{Ni} + jB_{Ni}$ by

$$Q_{Zi} = \frac{|X_{Ni}|}{R_{Ni}} = \frac{|x_{Ni}|}{r_{Ni}} \qquad (3.19)$$

$$Q_{Yi} = \frac{|B_{Ni}|}{G_{Ni}} = \frac{|b_{Ni}|}{g_{Ni}} \qquad (3.20)$$

The $Q$ factor is a dimensionless ratio and can be in either regular component values or normalized ones, as expressed in (3.19) and (3.20). The regular forms are used in our discussion except in the Smith chart–related analysis. Furthermore, since $Z_{Ni} = 1/Y_{Ni}$, it is straightforward to show $Q_{Zi} = Q_{Yi}$; that is, the network $Q$ is a unique parameter at a given section of the network, whether expressed in the impedance or admittance elements.

### 3.4.2 Bandwidth Expressed in $Q$

With the network $Q$ factor defined, we now consider how it is related to the bandwidth characteristic of the matching network. Since the relationship of the $Q$ factor and the bandwidth of a resonator is well-known [see (3.16)], we apply the concept of the network $Q$ factor to a parallel resonant circuit shown in Figure 3.11(a). The purpose is to examine the similarities and the differences between the network and the resonance circuit in terms of their respective $Q$ factors. Following the sequence of network construction sketched in Figure 3.7, we can reconstruct the parallel resonant circuit in two steps: connecting $B_1$ to $G_0$ first, and then adding $B_2$. After the first step, according to (3.20) the network $Q$ factor is

$$Q_1 = \frac{B_1}{G_0} \qquad (3.21)$$

In the second step, $B_2$ is chosen to be $-B_1$ (only at the resonant frequency), resulting in a pure resistive element $G_0$, which is the condition for resonance. Clearly, the network $Q$ factor defined in (3.20) is the same as the resonance $Q$ factor defined in (3.15). The difference between the two circuits is in the connection of the second component $X_2$: For the matching network, it is connected to $G_1 \| B_1$ in series [see Figure 3.11(b)], as opposed to parallel as in the case of resonator. This section shows that the network $Q$ factor is indeed a measure of the bandwidth of a matching network, but in a much more complex manner than the function in (3.16).

We first examine the impedance transformation in terms of the network $Q$ factors for a multiple-section matching network at a single operation frequency. Figure 3.7 is redrawn in Figure 3.11(c) with two simplifications for our analysis. First, the components used in the network are either reactance (for series) denoted as $X_i$ or susceptance (for parallel) as $B_j$. Second, the initial admittance $Y_0$ is assumed to be pure conductive (i.e., $Y_0' = G_0$), and at each subsequent step of transformation the end admittance is also pure conductive, denoted as $G_i$. It is obvious enough that $G_i$ is a network parameter that

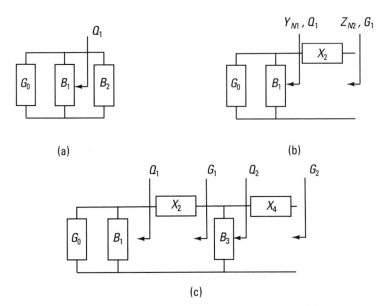

**Figure 3.11** $Q$ factors in three circuits: (a) parallel resonant circuit; (b) two-component matching network for $G_0 \rightarrow G_1$ transformation; and (c) four-component matching network for $G_0 \rightarrow G_2$ transformation.

the letter "$N$" is dropped in the index of $G_i$. In our notation used in Figure 3.11, the index for the component (or network section, $Z_{Ni}$) is twice that for the impedance transformation ($Q_j$ and $G_j$) because it takes two components to perform one transformation.

We start with the first step of transformation from $G_0$ to $G_1$ through the network of $B_1$ and $X_2$, as shown in Figure 3.11(b). Inspection of Figure 3.11(b) shows that the impedance after the connection of $X_2$ is given by

$$Z_{N2} = \frac{1}{G_0 + jB_1} + jX_2 \quad (3.22)$$
$$= \frac{G_0}{G_0^2 + B_1^2} + j\left(X_2 - \frac{B_1}{G_0^2 + B_1^2}\right)$$

To make $Z_{N2}$ pure resistive, $X_2$ must be chosen to cancel the imaginary part of $Z_{N2}$. Thus, we have

$$X_2 = \frac{B_1}{G_0^2 + B_1^2} \quad (3.23)$$

and

$$G_1 = \text{Re}\left(\frac{1}{Z_{N2}}\right) = \frac{G_0^2 + B_1^2}{G_0} = G_0\left(1 + Q_1^2\right) \quad (3.24)$$

where (3.21) is used. Equation (3.24) states that the original conductance $G_0$ is increased by a factor of $(1 + Q_1^2)$ through the matching network of $B_1$ and $X_2$. Note that the second component $X_2$ only affects the imaginary part (making it zero in this case). The real part of the final impedance is determined solely by $Q_1$(through $B_1$) and $G_0$. Also worth noting is that this network configuration can only increase the conductance. To transform in the opposite direction, the first element has to be in series. We can further transform the network impedance by connecting a second pair of $B_3$ and $X_4$ as shown in Figure 3.11(c). The network $Q$ factor after $B_3$ is

$$Q_2 = \frac{B_3}{G_1} = \frac{B_3}{G_0\left(1 + Q_1^2\right)} \quad (3.25)$$

Noting that $X_4$ is chosen only for cancelation of the imaginary part, the resulting conductance after $X_4$ is

$$G_2 = G_1\left(1 + Q_2^2\right) = G_0\left(1 + Q_1^2\right)\left(1 + Q_2^2\right) \quad (3.26)$$

For a network consisting of $2n$ sections the final conductance is simply

$$G_n = G_0 \prod_1^n \left(1 + Q_i^2\right) \quad (3.27)$$

Equation (3.27) suggests that the sections that have the highest $Q$ factors dominate the transformation ratio.

We now consider the frequency response of matching networks. Although many practical matching networks consist of multiple sections, the simple case in Figure 3.11(b) is illustrative enough to explain the basic concept. We use it in our analysis for the bandwidth of a network's frequency response. A thorough analytic treatment of the general case in Figure 3.11(c) is outside of the scope this book.

In the context of impedance matching, the network function in frequency response analysis is usually the input reflection coefficient (for RF amplifiers) or insertion loss (filter applications). For a lossless network, these two quantities are virtually the same because the insertion loss is completely due to the mismatch loss. The following analysis uses the reflection coefficient.

The network made of $B_1$ and $X_2$ in Figure 3.11(b) transforms a conductance $G_0$ to $G_1$ at a chosen frequency denoted as $\omega_0$. If a conductance $G_1$ were connected to this network from the $X_2$ side, we would have a two-port network that is terminated at both sides, which is the case considered in detail in Section 2.4. (If the transformation is to a complex impedance, the network should be terminated with its conjugate impedance.) According to the invariant property of lossless networks, the reflection coefficients at the two terminations must be equal. Since the $G_0$ side was taken as the circuit input where the reflection coefficient is measured, our analysis will be on this side. Evidently, the system is perfectly matched at $\omega_0$. Our interest is to derive a formula for the reflection coefficient when the frequency deviates from $\omega_0$.

Up to now, in our use of formulas, the frequency is implicitly included in the reactance and susceptance terms. To analyze the frequency response, the components need to be explicitly expressed as $L$ or $C$. For a transformation $G_0 \rightarrow G_1$ with a simple $L$-$C$ network, there are four possible configurations: $(C_p, L_s)$, $(L_p, C_s)$, $(C_s, L_p)$, and $(L_s, C_p)$, where the indices $p$ and $s$ stand for parallel and series. It is straightforward to verify that the first two configurations (the first element in parallel) is suitable for $G_1 > G_0$, and the other two for $G_1 < G_0$. Section 3.5.1 shows that the selection of matching network configurations is obvious on the Smith chart.

Figure 3.12(a) shows the specific network we are to analyze, where the configuration $(C_p, L_s)$ is used. The notations for the network terminations are changed from conductance ($G_0$ and $G_1$) to resistance ($R_0$ and $R_1$) in Figure 3.12 as the latter is more commonly seen in the literature. $R_0$ is chosen to be the reference impedance for $\Gamma$ in this analysis; thus, $\Gamma(\omega)|_{\omega=\omega_0}=0$.

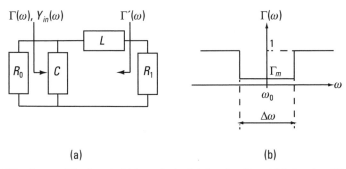

**Figure 3.12** Network for bandwidth analysis: (a) the circuit and (b) the simplified function of $\Gamma(\omega)$ for Bode-Fano criterion analysis.

We now derive the formula for $\Gamma(\omega)$.

In the conductance form, $\Gamma$ is given by

$$\Gamma(\omega) = \frac{G_0 - Y_{in}(\omega)}{G_0 + Y_{in}(\omega)} \tag{3.28}$$

$Y_{in}(\omega)$ can be derived from the circuit shown in Figure 3.12(a):

$$Y_{in}(\omega) = G_{in}(\omega) + jB_{in}(\omega)$$
$$= \frac{R_1}{R_1^2 + (\omega L)^2} + j\left(\omega C - \frac{\omega L}{R_1^2 + (\omega L)^2}\right) \tag{3.29}$$

The values of $R_1$, $C$, and $L$ in (3.29) are constrained by our assumption, in other words,

$$Y_{in}(\omega_0) = G_0 = \frac{1}{R_0} \tag{3.30}$$

Substituting (3.30) into (3.29) and equating the real and imaginary parts of both sides, we obtain two equations as

$$\frac{R_1}{R_1^2 + (\omega_0 L)^2} = \frac{1}{R_0} \tag{3.31}$$

$$\frac{L}{R_1^2 + (\omega_0 L)^2} = C \tag{3.32}$$

Then it is straightforward to show from the last two equations that

$$L = CR_0 R_1 \tag{3.33}$$

From (3.21), the $Q$ factor in this case is

$$Q = \omega_0 CR_0 \tag{3.34}$$

and (3.24) becomes

$$\frac{R_0}{R_1} = 1 + Q^2 \tag{3.35}$$

The index for $Q$ is dropped since it is the only $Q$ factor to be considered.

We are only interested in the change of $\Gamma(\omega)$ near $\omega_0$. For that purpose, we take a derivative of (3.28) at $\omega = \omega_0$:

$$d\Gamma(\omega)\Big|_{\omega=\omega_0} = -\frac{Y'_{in}(\omega_0)}{2G_0}d\omega \qquad (3.36)$$

where $Y'_{in}(\omega_0)$ is the derivative of $Y_{in}(\omega)$ at $\omega_0$. Its real and imaginary parts are evaluated from (3.29) separately:

$$G'_{in}(\omega_0) = -\frac{2R_1L^2\omega_0}{\left[R_1^2 + (\omega_0 L)^2\right]^2} \qquad (3.37)$$

$$= -2R_1C^2\omega_0$$

since (3.31) and (3.33). And

$$B'_{in}(\omega_0) = C - \frac{L}{R_1^2 + (\omega_0 L)^2} + \frac{2\omega_0^2 L^3}{\left[R_1^2 + (\omega_0 L)^2\right]^2} \qquad (3.38)$$

$$= 2R_0 R_1 \omega_0^2 C^3$$

We know from (3.32) that the first two terms in the last equation are canceled. Then, using (3.31) and (3.33) again, we obtain (3.38). Using (3.34) and (3.35) to eliminate $R_1$ and $C$ in (3.37) and (3.38) and then substituting them into (3.36), we reach the final result:

$$d\Gamma(\omega)\Big|_{\omega=\omega_0} = -\frac{G'_{in}(\omega_0) + jB'_{in}(\omega_0)}{2G_0}d\omega$$

$$= \frac{-Q^2 - jQ^3}{1+Q^2}\frac{d\omega}{\omega_0} \qquad (3.39)$$

Equation (3.39) indicates that the bandwidth of the frequency response is completely determined by the network $Q$ factor in this case. Two additional conclusions can be made: First, for a given relative bandwidth, $d\omega/\omega_0$, $d\Gamma$ can be infinitely small as $Q \to 0$, but the transformation function also diminishes as the transfer ratio $R_1/R_0 \to 1$. This implies that broadband matching can be achieved by breaking the matching network into small steps. Second, if $Q$ is significantly greater than 1, the bandwidth (for a fixed $|d\Gamma|$) is roughly proportional to 1/Q, just like in the case of resonant circuits. In addition, we state without mathematical proof that for a multiple-section matching network, $d\Gamma$ is mainly determined by the dominant $Q$ in the network [see (3.27)].

In the literature, the so-called Bode-Fano criterion [8] is often cited in the discussion of bandwidth limit [9, 10]. The Bode-Fano criterion deals with a general matching network like that shown in Figure 3.11(c) and considers the reflection coefficient at the output side [$\Gamma'(\omega)$ in Figure 3.12(a)]. This difference is inconsequential to our analysis since the reflection coefficients on two sides are the same if the network is lossless (invariant property). The Bode-Fano criterion states that the following inequality must hold for the reflection coefficient $\Gamma(\omega)$:

$$\int_0^\infty \ln\frac{1}{|\Gamma|} d\omega \leq \frac{\pi}{RC} \qquad (3.40)$$

where $R$ and $C$ are the first two parallel components in Figure 3.11(c) ($G_0$ and $B_1$). This inequality indicates that $\Gamma$ can be zero only at one frequency point. To see how (3.40) is related to the bandwidth of the network, we make a simplification on $\Gamma(\omega)$ as shown in Figure 3.12(b). Then (3.40) becomes

$$\Delta\omega \ln\frac{1}{|\Gamma_m|} \leq \frac{\pi}{RC} \qquad (3.41)$$

Although the functional forms are different, (3.39) and (3.41) convey the same message; that is, the bandwidth within which the reflection coefficient is smaller than a certain value is limited by the product of the first two components in the network, $R$ and $C$ [using $Q = \omega_0 CR$ to write (3.39) in $R$ and $C$].

### 3.4.3  Q Lines on the Smith Chart

We now turn our discussion to the topic of $Q$ lines on the Smith chart. It will be shown in Section 3.5.3.3 that incorporation of the $Q$ factor in the Smith chart allows the bandwidth consideration to be a part of matching network designs using the Smith chart.

The network $Q$ is defined in impedance in (3.19) [or admittance in (3.20)], which in turn corresponds to a point on the Smith chart. Thus, each point on the Smith chart has a unique $Q$ value. We wish to examine the constant $Q$ contours on the Smith chart. Consider a point at an impedance coordinate, $z = r + jx$, using the normalized form of (3.19) and the functions of $r(\Gamma_r, \Gamma_i)$ and $x(\Gamma_r, \Gamma_i)$ in (3.8) and (3.9), we have

$$Q = \frac{|x|}{r} = \frac{2|\Gamma_i|}{1-\Gamma_r^2-\Gamma_i^2} \qquad (3.42)$$

By the same technique used in Section 3.2 for constant circles, we obtain the equation for the constant $Q$ contours in the $\Gamma$ plane,

$$\Gamma_r^2 + \left(\Gamma_i \pm \frac{1}{Q}\right)^2 = \left(1+\frac{1}{Q^2}\right) \qquad (3.43)$$

where the $\pm$ sign implies it should be the same as that of $\Gamma_i$. (3.43) is a circle with the center at $(0, \pm 1/Q)$ and the radius of $\sqrt{1+1/Q^2}$. As is the case for (3.11), $Q = 0$ corresponds to the real axis, and the two groups of constant curves in the upper and lower halves are symmetric with respect to the real axis. Also, it can be seen that the two end points on the real axis ($\pm 1$, 0) are solutions to (3.43) regardless of $Q$ value. Therefore, all constant $Q$ curves go through these points. Figure 3.13 shows several constant $Q$ curves with the values at the lower half labeled.

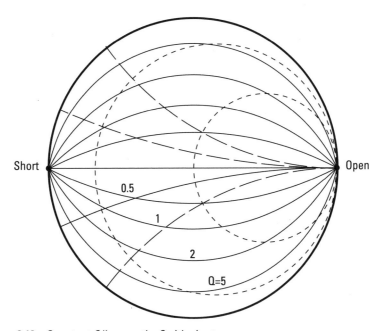

**Figure 3.13** Constant $Q$ lines on the Smith chart.

## 3.5 Matching Network Design Using the Smith Chart

We have finally arrived at the point where we can present how the Smith chart is utilized in a practical matching-network design. This section first examines the Smith chart operation using lumped elements and transmission lines. Our discussion assumes that readers have access to a computerized Smith chart. Section 3.5.2 presents a summary of some key functions and features that are desirable for a computerized Smith chart. Section 3.5.3 considers several important factors in the matching network design and shows that these design factors can be evaluated simultaneously on the Smith chart so that a balanced design can be achieved. In fact, this capability of the Smith chart is at the core of its effectiveness as a practical design tool.

### 3.5.1 Lumped Elements and Transmission Lines on the Smith Chart

Section 3.3 considers the Smith chart operations of reactance and susceptance without specifying whether they are capacitive or inductive. It also discusses the Smith chart representation of a transformation using a serial transmission line for a special case where the transmission line impedance is the same as the reference impedance. In actual circuit designs, however, we need to directly work with specific circuit elements as well as transmission lines with different impedances and configurations.

#### 3.5.1.1 Lumped Elements on the Smith Chart

If we start at the center of the Smith chart and add one of the three lumped elements, resistor $R$, inductor $L$, and capacitor $C$, there are six possibilities for the Smith chart operation as shown in Figure 3.14. The following are several general rules regarding the Smith chart operation of a lumped element:

- For a series connection, the transformation follows impedance coordinates; and for a parallel connection, it follows admittance coordinates.
- The reactance of an inductor $x_L = \omega L$ is positive, and its susceptance $b_L = -1/(\omega L)$ is negative. Section 3.2 showed that the upper half of the Smith chart has positive reactance and negative susceptance. Therefore, the upper half is the inductive region. By the same token, the lower half is the capacitive region.
- Regardless of the initial location, a series component always moves toward "open" (when $R_s$ and $L_s \rightarrow \infty$, and $C_s \rightarrow 0$). Similarly a parallel

component always moves toward "short" (when $R_p$ and $L_p \to 0$, and $C_p \to \infty$).

Section 3.4 concludes analytically that for a transformation $R_0 \to R_1$ using a two-component network, the configuration has to be series-parallel if $R_1 > R_0$. In fact, it is easy to see on the Smith chart that this conclusion is true for any point with the real part of its impedance bigger than $R_0$. Consider, for instance, point $A$ in Figure 3.14, which is inside the $r = 1$ circle (meaning $r_A > 1$); it is immediately clear from Figure 3.14 that the first component should not be in parallel, whether $C_p$ or $L_p$, because, if that were the case, the second series component could not bring the transformation to point $A$. (Their paths are outside the $r = 1$ circle and toward "open.") Of course, if more sections are allowed in the matching network, this restriction no longer applies. Generally,

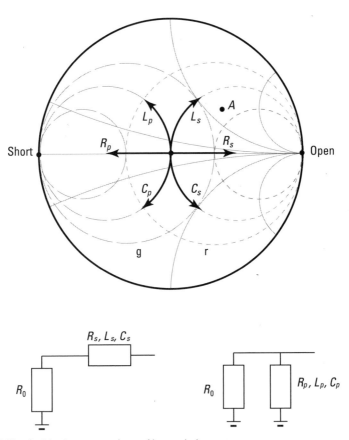

**Figure 3.14** Smith chart operations of lumped elements.

the Smith chart allows the user to easily visualize the possible transformation paths for a given pair of points on the chart.

### 3.5.1.2 Transmission Lines on the Smith Chart

Unlike lumped elements, a transmission line has two parameters, the characteristic impedance and electrical length. Section 3.3 shows that if the transmission line impedance is the same as the reference impedance, the transformation is a simple rotation with the rotation center at the center of the Smith chart. In such as a condition, when the impedance to be transformed is at the center, the transmission line has no effect. Physically, this observation implies that the length of a transmission line has no effect if the impedances are matched. This explains, from another angle, the importance of impedance standardization in industries.

When the two impedances are different, the Smith chart operation is no longer obvious. It is still possible in this case to manually perform the transformation on the Smith chart by taking extra conversions of reference impedances. With the computerized Smith chart widely available, however, the manual manipulation technique becomes unnecessary. We simply point out that the governing equation for this transformation is (1.20), where $Z_L$ and $Z'_L$ are the initial and final impedances of the transformation. As a matter of fact, the transmission line impedance is an effective parameter in manipulating the transformation path as illustrated in Figure 3.15(a), where the transformations using transmission lines with an electrical length of 45° and two impedances, 20Ω and 80Ω, are shown for three initial locations. Clearly, the transmission line impedance has a significant effect on the direction and length of the transformation path on the Smith chart.

In addition, a section of transmission can be used in parallel connection with the far end terminated by either a short circuit or an open one, which are denoted as $TL_{p\_short}$ and $TL_{p\_open}$, respectively in Figure 3.15(b). In either case the transmission line may be treated as an equivalent impedance transformed from an impedance $Z_L = 0$ (for short) or $Z_L = \infty$ (for open) as illustrated in Figure 3.15(b). Then from (1.20), we have the following:

- For $Z_L = 0$:

$$Z'_{L-S} = jZ_c \tan\beta\ell \qquad (3.44)$$

- For $Z_L = \infty$:

$$Z'_{L-O} = \frac{Z_c}{j\tan\beta\ell} \qquad (3.45)$$

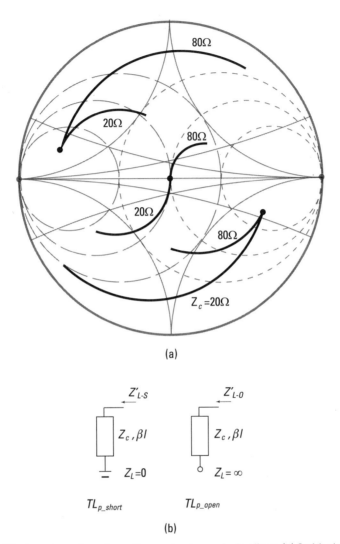

**Figure 3.15** Impedance transformations using transmission lines: (a) Smith chart operations using series transmission lines and (b) parallel transmission lines with short and open terminations.

Thus, the Smith chart operation of $TL_{p\_short}$ follows the same path (on a constant $g$ curve) as that for a parallel inductance while $TL_{p\_open}$ follows a parallel capacitance. The amount of change in susceptance is determined by (3.44) or (3.45). Again, the numerical computation in these cases is automatic in a computerized Smith chart.

## 3.5.2 Computerized Smith Chart

A computerized Smith chart can significantly ease the process of matching network design using the Smith chart. Various forms of computerized Smith charts are available from commercial publishers or as free online applications. Moreover, with the proliferation of programming languages, the development of a computer program that integrates the equation-based computation with a graphic user interface such as the computerized Smith chart is well within the scope of a do-it-yourself project. Figure 3.16 illustrates the user interface concept for a computerized Smith chart, and the following list summarizes some of the key requirements for a computerized Smith chart as a reference for the evaluation of available programs—and as a brief reference guide for readers who are interested in writing a code for the Smith chart application.

1. User input for the parameters for the chart, including the reference impedance and operation frequency: Both are required to convert the value of an actual circuit component to a point on the Smith chart.

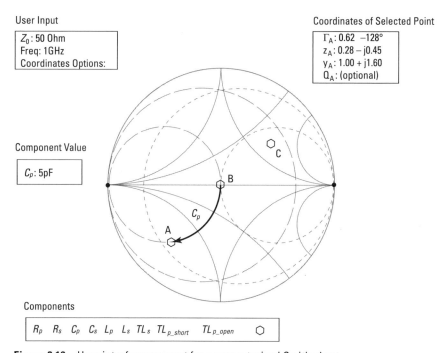

**Figure 3.16** User interface concept for a computerized Smith chart.

2. Display of coordinates and $Q$ lines: Since the parameters of a point on the Smith chart can be easily read out in a computerized chart, this display function is less critical for the computerized Smith chart than for the printed ones. Nevertheless, many users still find that the coordinates on the Smith chart provide a visual aid in selecting an appropriate component for the desired matching path. The program should provide options for "on" and "off" and the spacing of the coordinates for the display. In Figure 3.16, the display-control functions, "Coordinates Options," are grouped in the same window for "User Input."
3. Creation of as many points as needed on the Smith chart: This is accomplished in Figure 3.16 by clicking on the ○ symbol in the group of "Components." Once a point is created, the user can select and move it with a cursor in an interactive manner. For example, if point $A$ is selected, the program shows all three coordinates for the point as well as the $Q$ factor (which is optional). The program can also move the point to the target location based on the coordinates input by the user.
4. A collection of at least nine matching elements, including

$$R_p, R_s, C_p, C_s, L_p, L_s, TL_s, TL_{p\_short}, TL_{p\_open}.$$

Once a component is selected, an interactive window displays its value(s). For transmission lines there are two parameters.

5. The Smith chart operation: In this process, the user first selects a point (e.g., point $B$ in Figure 3.16), and then selects one of the components in the list. It is $C_p$ in Figure 3.16. A trace is displayed to indicate the path of the transformation for the chosen component value. The component values are in real units (using $Z_0$ for conversion). The displayed trace enables the user to quickly see if the selected component is suitable for the desired transformation. In Figure 3.16, $C_p$ is clearly a right choice for the transformation from $B$ to $A$ but is generally not appropriate for the transformation from $B$ to $C$ (although it is still possible, if there is a good reason for doing it that way). Again, the process is interactive in that the user can obtain the required component value from the display in the window of "Component Value" by dragging the end of the trace to the destination point, or the user can input the component value and let the program place the end of the trace at the corresponding location.

Each individual Smith chart program can certainly be implemented differently according to the programmer's preferences.

### 3.5.3 Design Considerations of Matching Networks

At this point, readers may have realized that the design task of a matching network using the Smith chart is essentially to construct a series of Smith chart operations that connect a starting point on the Smith chart ($\Gamma_0$ in Figure 3.1) to a destination point $\Gamma_S$. There are obviously endless paths for such a connection. As discussed in Chapter 2, the performance of an amplifier circuit is uniquely determined by the choice of $\Gamma_S$ and $\Gamma_L$. Therefore, if we assume that the matching networks are lossless, all $\Gamma_0 \rightarrow \Gamma_S$ transformations, regardless of the path, will yield the same performance at the chosen frequency point. However, in practical designs, there are a number of other factors that have to be taken into consideration. Different paths for the same $\Gamma_0 \rightarrow \Gamma_S$ transformation can have a significant impact on these design considerations. Most of them can be examined on the Smith chart, as will be outlined below.

#### 3.5.3.1 DC Bias Circuits and RF Matching Networks

Matching networks are about RF performance. However, if a matching network is part of an RF amplifier circuit, it usually interfaces with an active device. In that case, the DC bias circuits for the active device must be isolated from any external loading through the matching network unless the DC isolation is already implemented inside the active device as in some cases of RF ICs. Figure 3.17 shows a conceptual block diagram for a transistor-based RF amplifier. The separation of the DC bias circuits and the RF matching networks in the diagram highlights three requirements: (1) Both DC bias (voltage and current) and RF signals are applied to the same transistor ports; (2) DC bias must be prevented from being shunted to the ground through any external path; and (3) RF leakage to the DC supplies needs to be minimized. The third requirement is usually implemented in circuits with an inductor known as RF choke (RFC), which is detailed in Section 9.1. The second requirement imposes certain restrictions on the matching network, demonstrated as in the following two cases.

The first case is that a series capacitor is generally required between the RF input of the circuit (RF_in in Figure 3.17) and the device for DC blocking. (The same is true for the RF_out side.) The designer can decide whether or not to have this DC blocking capacitor be part of the matching network by

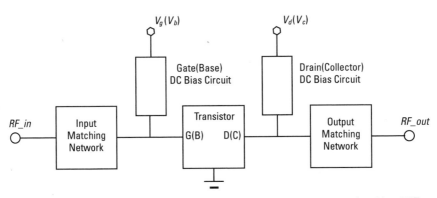

**Figure 3.17** Block diagram for a transistor amplifier. The transistor can be either FET or BJT.

selecting the capacitance value. If the series capacitor is the first component from RF_in, the corresponding Smith chart operation is shown for three values in Figure 3.18(a). It is evident from the chart that the end point hardly moves from the center for $C_s > 30$ pF (that is, it is transparent at the chosen frequency), implying that the capacitance in that range has negligible effect on the transformation. The advantage of this choice is that the matching network is not affected by the location of this capacitor, assuming a 50-$\Omega$ transmission line, which makes tuning more convenient as is demonstrated with some practical examples in Chapter 5. On the other hand, we will explain in a moment that a small capacitance can function as a low frequency rejection filter.

The second situation where the DC blocking function has to be considered is when a parallel inductor is used. Figure 3.18(b) shows two possible matching paths from the center to point $B$. The path through the midpoint $M$ is perfectly fine from the RF point of view but the parallel inductor would short the DC to ground. As a result, an alternative path such as the one going through the midpoint $M'$ has to be employed (a DC blocking capacitor may still be required at $Z_A$).

### 3.5.3.2 Filtering Functions of Matching Networks

For an RF amplifier used in a wireless communication system, in addition to specifications over the frequency band of operation (often known as the in-band in practice), the so-called out-of-band rejection is also frequently specified. This specification is a measure of the attenuation the amplifier can offer at specific frequencies outside of the operation band. Chapter 8 offers a more detailed discussion of out-of-band rejection from the perspective of system requirements. Here we explain how out-of-band specifications can be

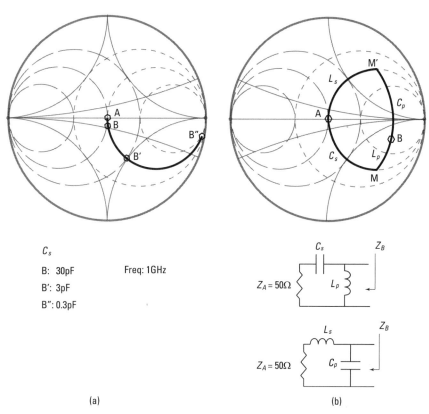

**Figure 3.18** DC blocking of matching networks: (a) Smith chart operation for three $C_s$ values, and (b) comparison of two matching network topologies.

incorporated into an in-band matching network design. That is, we consider the filtering functions of matching networks.

In most practical cases, if a system requires a sharp rejection at frequency bands near the in-band frequency, some dedicated bandpass filters such as surface acoustic wave (SAW) filters or integrated L-C filters will be required. From an analytic point of view, nothing prevents a matching network from being designed as a filter of higher orders that provides stringent filtering specifications. However, higher orders imply circuit complexity. The concern of manufacturability in volume production generally suggests that PCB-based circuits be designed as simple as possible. The goal here is to see how a matching network can be designed so that it offers some filtering functions without adding complexity to the circuit or compromise of in-band performance.

The Smith chart itself does not provide any information on the network performance at any frequencies other than the operation frequency set for the chart. Its utility in terms of the filtering function of a network, instead, is to allow the user to conveniently evaluate different circuit topologies and component values that have impacts on the out-of-band performance. For example, inspection of Figure 3.18(a) reveals that a change in DC blocking capacitance from 30 pF to 3 pF, which would increase the low-frequency rejection by roughly a factor of 10 (linear term), still allows a manageable network design in terms of network $Q$ factor. (See discussion below). Figure 3.18(a) also shows, from the location of the end point on the Smith chart, that if we want to use the same technique to further increase the low-frequency rejection by reducing the capacitance to 0.3 pF the in-band matching network design would be extremely difficult. The implication of this exercise is that the low-frequency rejection can be continuously improved by decreasing the DC blocking capacitance down to around 3 pF with little compromise to the in-band design.

Another utility of the Smith chart in this regard is evaluation of the circuit topology. We can use Figure 3.18(b) again to illustrate this point. Between the two paths for the transformation from point $A$ to $B$, the path through point $M$ ($C_S + L_p$) is a high-pass circuit while the other path ($L_S + C_p$) is a low-pass character. Putting aside the DC considerations, if the circuit realizations for the two paths turn out to be comparable, the designer then can pick the one that yields a desirable frequency characteristic. For certain conditions (frequency ranges and locations of the point $B$), the required component values may not be realistic. In that case, the filtering function of the circuit has to be implemented some other way.

### 3.5.3.3 The $Q$ Factor and Bandwidth of Matching Networks

Once the destination point is placed on the Smith chart, we can examine the $Q$ factor of the point. There are two scenarios. First, if the point has a low or moderate $Q$ factor and broadband matching is desirable, then the path for the matching should be kept in the low or moderate $Q$ areas. On the other hand, if some filtering effect is desirable (see Section 3.5.3.2), the $Q$ factor can be chosen to be moderately high. Second, when the destination point is at a high $Q$ area already, it is not possible to construct a low $Q$ matching network. If the $Q$ value of the final point is still judged to be acceptable, the matching should not further increase the total $Q$. In the case that the $Q$ factor is exceedingly high (close to the edge of the Smith chart), the design may need to be reconsidered.

We use an example to quantitatively demonstrate how the constant $Q$ lines on the Smith chart can be used to control the bandwidth of a matching network. Section 3.4 shows analytically that for a given impedance transformation ratio, increasing the number of matching sections lowers the $Q$ factor and consequently increases the bandwidth. The same matching problem discussed in Section 3.4 is reconsidered here using the Smith chart technique for a specific transformation, $r = 1 \rightarrow r = 0.2$.

Following the suggestion made in Section 3.5.2, the initial and final points are first placed on the chart. It is easy to see that a $C_p$-$L_s$ configuration can accomplish the required transformation. Figure 3.19 shows three matching processes with different $Q$ values. Their corresponding circuits are also shown. The reference impedance and the frequency are set at 50Ω and 1 GHz in this case. Circuit (a) in Figure 3.19 is a single section of $C_p$-$L_s$ with a $Q$ factor of 2 as predicted by (3.24).

In Circuit (b) of Figure 3.19, the transformation breaks into two steps, each having a $Q$ factor of about 1. Circuit (c) in Figure 3.19 shows a transformation that also consists of two sections but with an uneven split of the $Q$ factors between the two sections. (The $Q$ factor of the first step is about 1.5.) Figure 3.20 shows the plots of return loss versus frequency for Figure 3.19(a–c). As expected, the bandwidth increases as the $Q$ factor decreases. Quantitatively, for Circuit (a) in Figure 3.20, the relative bandwidth, $2d\omega/\omega_0$, measured between two points of RL = −20 dB (d$|\Gamma|$ = 0.1) is about 0.11, which is in good agreement with the calculated result using (3.39) for $Q = 2$. For matching networks with multiple sections, analytic results for bandwidth become considerably more complex. Furthermore, in our analysis the output side of the network is assumed to be terminated with a pure resistive element that is frequency-independent. In reality the termination is an active device (see Figure 2.10) that generally has strong frequency dependence. As a result, other than simulation, it is difficult to analyze the frequency response of the reflection coefficient at the circuit input. Nevertheless the $Q$ lines on the Smith chart still provide useful guidance for bandwidth control as evidenced in Figures 3.19 and 3.20.

This discussion also illustrates a significant advantage of the graphic method over the analytic approach in dealing with complex impedances. For the Smith chart operation, whether a point is a real or complex impedance makes no difference in the process, whereas in analytic analysis a complex number makes equation manipulation much more cumbersome, which was precisely the reason for the assumption that the initial and final impedances are pure resistive in our analysis in Section 3.4. This observation brings up another point: For the two-section matching networks [Circuit (b) and (c) in Figure 3.19], the midpoint does not need to be on the real axis, as assumed in the analysis. Without this restriction, the selection of component values becomes flexible. For example, if the 3.73-nH inductor called for in Circuit (c)

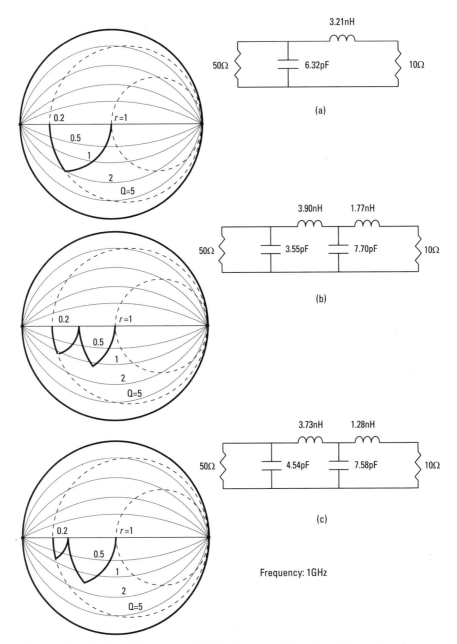

**Figure 3.19** Circuit realizations and Smith chart operations of matching networks with different $Q$ values: Circuit (a), one-step transformation, $Q=2$. Circuit (b), two-step transformation with two equal Q factors. Circuit (c), two-step transformation with different Q factors.

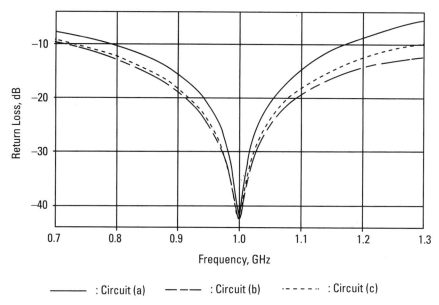

**Figure 3.20** $S_{11}$ versus frequency for the networks in Figure 3.19.

of Figure 3.19 is not practical, a different value can be used with little impact on the matching network, provided that some adjustments are made on the other components to compensate for the change.

### 3.5.3.3 Lumped Elements Versus Transmission Lines

As noted in Chapters 1 and 2, transmission lines can be used as circuit elements in matching networks. Figure 3.21(a) shows an example of using transmission lines for the same matching problem discussed in Figure 3.19 Circuit (a), which is repeated as Figure 3.21(b) for comparison. It is easy to visualize that a hybrid matching circuit made of lumped elements and transmission lines such as $C_p + TL_2$ or $TL_1 + L_s$ in this case (with adjustments in the component values) can also achieve the required transformation. The key to the decision on lumped elements versus transmission lines is the frequency. Some other factors such as board space constraint, cost, and power loss can be part of the considerations also. Note that, with the exception of the transmission line impedance, the component values for all nine components listed in Section 3.5.2 are scaled with frequency in a Smith chart operation. This feature allows the user to quickly estimate the suitability of certain matching configurations for a given frequency. Consider the two transformations shown in Figure 3.21. The component values shown in Figure 3.21 are for the frequency of 1 GHz.

We can see that the chosen capacitance and inductance are in reasonable ranges for practical circuits. In contrast, for a typical PCB substrate of $\varepsilon_r = 4$ and a dielectric thickness of 0.25 mm (10 mil), the required physical lengths for the two transmission lines are 22 mm and 12.4 mm, respectively. Both could be unacceptable for practical products due to the size constraints. Therefore, at 1 GHz or lower frequencies, matching circuits are realized mainly using lumped elements with transmission lines used for fine tuning purposes in some situations. On the other hand, if the frequency is 10 GHz, the same lumped element circuit would require the capacitance and inductance to be 10 times smaller, which is unrealistic in practice. At this frequency, the transmission-line solution becomes more convenient. For a practical product design with operation frequency in the several gigahertz range, a mix of lumped elements and transmission lines is commonly employed for matching circuits.

In summary, this chapter considers separately several factors in matching network designs. A complete design should take all these factors into account at once. These factors are sometimes in conflict. Should that be the case, some compromises may need to be made based on the priorities of specification requirements. The Smith chart technique allows designers to examine these design requirements simultaneously, and thereby obtain a satisfactory solution. Chapter 5 presents a number of practical examples.

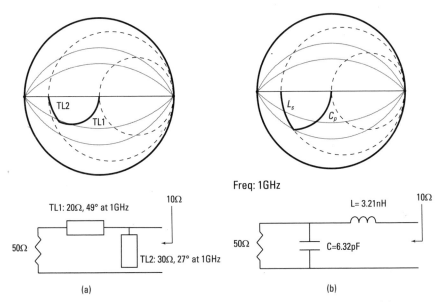

**Figure 3.21** Transmission line implementation for the network in Figure 3.19(a).

# References

[1]  Smith, P. H., *Electronic Applications of the Smith Chart* (Second Edition), Noble Publishing, 1995, Atlanta.

[2]  Daryanani, G., *Principles of Active Network Synthesis and Design*, John Wiley and Sons, 1976.

[3]  Anderson, B. D. O., and S. Vongpanitlerd, *Network Analysis and Synthesis*, Dover, 2006.

[4]  Dorf, R. C., and R. H. Bishop, *Modern Control Systems* (Seventh Edition), Addison-Wesley, 1995 (and later editions).

[5]  Riley, K. F., M. P. Hobson, and S. J. Bence, *Mathematical Methods for Physics and Engineering*, Third Edition, Cambridge University Press, 2006.

[6]  Chen, W. K., *Broadband Matching Theory and Implementations*, Third Edition, World Scientific, 2016.

[7]  Alexander, C. K., and M. N. O. Sadiku, *Fundamentals of Electric Circuits*, McGraw-Hill, 2000.

[8]  Bode, H. W., *Network Analysis and Feedback Amplifier Design*, Van Nostrand, 1945.

[9]  Pozar, D. M., *Microwave Engineering*, Second Edition, John Wiley and Sons, 1998.

[10]  Misra, D. K., *Radio-Frequency and Microwave Communication Circuits: Analysis and Design*, John Wiley and Sons, 2001.

# 4

# Noise and Its Characterization in RF Systems

In today's wireless communication system designs, noise reduction and control are often some of the critical engineering considerations. If noise is understood as unwanted signals, RF engineers generally need to deal with the noise issue on two fronts. One is noise emission control, which is mandated by various regulatory agencies (along with other possible requirements). The second major design area is optimization of noise performance in receivers.

Noise emission is an EM compatibility (EMC) subject and is not systematically covered in this book, except in the section on PCB layout for signal integrity (Section 8.6). Interested readers can consult the more complete treatments available in a number of texts on the subject (e.g., [1, 2]).

From the perspective of a wireless receiving system, there are two categories of electrical noise sources: intrinsic and external. The intrinsic noise source is the noise generated by the components such as semiconductor devices and resistors used in the signal receiving circuit. The external noise can be a broadband background noise from the environment, from other parts of the circuit, or from an external signal that is somehow coupled into the receiving path, causing interference with the intended signal. Strictly speaking, interfering signals are not noise in the sense that they are not originated from a random process. While in principle they can be completely eliminated if the circuit is designed perfectly, in practice, interference, particularly on-board interference (from another circuit element on the same board, such as an oscillator or a transmitter power amplifier), can be difficult to cope with in certain conditions. Chapter 8 reviews several techniques for the control of external noise. This chapter focuses exclusively on the intrinsic noise with the exception of a brief discussion of background thermal noise in the context of satellite communication systems. Sections 4.1 and 4.2 discuss the characteristics of several common types of noise, with an emphasis on thermal noise. Then, Section 4.3 introduces the equivalent noise circuit. Finally, Sections 4.4 and 4.5 discuss the noise figure, an especially important concept in low-noise RF circuit designs, in great depth. Chapter 5 covers practical design techniques for LNAs.

## 4.1 Noise Sources

For practical purposes, noise in electronic systems represents simply fluctuations of certain measurable quantities such as voltage, current, and power. At a more fundamental level of understanding, electrical noise originates from the random nature of the motion of electrical charge carriers (electrons in most cases but other charged particles as well). The randomness in the motion of charge carriers at a microscopic level causes the corresponding macroscopic quantities to fluctuate. There is rich literature on the microscopic mechanisms of various noise sources. References [3–6] discuss this topic in varying degrees of depth. Please note that this chapter uses some common terms in the statistical analysis of random processes are without any further explanation. Interested readers can find detailed discussions on these concepts in many textbooks on communication theory or signal processing [7, 8].

In terms of RF circuit designs, one of the most important attributes of a noise source is its frequency characteristic. There are two common types of noise in this regard. One is white noise, which has a noise power spectral density that is independent of frequency in the RF/microwave range. The two most common examples of white noise are (1) thermal noise, which is generated by random thermal motion, and (2) shot noise, which is caused by the discrete nature of charge carriers. The other noise category is $1/f$ noise. As the name implies, its spectral density decreases inversely with frequency up to a certain frequency point called the corner frequency. Above the corner frequency, the frequency spectral density gradually decreases to the white noise floor. The corner frequency of $1/f$ noise varies depending on the device but is usually on the order of megahertz. Hence $1/f$ noise is essentially a low-frequency noise and generally has no effect on the noise performance of a device in an RF range. However, the $1/f$ noise of an active device is critical in RF oscillator design, because it determines the phase noise characteristics of the oscillator. Chapter 9 discusses this topic. The remainder of this chapter considers only white noise, particularly thermal noise.

## 4.2 Thermal Noise

Thermal noise is unique, among various types of noise, in that it is associated with temperature and as such, is always present in any electronic system as long as the operating temperature is not at absolute zero. Also, the thermal noise of a circuit element is fully characterized by its resistive component at a given temperature. In contrast, the shot noise and the $1/f$ noise of a device are generally dependent on bias conditions. For these reasons, it is common practice in noise characterization that the thermal noise of a resistance is used as a reference and that other types of noise are represented with an equivalent temperature or an equivalent noise resistance. This method will be repeatedly used later in this chapter (Section 4.4 and 4.5).

A complete treatment of thermal noise, also referred to as Johnson noise or Johnson-Nyquist noise, requires thermodynamic physics [3, 6], which is outside the scope of this book. Here we simply outline the relevant theories with an emphasis on their implications in practical circuit designs. We start with Nyquist's theorem, which states that the available thermal noise power from a resistance at temperature $T$ in Kelvin is given by

$$N = kTB \qquad (4.1)$$

where $k = 1.38 \cdot 10^{-23}$ J/K is the Boltzmann constant, and $B$ is the bandwidth of the measurement system. Equation (4.1) indicates that the available thermal noise power is only determined by the bandwidth and temperature and is independent of the resistance value and that $kT$ is the power spectral density for thermal noise.

While in practice a measurement system always has limited bandwidth, (4.1) does impose a difficulty mathematically because it implies an infinite power when $B$ approaches infinity. This difficulty arises from (4.1) being only a low frequency approximation of a more general form of noise power spectral density known as the Planck distribution[3]:

$$P(f) = \frac{hf}{e^{hf/kT} - 1} \tag{4.2}$$

where $h$ is Planck's constant ($6.626 \times 10^{-34}$ J·s). The integration of $\int_0^\infty P(f) df$ is finite; thus the Planck distribution does not have the same divergence problem as $f \to \infty$. For the condition $hf \ll kT$, $P(f) = kT$ leading to (4.1). At room temperature $T = 300$K, the condition of $hf = kT$ occurs at $f = 6$ THz ($\lambda = 0.05$ mm in the air). Therefore, up to millimeter waves, (4.1) is valid.

Readers who learned about blackbody radiation in college physics (e.g., [9]) may wonder how the Planck distribution, which was first proposed to explain blackbody radiation, is also applicable to the noise power distribution of electrical resistance, seemingly a very different physical system. Nyquist was the first to establish the equivalence between these two systems. In his original paper [10], he made use of a thought experiment in which a lossless transmission line connecting two resistors was modeled as a one-dimensional cavity filled with EM standing waves. Then, it became clear that the Planck power spectral distribution, which was originally derived after the quantum correction, from an analysis of three-dimensional cavity radiation, can be applied to the case of the thermal noise power density of the resistor. Readers are referred to [3, 6] for more detailed discussions.

For engineers, it is more relevant to realize that the thermal noise described in (4.1) is proportional to the temperature and bandwidth of the system. Both have practical importance in system design and characterization, as discussed in Sections 4.5 and 8.5.

Note that this section also uses the term available power (introduced in Chapter 1 for a signal source) for thermal noise. The basic concept of this term regarding the power transfer between a source and a load is the same for both a signal source and a thermal noise source. However, there is one significant difference between the two cases. Recall that [see Figure 4.1(a)] the available power of a signal source is the power transferred from the source to the load when they are matched. The concept of power transfer from a thermal noise source is less straightforward. Consider the thought experiment shown in Figure 4.1(b), where a resistance $R_s$ as a source at temperature $T_s$ is connected to a load resistance at temperature $T_L$. To be consistent with the convention in Chapter 2, the arrow in the diagram is for the direction of the source that provides available power rather than the direction of power flow. If $R_L$ is at a temperature of $T_L = 0$K, the power dissipated in $R_L$ is the available thermal noise power of $R_s$. When $T_L$ rises above 0K, $R_L$ starts to deliver thermal noise power to $R_s$, resulting in a reduction in net power transfer. This process continues until $T_s = T_L$ when there is no net power transfer between $R_s$ and $R_L$, the so-called thermodynamic equilibrium condition. Therefore, unlike a signal power source, there can be no net thermal noise power transfer between the source and the load if they are at the same temperature. This line of thinking also leads to another important conclusion: that only a circuit component that is capable of dissipating power can generate noise power. The proof is straightforward: In Figure 4.1(b), consider a case where $R_L$ is replaced with a pure reactive component (and therefore cannot dissipate power from the source) and $T_L = T_s$; if this component could generate thermal noise power, there would be a net power transfer from the load to the source, in a direct violation of the thermodynamic equilibrium condition. In fact, it can be generally proved [3] that if a complex electrical component or a circuit can be represented by an equivalent impedance, its thermal noise power is determined only by the real part of the impedance. This concept is utilized in the discussions in Sections 4.4 and 4.5.

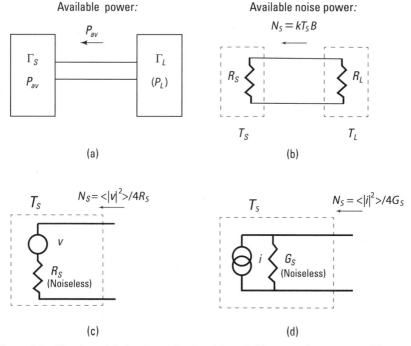

**Figure 4.1** Circuit models for thermal noise: (a) available power from source, (b) resistance as a thermal noise power source, (c) equivalent noise voltage source, and (d) equivalent noise current source. The arrows in the diagram point to the sources that provide available powers.

## 4.3 Equivalent Noise Circuits

Chapter 1 mentions that the design and analysis of a linear RF circuit can be performed using the S-parameters (which are defined in power waves) without dealing with linear parameters such as voltage, current, and impedance (or admittance). For noise, it is still true that noise parameters for practical use in RF circuitry are also defined in power terms. However, in our attempt to derive the formulas for these noise parameters, noise voltage and noise current turn out to be much easier to use. This is because analytic techniques for linear networks can be fully utilized when these linear parameters are used in circuit analysis. In fact, two alternative forms of Nyquist's law are expressed in terms of equivalent noise voltage $v$ and noiseless resistance $R_s$ and noise current $i$ and noiseless conductance $G_s = 1/R_s$, as illustrated in Figure 4.1(c, d), respectively:

$$\langle |v|^2 \rangle = 4kTBR_s \qquad (4.3)$$

$$\langle |i|^2 \rangle = 4kTBG_s \qquad (4.4)$$

By the definition of available power of a source, we can see that (4.3) and (4.4) are equivalent to (4.1). The terms $v$ and $i$ in (4.3) and (4.4) are random complex variables, and the symbol $\langle \ldots \rangle$ denotes the average, which can be either an ensemble average or a time average since thermal noise is generally considered ergodic [6, 7]. Table 4.1 lists some numbers based on (4.1, 4.3, and 4.4) for thermal noise at $R_s = 50\Omega$ ($G_s = 0.02S$) and $T = 290K$, (the standard temperature for noise characterization—see Section 4.5).

Note that a standard formula for $50\Omega$ load: value in dBV = value in dBm − 13 dB, which is often cited in the literature, seems inconsistent with the numbers in Table 4.1. The reason is that the 13-dB difference between dBV and dBm is for a case where both power and voltage are on a specific resistance, whereas, in our calculation, the power and voltage are the available power and the equivalent source voltage (or open-circuit voltage), respectively,

**Table 4.1**
Thermal Noise Floor in Power, Voltage, and Current, 50Ω, 290K

| Thermal Noise Power | Noise Voltage Source | Noise Current Source |
|---|---|---|
| $4.0 \dfrac{10^{-21} W}{Hz} = -174 \dfrac{dBm}{Hz}$ | $\dfrac{0.89 nV}{\sqrt{Hz}} = -181 \dfrac{dBV}{\sqrt{Hz}}$ | $\dfrac{18 pA}{\sqrt{Hz}} = -215 \dfrac{dBA}{\sqrt{Hz}}$ |

of a source. After subtraction of 6 dB (the factor of 2), the two cases are equivalent. As shown in Table 4.1, thermal noise is small at 300K. It is generally a challenging task to design a system that has its noise performance close to the thermal noise floor.

We can extend the notion of equivalent noise generators to a general one-port network where the mechanisms for noise generation are not limited to resistive elements. Then the noise characteristics of a one-port network can be represented by either an equivalent noise temperature or an equivalent resistance (conductance) without any knowledge of the actual noise generation mechanism inside the network. Furthermore, the corresponding noise voltage and current sources defined in (4.3) and (4.4), $v$ and $i$, can be interpreted as respective Thevenin's and Norton's equivalent sources for noise power.

It is straightforward to apply the above described technique of equivalent noise representation to a two-port network. The basic concept in this case was first proposed by Rothe and Dahlke in 1956 [11]. The essence of this technique is to move the noise generators outside the network and treat the network as noise-free. Since there are two ports, two noise generators are generally required (this will be proven in a moment). Figure 4.2(a) shows a possible equivalent circuit where a noisy two-port network is represented by a noise-free network plus two noise generators, $i_{n1}$ at port 1 (always taken as the input if the device is an amplifier) and $i_{n2}$ at port 2 (output). These noise generators are always treated as an ideal source in a circuit analysis; thus they do not alter the impedance condition. Then, if the network is characterized by the $Y$ parameters, we have the following terminal equations:

$$I_1 = Y_{11}U_1 + Y_{12}U_2 + i_{n1} \tag{4.5}$$

$$I_2 = Y_{21}U_1 + Y_{22}U_2 + i_{n2} \tag{4.6}$$

From this set of equations, we can see why two noise generators are required in order to cover all circuit scenarios. Consider a case of $i_{n2} = 0$. If port 1 is shorted, that is, $U_1 = 0$, then from (4.6), $I_2 = Y_{22}U_2$, which implies that no noise from the device would appear at port 2. In reality, a short-circuited port 1 certainly could not prevent the presence of noise in port 2. For the completeness of our argument, if port 1 is terminated with a finite load, $U_1$ is no longer vanished and can be expressed in terms of $i_{n1}$, according to (4.5). Then $I_2$ has a component that is associated with $i_{n1}$, although quantitatively, it is generally incorrect. In conclusion, we can make a general statement that two noise generators are necessary and sufficient to represent the noise characteristics of a noisy two-port network.

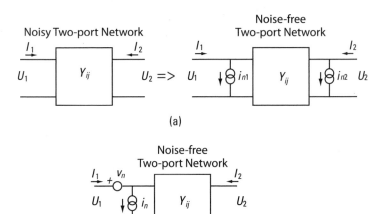

**Figure 4.2** Equivalent noise circuits for a noisy two-port network, represented by (a) noise current generators $i_{n1}$ and $i_{n2}$. (b) noise voltage generator $v_n$ and noise current generator $i_n$.

As in the case of parameters for two-port network characterization, where multiple options are possible, noise generators can be either voltage sources or current sources and can be at the device input or output. In addition to the representation by $i_{n1}$ and $i_{n2}$ in Figure 4.2(a), another configuration commonly used in noise analysis is shown in Figure 4.2(b) where two noise sources, $i_n$ and $v_n$, are placed at the device input. These two sets of noise generators must be related since they are for the same device. It is an easy exercise to show [11]

$$i_n = i_{n1} + Y_{11}v_n = i_{n1} - \frac{Y_{11}}{Y_{21}}i_{n2} \tag{4.7}$$

$$v_n = -\frac{1}{Y_{21}}i_{n2} \tag{4.8}$$

Generally, the configuration of Figure 4.2(a) is convenient for device-physics–based analytic works [3, 12, 13] while the configuration in Figure 4.2(b) is mostly used in circuit analysis.

From the device-physics point of view, the noise current at the output side, $i_{n2}$, such as the noise current between the drain and source of an FET device, is always the dominant noise source for any transistor. Quantitative analyses for $i_{n2}$ can be exceedingly complicated due to the complexity of the device structure of modern RF transistors and the difficulty in accurate modeling the microscopic random processes that contribute to the noise under consideration. For circuit designers, the mechanism of $i_{n2}$ is of little importance. What is relevant in (4.7) and (4.8) to RF circuit design is that both $i_n$ and $v_n$, which are random variables, are related to the same variable $i_{n1}$, indicating that they are partially correlated, a term commonly used in random variable analysis [7, 8]. In a typical analog circuit analysis of a transistor, $i_n$ and $v_n$ are often assumed to be independent (no correlation), which can be justified by large gains in the frequency range. As the frequency moves into the RF region, the coupling between the input and the output increases, which not only lowers the RF gain but also increases the correlation between the noise generation at the input and output. As a result, for RF circuit analysis, the correlation between $i_n$ and $v_n$ generally cannot be ignored.

To quantitatively characterize this correlation, we note that for a pair of complex voltage and current, $i_n$ and $v_n$, on a specific load the power dissipation is given by $v^*_n \cdot i_n$. The complex conjugate in this expression accounts for the fact that only the in-phase components contribute to the power dissipation [14]. Then the correlation coefficient that is relevant to our discussion is [7, 8]:

$$\rho = \frac{\langle v^*_n i_n \rangle}{\sqrt{\langle |v_n|^2 \rangle \langle |i_n|^2 \rangle}} \tag{4.9}$$

Here it is assumed that the means of $i_n$ and $v_n$ are zero. Note that $\rho$ is also a complex number. Now we can write $i_n$ as

$$i_n = i_{nu} + i_{nc} \tag{4.10}$$

where $i_{nu}$ and $i_{nc}$ are uncorrelated and correlated to $v_n$ respectively. Further we introduce a new parameter $Y_c$ as

$$i_{nc} = Y_c v_n \tag{4.11}$$

It follows that $Y_c$ is related to $\rho$ by

$$Y_c = \rho \sqrt{\frac{\langle |i_n|^2 \rangle}{\langle |v_n|^2 \rangle}} \tag{4.12}$$

Since $i_{nu}$ and $i_{nc}$ are uncorrelated, we have

$$\langle |i_n|^2 \rangle = \langle |i_{nu}|^2 \rangle + |Y_c|^2 \langle |v_n|^2 \rangle \tag{4.13}$$

That is, $\langle|i_n|^2\rangle$ can be calculated from $\langle|i_{nu}|^2\rangle$, $\langle|v_n|^2\rangle$ and $Y_c$. The last four parameters ($Y_c$ counts for two since it is a complex number) fully describe the noise characteristics of a two-port network.

Both equivalent current and noise sources, $\langle|i_n|^2\rangle$ and $\langle|v_n|^2\rangle$, are measureable quantities [3] at analog frequencies and are actually used as device parameters on data sheets in some cases. However, they are not suitable as practical parameters for RF devices and circuits for the reason explained in Chapter 1. In Section 4.4, we derive a different set of noise parameters that are defined in terms of power and are commonly used in RF circuitry. Nevertheless, $i_n$ and $v_n$ are well-understood conceptually and are easy to manipulate in the circuit analysis. We will use them as a starting point in our formula deduction, which eventually leads to the desired results at the end of Section 4.4.

## 4.4 The Noise Figure of Linear Two-Port Networks

The noise performance of an RF device/circuit is most often characterized by the noise figure. There are several versions of the definition of noise figure in the literature. For most practical conditions, these different definitions are equivalent, but they are not identical in all circumstances. This section uses a definition that provides a precise and convenient starting point for our analysis.

Although the concept of noise figure is applicable to any $n$-port network [4, 15], this chapter considers only the two-port network case. The basic idea of noise figure is to characterize the noise performance of a device by comparing its noise contribution with the thermal noise from the source. To quantitatively calculate the noise figure of a device, consider the circuit shown in Figure 4.3 where the device, represented by a noiseless device and two noise generators ($i_n$ and $v_n$), is connected to a source characterized by its admittance $Y_s$ (with the choice of admittance instead of impedance entirely for convenience) and the corresponding equivalent noise current $i_{sn}$ defined in (4.4). Noise figure then is defined as

$$F = \left.\frac{\text{Available noise power of noisy device}}{\text{Available noise power of noiseless device}}\right|_{\text{Output}} \quad (4.14)$$

In consideration of Figure 4.3, the noise figure in (4.14) can also be interpreted as the ratio of the total noise power in the presence of the noise generators to the noise power without the noise generators. Since both noise generators are assumed to be ideal, their presence does not affect the source impedance (or $\Gamma_S$) presented to the device. As a result, the available power

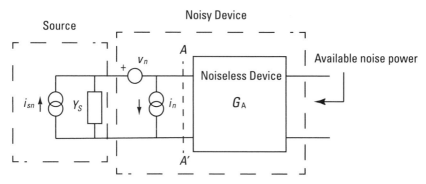

**Figure 4.3** Circuit model for noise figure calculation.

gain of the device is unchanged from the case with generators to that without. Then, according to the discussion of available gain in Section 2.1, we can see that $F$ in (4.14), can also be calculated as the same ratio at the input of the noiseless device (at $A - A'$ terminals). Let $N_s$ and $N_d$ be the available noise power at the input attributable to the source and the device respectively; then $F$ is expressed as

$$F = \frac{N_s + N_d}{N_s} = 1 + \frac{N_d}{N_s} \qquad (4.15)$$

If in (4.15), both $N_d$ and $N_s$ are assumed to be proportional to the bandwidth (which is termed as a narrowband approximation in some texts), then $F$ is independent of the bandwidth. This is important because the noise figure then can be used as a figure of merit for noise performance without specifying the bandwidth of the systems to be considered. This condition is assumed throughout our analysis in this chapter.

Before proceeding further with the analysis, let's review the terms used for noise figure in practice. It is common in the literature that $F$ defined in (4.14) is called the noise factor, and its decibel form, $10\log F$ is referred to as the noise figure. There is nothing wrong with this practice. On the other hand, no such distinction is deemed necessary for other parameters such as gains and S-parameters, which can also be in either linear or logarithmic scales. A single term, noise figure, is preferred in this book regardless of the scale used for the quantity. Of course, the correct scale has to be used in numerical calculations for noise figure.

Now we return to the noise figure expressed in (4.15). Unlike $i_n$ and $v_n$, which are parameters of the device only, noise figure, in principle, depends on the choice of source (both $i_{sn}$ and $Y_s$). In practice, however, the source is standardized, at least in the RF industry (see Section 4.5.2), so noise figure is a good figure of merit for comparison among different devices. In addition, thermal noise is generally unavoidable (although this may not apply to cryoelectronics, discussed in Section 4.5.2), and it sets the noise floor of a receiving system. Noise figure, which is defined with the thermal noise of the source as a reference, provides a sense of the noise contribution from the device with respect to that noise floor. From (4.15), we have some numbers as follows:

- For a noiseless device ($N_d = 0$), $F = 1$ (or 0 dB). This is the floor for $F$.
- When $N_d = N_s$ (noise contributions from the source and device being equal), $F = 2$ (or 3 dB).
- For a noise figure of 0.5 dB, a spec commonly achieved with the current state of the art technologies, $N_d/N_s = 0.12$, which means that the noise contribution from the device is only about 10% of the total noise power. Thus, from this level (0.5 dB), any further effort in reduction of $N_d$ lands us in the territory of diminishing returns. (Satellite communication systems can be an exception—see Section 4.5.)

We begin the derivation of the noise figure $F$ by noting that the circuits to the left of the noiseless device input in Figure 4.3 can be treated as equivalent Norton current sources and that their values are the short-circuit currents at the $A$-$A'$ terminals. Let $i_{AN}$ and $i_{A0}$ be the Norton equivalent current sources for the cases with and without noise generators; then we have

$$i_{AN} = i_{sn} - (Y_s v_n + i_n); \text{ and } i_{A0} = i_{sn}$$

Since the source admittance is unaffected by the noise generators, the ratio of the available powers in (4.15) can be expressed in Norton currents. That is, $F$ can be written as

$$F = \frac{\langle |i_{AN}|^2 \rangle}{\langle |i_{A0}|^2 \rangle} = \frac{\langle |i_{sn} - (Y_s v_n + i_n)|^2 \rangle}{\langle |i_{sn}|^2 \rangle} \quad (4.16)$$

The numerator of (4.16) may be simplified to

$$\langle |i_{sn} - (Y_s v_n + i_n)|^2 \rangle = \langle |i_{sn}|^2 \rangle + \langle |(Y_s v_n + i_n)|^2 \rangle \quad (4.17)$$

Here $\langle i_{sn}(Y_s v_n + i_n)^* \rangle = 0$ is used, since the noises from the source and the device are obviously uncorrelated. Using (4.10) and (4.11) for $i_n$ and noting that $i_{nu}$ is uncorrelated with $v_n$, the second term in (4.17) is

$$\langle |Y_s v_n + i_n|^2 \rangle = \langle |i_{nu}|^2 \rangle + |Y_s + Y_c|^2 \langle |v_n|^2 \rangle \tag{4.18}$$

Writing $Y_c$ and $Y_s$ as

$$Y_c = G_c + jB_c \tag{4.19}$$

and

$$Y_s = G_s + jB_s \tag{4.20}$$

and using (4.18), the right-hand side of (4.17) becomes

$$\langle |i_{sn}|^2 \rangle + \langle |i_{nu}|^2 \rangle + \left((G_c + G_s)^2 + (B_c + B_s)^2\right)\langle |v_n|^2 \rangle$$

Then (4.16) can be expressed as

$$F = 1 + \frac{\langle |i_{nu}|^2 \rangle + \langle |v_n|^2 \rangle \left((G_c + G_s)^2 + (B_c + B_s)^2\right)}{\langle |i_{sn}|^2 \rangle}$$

By Nyquist's law, the source noise current and source conductance are related by

$$\langle |i_{sn}|^2 \rangle = 4kT_s G_s B \tag{4.21}$$

In a similar manner, we introduce equivalent noise resistance $R_n$ and an equivalent noise conductance $G_{nu}$ such that

$$\langle |v_n|^2 \rangle = 4kT_d R_n B \tag{4.22}$$

$$\langle |i_{nu}|^2 \rangle = 4kT_d G_{nu} B \tag{4.23}$$

where $T_d$ is the temperature of the device. Note that $R_n$ and $G_{nu}$ are not associated with any physical entity. The terms $\langle |v_n|^2 \rangle$ and $\langle |i_{nu}|^2 \rangle$ have both thermal and nonthermal components; hence they are generally complex functions of $T_d$.

With these notations, we write $F$ as

$$F = 1 + \frac{T_d}{T_s} \frac{G_{nu} + R_n \left[ (G_c + G_s)^2 + (B_c + B_s)^2 \right]}{G_s} \quad (4.24)$$

In most practical conditions, $T_d = T_s$; hence $F$ is usually seen as

$$F = 1 + \frac{G_{nu} + R_n \left[ (G_c + G_s)^2 + (B_c + B_s)^2 \right]}{G_s} \quad (4.25)$$

Equation (4.25) indicates that for a given device (i.e., $G_{nu}$, $R_n$, $G_c$ and $B_c$ are given), $F$ is a function of the source admittance (or impedance) $Y_s = G_s + jB_s$. Furthermore the functional form of (4.25) implies that there is a minimum of $F(Y_s)$. Let $Y_{opt} = G_{opt} + jB_{opt}$ be the value of $Y_s$ for the minimum of $F$ that will be denoted as $F_{min}$. Since $B_s$ only appears in the numerator in (4.25), $F_{min}$ must occur at

$$B_{opt} = -B_c \quad (4.26)$$

$G_{opt}$ is determined by $\partial F / \partial G_s = 0$. The result is

$$G_{opt} = \sqrt{G_c^2 + \frac{G_{nu}}{R_n}} \quad (4.27)$$

Thus, we have

$$\begin{aligned} F_{min} &= 1 + \frac{G_{nu} + R_n (G_c + G_{opt})^2}{G_{opt}} \\ &= 1 + 2R_n (G_{opt} + G_c) \end{aligned} \quad (4.28)$$

By expanding the expression of $F - F_{min}$, $F$ can be expressed as

$$F = F_{min} + \frac{R_n}{G_s} |Y_s - Y_{opt}|^2$$

Finally, using the relationships (see (1.13))

$$\Gamma_s = \frac{Y_0 - Y_s}{Y_0 + Y_s}$$

and

$$\Gamma_{opt} = \frac{Y_0 - Y_{opt}}{Y_0 + Y_{opt}}$$

and after some equation manipulation [note that $G_s = (Y_s + Y_s^*)/2$], we deduce the desired expression for $F$:

$$F = F_{min} + \frac{4r_n |\Gamma_{opt} - \Gamma_s|^2}{\left(1 - |\Gamma_s|^2\right)\left|1 + \Gamma_{opt}\right|^2} \qquad (4.29)$$

Here $r_n = R_n Y_0$ and is referred to as normalized equivalent noise resistance.

The noise figure in (4.29) is expressed in a new set of four noise parameters, $F_{min}$, $\Gamma_{opt}$ (magnitude and phase) and $r_n$, and the dependence of $F$ on the source impedance is through $\Gamma_s$. This formula for noise figure is standard in the RF business. The noise parameters (also called noise data), if available, are usually provided by manufacturers for a certain frequency range in the s2p files. (See an example in Section 1.5.) Because of the numerical complexity seen in (4.29), actual LNA design is ideally carried out using an EDA simulator. (See Section 5.4 for a detailed discussion of this topic.) Nevertheless, a designer still benefits from some general understanding of the significance of these parameters in terms of circuit performance. These parameters are briefly summarized as follows.

- $\Gamma_{opt}$: This parameter indicates the location of the source impedance for achieving $F_{min}$. Strictly speaking, it is not a figure of merit by itself. Ideally it should be close to $S_{11}$ so that the optimal noise figure and input return loss can be achieved simultaneously. However, that is normally not the case in practice. $\Gamma_{opt}$ is a function of all four original noise parameters, $i_n$, $v_n$ and their correlation coefficient.
- $r_n$: From the circuit design point of view, $r_n$ determines how sensitive the noise figure is to the mismatch from $\Gamma_{opt}$. The smaller it is, the less sensitive. (See Section 5.3.) According to (4.22), $r_n$ is only related to $v_n$.
- $F_{min}$: Its value sets the lower limit on the noise figure and is obviously the most critical parameter regarding the noise performance. One important aspect of $F_{min}$ in a circuit design is its dependence on bias current. As indicated in (4.28), $F_{min}$ is determined by a product of two factors, $R_n$ and $G_{opt} + G_c$. Since the internal noise of the device

is partially related to the bias current, some current dependence of $F_{min}$ is expected. The section will now briefly review this dependence.

Analysis of $R_n$ is relatively straightforward. From (4.22) and (4.8), $R_n$ can be written as

$$R_n = \frac{\langle |i_{n2}|^2 \rangle}{4kTB|Y_{21}|^2}$$

There are two conflicting factors in the function of $R_n$ versus bias current. On one hand, as explained earlier, $i_{n2}$ has a component that is directly related to the bias current; therefore $i_{n2}$ increases with the bias current. On the other hand, $Y_{21}$ is associated with the transistor gain, and it generally increases with the bias current until the saturation sets in. Thus, $R_n$ is expected to reach a minimum value at a certain bias current. Analysis of the other term ($G_{opt} + G_c$) is more complicated, even qualitatively, and we do not attempt it here. For all practical purposes, the behavior of $F_{min}$ versus bias current is observed to be a U-shaped curve as illustrated by a measurement of $F_{min}$ versus $I_d$ on an FET microwave transistor in Figure 4.4. Figure 4.4 also plots the associated gain $G_a$ vs. $I_d$. (See Section 5.1 for the exact definition.) It can be seen that the minimum of $F_{min}$ indeed occurs where the gain curve starts to saturate. Details of $F_{min}$ as a function of bias current varies greatly depending on the type of device. In general, the noise figure on the datasheet is specified at, or close to, the optimal bias condition. Sometimes manufacturers provide additional noise data at different bias points. In practice if the bias point is chosen considerably different from the datasheet condition for the noise figure specification, some degradation in the noise figure should be expected.

For specification purposes, manufacturers normally only provide the noise figure rather than the full set of noise parameters. For transistors, the datasheet noise figure specs are usually the values of $F_{min}$ (measured with matching circuit), whereas for low-noise IC amplifiers, the noise figure specs are often the measured values under the 50-$\Omega$ condition.

In the derivations from (4.16) to (4.29), we simply introduce a new set of parameters to replace the original ones. Therefore, just like the noise current and voltage, $i_n$ and $v_n$, the noise parameters in (4.29) do not correspond to any physical quantities of the device and are not directly measurable (i.e., they cannot be determined with a single measurement). In practice, the noise parameters are usually obtained through a parameter extraction process in which numerical fitting is performed over a large enough set of data points

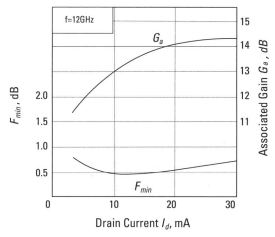

**Figure 4.4** $F_{min}$ and $G_a$ versus drain current $I_d$ for a 12GHz FET.

measured with an impedance tuner. While commercial equipment for noise parameter extraction became available in recent years, the noise characterization of a device is still largely done by the manufacturer or specialists in the field.

## 4.5 The Noise Figure in Practical Applications

For practical LNA designs, (4.29) seems to offer straightforward guidance: Select $\Gamma_s = \Gamma_{opt}$ to achieve $F_{min}$. However, both the notion of $F_{min}$ and the condition $\Gamma_s = \Gamma_{opt}$ actually need to be examined more closely in terms of actual circuit performance. Specifically, this section considers two factors—source impedance and temperature—that affect the noise performance of an amplifier.

### 4.5.1 Noise Figure and Source Impedance

As mentioned in Section 4.4, the noise figure is not only dependent on the device noise parameters but also on the source impedance (or $\Gamma_s$). In analog circuitry, the source impedance does come with a wide range depending on the circuit functions and other factors; a perfect current source and a voltage source are examples at two extremes (infinite and zero respectively). In that case, it is practically possible to select a source with a more favorable source impedance for noise performance. For RF circuits, on the other hand, the impedance of a signal source is almost always 50Ω. Then the question for the designer is how to transform 50Ω to a desired source impedance for optimal noise performance.

To underscore the basic concept of our discussion in this section, we reformulate the noise figure in a simpler form. The simplified expression for noise figure will make the algebra more manageable in our analysis, and in fact, it is also commonly used in noise analysis of analog circuits. Specifically, we make two simplifications: (1) that the source impedance (again using the term impedance loosely here, whether impedance $Z$ or admittance $Y$ is used in the analysis) is real (i.e., $Y_s = G_s$) and (b) that the noise generators, $i_n$ and $v_n$, are uncorrelated. Then (4.29) can be simplified as

$$F = F_{min} + \frac{\langle |v_n|^2 \rangle (G_{so} - G_s)^2}{4kTBG_s} \quad (4.30)$$

where

$$F_{min} = 1 + \frac{\sqrt{\langle |i_n|^2 \rangle \langle |v_n|^2 \rangle}}{2kTB} \quad (4.31)$$

and

$$G_{so} = \sqrt{\frac{\langle |i_n|^2 \rangle}{\langle |v_n|^2 \rangle}} \quad (4.32)$$

As expected, (4.30) offers the same conclusion as (4.29) regarding the optimal noise performance, that is, making $G_s$ equal to $G_{s0}$. Consider a case of $G_s < G_{s0}$; an easy solution seems to add a conductance $G'_s$ such that $G_{s0} = G_s + G'_s$. Considering that a resistive element (a shunt resistance in this case) has been treated as a noise generator, it seems counterintuitive that adding $G'_s$ would improve the noise performance of an amplifier. Upon examination of (4.15), we realize that the available noise power of the source $N_s$ is from thermal noise and independent of the source impedance. Thus, adding $G'_s$ has no effect on $N_s$. On the contrary, the available noise power from the device $N_d$ is determined by the equivalent noise voltage and current sources, $v_n$ and $i_n$, and is dependent on the source impedance. (Readers may work out the expression for $N_d$ as a function of $G_s$ using the circuit in Figure 4.3.) The function $N_d(G_s)$ reaches a minimum value at $G_{s0}$, defined in (4.32), which corresponds to the minimum of $F$, $F_{min}$. The same argument is valid for the more complex analysis in Section 4.4, which led to the expression for $\Gamma_{opt}$. Therefore, if only noise power is concerned, in the case of $G_s < G_{s0}$, increasing the source conductance by adding $G'_s$ indeed reduces the total noise power at the device input. However, amplifiers are generally for signal amplification. Once signals are taken into account, the situation is different. This leads us to examine another form of noise figure that is defined in terms of the signal-to-noise ratio (SNR), denoted as $S/N$:

$$F = \frac{\left(\dfrac{S}{N}\right)_{in}}{\left(\dfrac{S}{N}\right)_{out}} \qquad (4.33)$$

where both $S$ and $N$ are available powers from the signal and noise sources and the subscripts "in" and "out" are for the device input and output. Figure 4.5(a) illustrates the condition that we consider. It was mentioned at the beginning of

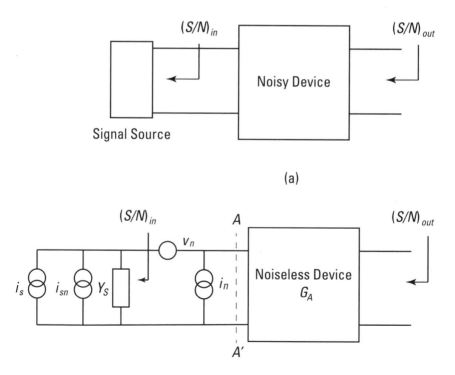

**Figure 4.5** (a) SNRs in the noise figure definition and (b) equivalent circuit for noise figure.

Section 4.4 that noise figure is essentially a measure of added noise power by the device with the thermal noise as a reference. Therefore, the noise power in the term $(S/N)_{in}$ should be understood as the thermal noise power only. This point is important when considering the cascaded noise figure in Chapter 8.

As discussed in Chapter 8, $(S/N)_{out}$ of the first-stage LNA is most critical in terms of the noise performance of a wireless receiving system, because it determines the sensitivity floor. As a matter of fact, the noise figure definition in (4.33) is more common in the literature than that in (4.14). The latter, as mentioned earlier, is more convenient as a starting point for circuit analysis.

A practically useful case is when the noisy device in Figure 4.5(a) is an attenuator. Since the noise power at the output is the same as that at the input (both are simply the thermal noise, assuming the source and the attenuator at the same temperature) and the signal is reduced by the attenuation at the output, the noise figure is simply the attenuation according to (4.33).

With the noise figure defined in (4.33), we will seek a condition that maximizes $(S/N)_{out}$ as the design goal. In the scheme shown in Figure 4.5(a), the device provides the same gain for the signal and the noise; then, we can use the same technique of equivalent noise generators shown in Figure 4.3 and move $(S/N)_{out}$ to the input of the noiseless device (at $A\text{-}A'$ terminal) as shown in Figure 4.5(b). With this equivalent circuit, utilizing the Norton current source again, we obtain

$$\left(\frac{S}{N}\right)_{in} = \frac{\langle |i_s|^2 \rangle}{\langle |i_{sn}|^2 \rangle} \tag{4.34}$$

and

$$\left(\frac{S}{N}\right)_{out} = \left(\frac{S}{N}\right)_{A\text{-}A'} = \frac{\langle |i_s|^2 \rangle}{\langle |i_{sn}|^2 \rangle + \langle |i_n|^2 \rangle + G_s^2 \langle |v_n|^2 \rangle} \tag{4.35}$$

where $i_s$ is the equivalent source current for signal. Thus, from (4.34) and (4.35) we can see that $F$ defined in (4.33) is the same as that in (4.16) (after simplifications made in this section), proving the equivalence of the two definitions. However, with the noise figure defined in the SNR, we can analyze the effects of a matching circuit on both signal and noise, as demonstrated in Sections 4.5.1.1 and 4.5.1.2.

Figure 4.6 shows two possible matching schemes, namely, (1) a conductance $G'_s$ or (2) a transformer $T_r$ that transforms the source conductance from $G_s$ to the desired value $G_{s0}$. It is important to recognize that $(S/N)_{in}$ remains the same in these schemes, and the difference is in $(S/N)_{out}$.

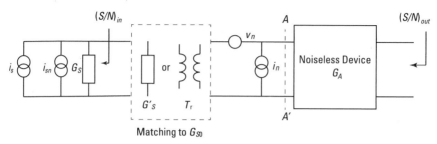

**Figure 4.6** Two matching circuits for achieving $F_{min}$.

### 4.5.1.1 Case 1: Adding $G'_s$

Although the addition of $G'_s$ changes the matching condition between the source and the device, we only need to deal with the SNR; therefore the effect of the gain is always canceled regardless of its actual value.

$(S/N)_{out}$ now becomes

$$\left(\frac{S}{N}\right)'_{out} = \frac{\langle |i_s|^2 \rangle}{\langle |i'_{sn}|^2 \rangle + (G_s + G'_s)^2 \langle |v_n|^2 \rangle + \langle |i_n|^2 \rangle} \quad (4.36)$$

Here $\langle |i'_{sn}|^2 \rangle = 4kTB(G_s + G'_s)$. Comparing with $(S/N)_{out}$ in (4.35), $(S/N)'_{out}$ has extra terms in the denominator: $\langle |v_n|^2 \rangle [(G'_s)^2 + 2G_s G'_s] + 4kTBG'_s > 0$. Thus, we have proved that adding $G'_s$ decreases $(S/N)_{out}$ and hence is not a desirable solution. Since $(S/N)_{out} = (S/N)_{in}/F$, a decrease in $(S/N)_{out}$ can also be interpreted as a decrease in $(S/N)_{in}$ or an increase in $F$, which is the reason that different explanations can be found in the literature as to why adding a resistive element is not a desirable solution.

### 4.5.1.2 Case 2: Using a Lossless Transformer to Increase $G_s$ to $G_s + G'_s$

Define transformer ratio $T_r$ as

$$T_r = \frac{G_s + G'_s}{G_s} \quad (4.37)$$

$T_r$ is related to the primary and secondary turns, $N_{pri}$ and $N_{sec}$, by $T_r = (N_{pri}/N_{sec})^2$. We will show that the value of $T_r$ that yields an optimal $(S/N)_{out}$ leads to the same conclusion as that in (4.32).

The secondary current of a transformer $I_{sec}$ is related to the primary current $I_{pri}$ by $I_{sec} = I_{pri}(N_{pri}/N_{sec}) = \sqrt{T_r} i_{pri}$, which holds for any secondary load condition including short circuit. Then in Figure 4.6, the respective Norton equivalent currents for $i_s$ and $i_{sn}$ at the terminal $A$-$A'$ become $\sqrt{T_r}i_s$ and $\sqrt{T_r}i_{sn}$. The transformer has no effect on the noise current $i_n$, and the contribution to the noise current at the $A$-$A'$ terminal from the noise voltage $v_n$ changes to $(G_s + G_s')v_n = T_r G_s v_n$. Thus $(S/N)_{out}''$ in this case is

$$\left(\frac{S}{N}\right)_{out}'' = \frac{T_r\langle|i_s|^2\rangle}{T_r\langle|i_{sn}|^2\rangle+(T_r G_s)^2\langle|v_n|^2\rangle+\langle|i_n|^2\rangle} = \frac{\langle|i_s|^2\rangle}{\langle|i_{sn}|^2\rangle+T_r G_s^2\langle|v_n|^2\rangle+\dfrac{\langle|i_n|^2\rangle}{T_r}}$$

(4.38)

Let $T_{r0}$ be the value of $T_r$ when $(S/N)_{out}''$ reaches a maximum in (4.38). Since only two terms in the denominator have $T_r$, the maximum of $(S/N)_{out}''$ occurs at:

$$\frac{\partial}{\partial T_r}\left(T_r G_s^2\langle|v_n|^2\rangle + \frac{\langle|i_n|^2\rangle}{T_r}\right) = 0$$

The solution is

$$T_{r0} = \frac{1}{G_s}\sqrt{\frac{\langle|i_n|^2\rangle}{\langle|v_n|^2\rangle}}$$

Then the corresponding source conductance after the transformer is

$$G_{s0} = G_s T_{r0} = \sqrt{\frac{\langle|i_n|^2\rangle}{\langle|v_n|^2\rangle}}$$

which is the same result as that in (4.32). Furthermore, the noise figure in this case is

$$F''(T_r) = \frac{\left(\dfrac{S}{N}\right)_{in}}{\left(\dfrac{S}{N}\right)_{out}''} = \frac{\langle|i_{sn}|^2\rangle+T_r G_s^2\langle|v_n|^2\rangle+\dfrac{\langle|i_n|^2\rangle}{T_r}}{\langle|i_{sn}|^2\rangle}$$

At $T_r = T_{r0}$, we have

$$F''(T_{r0}) = 1 + \frac{\sqrt{\langle |i_n|^2 \rangle \langle |v_n|^2 \rangle}}{2kTB} \qquad (4.39)$$

$F$ in (4.39) is the same as $F_{min}$ given in (4.31). Thus, we have proved that by using a transformer to transform $G_s$ to the optimal value $G_{so}$ we can obtain the minimum noise figure $F_{min}$. This concept can be extended to a more general case where the noise figure is given in (4.29). In that case, $F_{min}$ can be achieved by matching $\Gamma_s$ to $\Gamma_{opt}$ with a lossless network.

These analyses and conclusions can be easily applied to the case of $G_s > G_{so}$. We only need to replace the equivalent noise current source and the parallel source conductance with an equivalent noise voltage source and a series source resistance in Figure 4.6. The process will not be repeated here.

In conclusion, the noise figure is a function of both the noise characteristics of the device ($\Gamma_{opt}$, $F_{min}$ and $r_n$) and the source impedance. Only with a commonly defined source impedance is the noise figure a good measure for noise performance when comparing different devices. In the RF business, 50Ω is generally assumed for noise figure specifications.

The SNR at the device output is what matters in terms of system performance. It is this parameter that should be used in analyses of the noise performance of a circuit.

When the source impedance is transformed to $\Gamma_{opt}$ with a lossless matching network, the noise figures defined in noise power and $(S/N)_{out}$ are optimized at the same time. This is a key concept in low noise RF amplifier designs.

### 4.5.2 Temperature Effects on Noise Performance and Characterization

It is important to note that the unit for temperature in all the noise related formulas is kelvin (K). The average room temperature (or the average temperature at the Earth's surface for outdoor applications) is usually taken as 298K (25°C). On the Kelvin scale, a temperature range that is considered sufficiently wide by typical application requirements, say, −40 to 85°C, is still only a ±20% variation from the average. For this reason, in practical designs, the focus is usually on the performance at room temperature, with limited effort given to extreme temperatures other than ensuring that the system is functional at these temperatures. However, for a more precise system characterization, temperature is still a critical factor. This section discusses several temperature related topics.

#### 4.5.2.1 Standard Temperature for the Noise Figure

In (4.24), the noise figure is explicitly written as a function of the source and device temperatures, $T_s$ and $T_d$. If noise figure is used as a differentiating factor in device or system selection, both temperatures should be standardized. In practice, the device temperature is nominally specified at 25°C. The IEEE specification for the source temperature in noise figure measurements is 290K (or 16.85°C) [15, 16]. While in principle the actual device temperature in the noise figure measurement should be at the specification temperature, no such condition is required for the source temperature if the Y method is employed. (See Chapter 10 for a detailed description.) The source temperature standard is for calculation only. The difference between 25°C and 17°C is small enough that the formulas derived in the last section under the assumption $T_s = T_d$ are accurate for virtually all practical purposes. To use the noise figure of an amplifier measured at room temperature for estimating the noise performance in a real-world environment, both the actual source temperature and the amplifier operation temperature need to be reasonably close to 25°C. If either one significantly deviates from the standard condition, the noise figure should be modified accordingly. We consider each case separately in the following section.

#### 4.5.2.2 Antenna Noise Temperature and Noise Figure

Inspection of (4.16) reveals that the source impedance (represented by $T_s$ in our analysis) has two roles in noise figure analysis: (1) it is responsible for the source noise generator $i_{sn}$ and (2) it presents a load to the device noise generators, $i_n$ and $v_n$. In a wireless system, signals are from an antenna. Since there are no physical resistive elements inside the antenna, questions arise as to (1) what the impedance is in place of $Y_s$ in noise calculation using (4.16) and (2) where the source noise originates. To the first question, the answer is the input impedance of the antenna, which is almost always 50Ω in practice. As a load, an antenna has the same effect as a physical resistance on the generators. The difference is that the antenna radiates the power instead of absorbing it. Also, antennas are generally narrowband.

The answer to the second question is less obvious. Noise power at the output of an antenna into the next stage amplifier is, just like signals, from the background radiation captured by the antenna. The underlying principle for this radiation is the physics of thermal radiation, also known as blackbody radiation. Detailed discussions on this topic can be found in the literature [6, 17]. For the purposes of this book, note that the amount of noise power from the radiation depends on the temperature of the object that emits the radiation. In practice, antenna noise temperature is used to characterize the antenna in noise analysis. Conceptually, antenna noise temperature, denoted as $T_{ant}$, can be loosely interpreted as the temperature of the objects that the antenna sees, and the noise power from an antenna $N_{ant}$ is related to $T_{ant}$ rather than to the physical temperature of the antenna according to Nyquist's law:

$$N_{ant} = kBT_{ant} \qquad (4.40)$$

Therefore, $T_{ant}$ is the temperature for $T_s$ in the noise figure calculation.

For typical terrestrial wireless applications such as cellular and Wi-Fi systems, $T_{ant}$ is close to room temperature, because most objects to which antennas are aimed are at the Earth's surface temperature. Therefore, in these applications, a noise figure measured in the lab is usually used for the system analysis without any further questions. On the other hand, for satellite communication systems, antennas are aimed at the sky, which can be much colder than the Earth's surface. Then the corresponding antenna noise temperature can be significantly lower than the room temperature. An exception is an event, known as a Sun outage or Sun transit, which occurs when the Sun is directly aligned with the satellite and the receiver. During this event, the receiver antenna directly receives the thermal radiation from the sun (noise), causing a temporary loss of reception.

The actual value of $T_{ant}$ varies greatly depending on various factors and can be estimated using measurement data under the specific condition [17, 18]. We are only concerned here with how the noise figure is dependent on $T_{ant}$. To this end, we write (4.24) as

$$F_\alpha = 1 + \alpha A(T_d) \qquad (4.41)$$

where $\alpha = T_d/T_s$ and $A(T_d)$ is given by:

$$A(T_d) = \frac{G_{nc} + R_n\left[(G_c + G_s)^2 + (B_c + B_s)^2\right]}{G_s}$$

$F_\alpha$ is related to the regular noise figure $F(\alpha = 1)$ by:

$$F_\alpha - 1 = \alpha(F - 1) \qquad (4.42)$$

From (4.15), $F - 1$ is the ratio of the noise from the device to that of the source; then (4.42) implies that when the source temperature is lower than the device temperature ($\alpha > 1$) the noise contribution from the device, relative to the source noise, is increased by a factor of $\alpha$. Take $F = 1.12$ (0.5 dB) as an example, for $T_{ant} = 100K$ ($\alpha = 3$, if $T_d$ is taken to be 300K), $F_{\alpha=3} = 1.36$, which means that the noise from the device is still 36% of the source noise, as opposed to 12% when $T_s = T_d$. This is why engineers working on satellite communication systems are often more willing to go the extra mile for noise figure reduction.

### 4.5.2.3 Noise Temperature

The preceding discussion leads us to the concept of noise temperature, a term more commonly used in satellite communication systems than noise figure. Figure 4.7 illustrates how this term is defined. In this scheme the noise contribution from the device ($N_d$ in (4.15)) is represented by an equivalent input noise temperature, (usually simply called noise temperature) denoted by $T_e$, that is, an increase in source temperature by $T_e$ generates extra noise power equal to $N_d$:

$$N_d = kBT_e \qquad (4.43)$$

Assuming the physical temperature of the source is $T_s$, then the total noise power at the device input is:

$$N_d + N_s = kB(T_e + T_s) \qquad (4.44)$$

By letting $T_s$ be the reference temperature $T_0$, we can relate $T_e$ to noise figure $F$ as

$$T_e = T_0(F - 1) \qquad (4.45)$$

Unlike noise figure, which depends on the source temperature, the noise temperature of a device is strictly a noise characteristic of the device, as indicated in (4.43). Therefore, for applications where the source temperature is unspecified, noise temperature is more convenient than noise figure as a metric for device comparison. Using the same example of $F = 1.12$, according

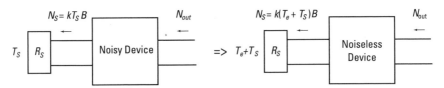

**Figure 4.7** Concept of equivalent input noise temperature.

to (4.45), $T_e = 35K$ in this case. Then for $T_s = 100K$, we immediately reach the same conclusion as that obtained from (4.42).

#### 4.5.2.4 Effect of Device Temperature on Noise Figure

Next we turn our attention to the effect of device temperature on noise figure. In some applications system designers need to be concerned with the noise figure degradation at certain higher temperatures, say 85°C for commercial applications and 125°C for automotive (grade 1) applications. Most vendors do not specify the noise figure at higher temperatures and only offer measured high temperature data in special situations. Analytically, accurate numerical estimates turn out to be difficult due to the factor $A(T_d)$ in (4.41), which generally has a complex dependence on temperature. Nevertheless, as a rough approximation, we assume $A(T_d)$ to be constant. Then (4.41) can be used to estimate the noise figure at higher temperatures. Table 4.2 shows the numerical results for two temperatures, 85°C and 125°C.

Generally, the degradation in noise figure in these temperatures could be considered secondary but not completely insignificant. In reality, measured data can be somewhat higher or lower than the values in the table.

#### 4.5.2.5 Cryoelectronics

For the completeness of our discussion we briefly mention a case where both $T_s$ and $T_d$ are significantly lower than the reference temperature, which is called cryoelectronics. In this condition, noise figure is no longer a useful figure of merit because the design consideration is the absolute noise level (noise from the source plus noise from the device) instead of the ratio of them.

Cryoelectronics is a highly specialized engineering field. For the purposes of this book, it is sufficient to mention that GaAs FETs are usually a good choice for the devices used in this application.

**Table 4.2**
Noise Figure $F_\alpha$ at Higher Device Temperatures, $F_0$ is the Noise Figure at $T_s = T_d = 25°C$

| | | $F_\alpha$ | | $F_\alpha$ (dB) | |
|---|---|---|---|---|---|
| $F_0$(dB) | $F_0$ | 85°C | 125°C | 85°C | 125°C |
| 1 | 1.26 | 1.31 | 1.35 | 1.18 | 1.29 |
| 0.5 | 1.12 | 1.15 | 1.16 | 0.59 | 0.66 |

# References

[1] Weston, D. A., *Electromagnetic Compatibility*, Marcel Dekker, 2001.

[2] Montrose, M. I., *EMC and the Printed Circuit Board*, IEEE Press, 1999.

[3] van der Ziel, A., *Noise*, Prentice-Hall, 1954.

[4] Engberg, J., and T. Larsen, *Noise Theory of Linear and Nonlinear Circuits*, John Wiley and Sons, 1995.

[5] Ott, H. W., *Noise Reduction Techniques in Electronic Systems*, Second Edition, John Wiley and Sons, 1988.

[6] Reif, F., *Fundamentals of Statistical and Thermal Physics*, Waveland Press, 2009.

[7] Haykin, S., *Communication Systems*, Third Edition, John Wiley and Sons, 1994 (and later editions).

[8] Ziemer, R. E., and W. H. Tranter, *Principles of Communications*, Seventh Edition, Wiley, 2015.

[9] Giancoli, D. C., *Physics for Scientists and Engineers*, Third Edition, Prentice Hall, 2000.

[10] Nyquist, H., "Thermal Agitation of Electric Charge in Conductors," *Phys. Rev.*, Vol. 32, No. 1, 1928.

[11] Rothe, H., and W. Dahlke, "Theory of Noisy Foupoles," *Proc. of the IRE*, Vol. 44, No. 6, 1956.

[12] Schwierz, F., and J. J. Liou, *Modern Microwave Transistors, Theory, Design and Performance*, John Wiley and Sons, 2003.

[13] Sze, S. M., and K. K. Ng, *Physics of Semiconductor Devices*, Third Edition, John Wiley and Sons, 2007.

[14] Alexander, C. K., and M. N. O. Sadiku, *Fundamentals of Electric Circuits*, McGraw-Hill, 2000.

[15] Adler, A., et al., "Description of the Noise Performance of Amplifiers and Receiving Systems," *Proc. of the IEEE*, Vol. 436, 1963.

[16] *The New IEEE Standard Dictionary of Electrical and Electronics Terms*, Fifth Edition, IEEE, 1993.

[17] Maral, G., and M. Bousquet, *Satellite Communications Systems*, Fifth Edition, John Wiley and Sons, 2009.

[18] Balanis, C. A., *Antenna Theory*, Third Edition, John Wiley and Sons, 2005.

# 5

# RF Amplifier Designs in Practice

Chapters 1–4 introduce several important concepts and principles in RF device characterization and circuit design. This chapter discusses various practical design issues with emphasis on PCB-based RF amplifier circuits. Materials in this chapter are mainly, but not completely, limited to linear circuits. Chapters 6 and 7 cover power RF amplifiers.

## 5.1 Circuit Specifications and Transistor Selection

### 5.1.1 Circuit Specifications of Linear Amplifiers

The first step of an amplifier design usually involves the definition of specifications. For linear amplifiers, they generally include operation frequency, gain, input and output return losses, and noise figure in the case of low noise applications. In addition, in many wireless communications systems, various linearity specifications are also required for out-of-band interference rejections, even when the in-band signal is well within the linear region of an amplifier. This topic is left to Chapter 8. This chapter focuses on the linear parameters.

The frequency characteristics of an amplifier are usually the first specification to be considered. Most wireless protocols operate in a relatively narrow frequency band, simply because of spectrum restrictions by regulatory agencies.

For these applications, other than avoiding extremely high-Q matching, which were discussed in Chapter 3, a design for the midpoint of the frequency band is usually sufficient. On the other hand, if a broadband frequency response in gain or return loss is demanded, some special circuit configurations might be needed. Two such cases are discussed in Section 5.4.

The fundamental function of an RF amplifier is to provide power gain. Generally, a gain considerably lower than 10 dB is too low to be worth one stage of amplification, unless a special situation justifies such an implementation. At the same time, a single-stage RF amplifier with a gain considerably more than 20 dB is difficult to practically implement, due to issues of stability and consistency in matching. For these reasons, most single-stage RF amplifiers in practice have a gain between 10 and 20 dB. In addition, at higher frequencies and high power levels, the availability of transistors has to be taken into account in defining the gain specification of an amplifier.

The consideration of return loss is slightly more complicated. The reason for seeking a good return loss is often cited in the literature as to improve the gain, which, in principle, is correct as indicated by (2.22). But numerically, unless the return loss is very poor (e.g., worse than 5 dB), the increase in gain from an improvement in return loss may not be significant enough to justify either a degradation in noise figure (see Section 5.3 for a detailed discussion) or the extra components required for a network that yields a better match. In practice, the specification for return loss of an RF amplifier often stems from concerns over how the amplifier interfaces with other components. One issue is the interference or loading effect (i.e., changing the operation conditions) due to excessive power reflection to the source that drives the amplifier. The other issue is the potential large variation in performance specification caused by a poor return loss. This second issue can be especially problematic for a high volume and multisite production environments because of the yield requirements. To quantitatively analyze this problem, we revisit the mismatch factor defined in (1.32)

$$M = \frac{\left(1-|\Gamma_S|^2\right)\left(1-|\Gamma_L|^2\right)}{|1-\Gamma_S\Gamma_L|^2} \quad (5.1)$$

To be specific in our discussion, we assume that the amplifier is the load [i.e., its input return loss is related to $\Gamma_L$ in (5.1)] and the source to the amplifier is characterized by $\Gamma_S$. The premise for our consideration is that $\Gamma_S$ is uncertain, particularly its phase (recall the phase rotation due to a transmission line). Then we notice that $M$ in (5.1) is unpredictable even if $|\Gamma_S|$ and $|\Gamma_L|$

are given because of the uncertain term $|1 - \Gamma_S\Gamma_L|$ in the denominator, which has two extremes at $1 - |\Gamma_S||\Gamma_L|$ ($\Gamma_S$ and $\Gamma_L$ are in phase) and $1 + |\Gamma_S||\Gamma_L|$ (out of phase). Figure 5.1 shows the difference in decibels of the $M$ values at these two extremes, $\Delta M = 10\log(1 + |\Gamma_S||\Gamma_L|) - 10\log(1 - |\Gamma_S||\Gamma_L|)$, as a function of $|\Gamma_L|$ for two $|\Gamma_S|$ values. It can be seen from Figure 5.1 that for a moderate value of $|\Gamma_S| = 0.3$ (10-dB return loss), there is a 2.5-dB uncertainty in mismatch factor at $\Gamma_L = 0.5$ (6-dB return loss). Considering that the system tolerance is an aggregation of multiple factors, a potential variation of 2.5 dB from a single stage can be unacceptable. Of course, $\Delta M$ in the figure is the worst-case scenario. Nevertheless, this analysis reveals why a good return loss is generally desirable in volume production; otherwise, an excessively large tolerance would be required to achieve an admissible production yield. In practice, 10 dB is generally considered reasonable for a return loss specification, while special efforts must be made, in selection of circuit topology and component tolerance, to maintain a 20-dB return loss specification for amplifiers operating in the gigahertz range in production.

### 5.1.2 Transistors for RF Applications

The semiconductor device technology for RF applications is constantly evolving. Since the first generation (1G) mobile phone system in the 1980s (generally considered the start of the boom of RF technology in wireless communication applications), the most noticeable advances are perhaps in RF complementary

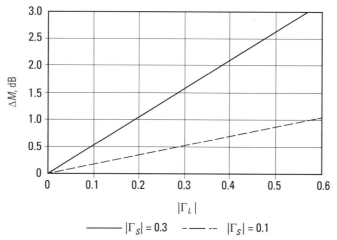

**Figure 5.1** Maximum variation in mismatch factor due to phase uncertainty.

metal-oxide-semiconductor (CMOS) technology. It is a long-held understanding that silicon is inferior to compound semiconductors, such as GaAs and GaN, for RF devices due to its relatively low electron mobility (which determines the saturation speed of electrons in the channel). However, the upper frequency limit of a field-effect transistor (FET) is roughly inversely proportional to the gate size. Therefore, the low electron mobility in Si is compensated for by the continued reduction in the channel length along with other technology advances. Currently, the operation frequency of Si-based CMOS is well into the tens of gigahertz and beyond. As a result, highly integrated RF integrated circuits (ICs) are almost all CMOS-based. [In the microwave business, it is customary to refer to ICs as monolithic microwave ICs (MMICs).] On the other hand, various discrete RF transistors are still widely used in practice because of their unmatched performance in high-power, low-noise, and high-frequency (>10-GHz) applications. For discrete transistors, flexibility in frequency and configuration selections, and in some cases, lower cost, are also advantageous versus IC solutions in many situations. Moreover, for applications of moderate market demand, developments of an IC or an integrated module may simply not be a profitable solution.

There are two major categories of RF transistors, bipolar junction transistors (BJTs) and FETs. Each has two main subcategories: For BJTs, there are NPN and PNP transistors, and for FETs, there are depletion and enhancement mode transistors. The PNP transistor has a relatively low-frequency range and is normally used with an NPN BJT in a complimentary configuration, such as a push-pull amplifier [1, 2]. In practical RF amplifier designs using a single transistor, the NPN and the two FETs are normally considered, based on the circuit specifications. For general principles of RF transistors, two texts [3, 4] written in the 2000s still provide sufficient information, although some performance data cited there may be outdated. In terms of device selection for a specific application in practice, it is difficult to offer accurate guidance because of the constant changing nature of the market both in device performance and availability. Instead, it is advised to thoroughly research the market to determine the best options for a given design task.

From a circuit design perspective, the design techniques in terms of matching for RF performance, whether it is a linear or noise specification (discussed in Chapters 3 and 4) or a power one (discussed in Chapter 6), essentially remain the same regardless of the device types. The main difference is in the input DC bias circuit (base for BJTs and gate for FETs). A BJT is a current-controlled device, and its input port (base-emitter) behaves like a diode. In comparison, a FET is a voltage-controlled device with very high input (gate-source) DC resistance, which can be virtually treated as an open circuit in most practical cases. Since the power supply for a circuit board is always a voltage source, the base bias circuit of a BJT needs to be carefully designed so that the base current is set at the target value. The bias circuit for the gate of an FET is much simpler because the gate current is usually small enough to be ignored. The difference in gate bias circuits for enhancement and depletion modes FETs [1, 2] is more significant in design considerations. The enhancement-mode transistors include Si-based laterally diffused metal-oxide semiconductors (LDMOSs) and some types of GaAs FETs, while most GaAs pseudomorphic high-electron mobility transistors (pHEMTs) operate in the depletion mode. A depletion-mode transistor requires a negative supply voltage, which increases cost and circuit complexity. For this reason, depletion-mode FETs are used predominantly in applications at higher frequencies (>10 GHz) where their RF performance currently still offers considerable advantages over other types of RF transistors.

Bias circuits are an important part of an RF circuit design. References [5–7] provide detailed discussions on various bias circuit schemes for both BJTs and FETs. A bias circuit recommended by the manufacturer for a specific transistor is also a good starting point in practice. This book's discussion of this topic is limited to Section 5.2, which describes the stability associated with bias circuits.

### 5.1.3 Datasheet Specifications for RF Transistors

Since device selection usually starts with datasheets, let's briefly consider the datasheet specifications for typical RF transistors, focusing on several items that are practically important and sometimes confusing due to differences in terminology and methods by manufacturers.

We start with a note on transistor configurations. A transistor has three terminals. In analog circuits, transistors can be connected in different ways for various circuit functionalities. For RF transistors, the RF performance is generally specified in the amplifier configuration where one terminal is the common (emitter for BJTs and source for FETs) terminal, and the other two are assigned to be the input (base and gate) and output (collector and drain) respectively.

#### 5.1.3.1 Usable Frequency Range and Transition Frequency $f_T$

The first question in transistor selection for an RF application is whether the device is usable at the specified frequency. Traditionally, this question is answered by the transition frequency, $f_T$, which also goes by the names of gain bandwidth product (or simply gain bandwidth), unity gain frequency, and cutoff frequency. Regardless of the name, $f_T$ is the frequency where the short-circuit current gain, denoted as $h_{21}$ (one of the $h$ parameters), becomes unity.

$f_T$ is a well established concept for the frequency response of a transistor in analog circuit theory, and its value is determined from the function of $h_{21}(f)$, which can be derived based on the small-signal equivalent circuit models for BJTs or FETs [1, 8]. At sufficiently high frequencies (beyond the −3-dB roll-off point [1, 2]), we can use the much simplified FET equivalent circuit shown in Figure 5.2 for illustration. From Figure 5.2, we have

$$h_{21} = \frac{i_2}{i_1} = \frac{g_m}{2\pi C_1} \frac{1}{f} \qquad (5.2)$$

$f_T$ is the frequency point when $h_{21} = 1$. Then (5.2) indicates that the current gain at any frequency $f$ can be calculated by

$$h_{21}(f) = \frac{f_T}{f} \qquad (5.3)$$

Equation (5.3) is the basis for $f_T$ being used as guidance for the usable frequency range of a transistor.

Although $f_T$ was originally used to characterize the frequency response at analog and low-RF ranges, it is still a standard figure of merit for modern microwave transistors when the frequencies are well into the domain of hundreds of gigahertz or even higher. The use of $f_T$ is valid only if the inverse proportionality in (5.3) holds, which is generally true at the device level in spite of increased complexity in device structures. However, in most commercial applications, packaged devices are used. At high enough frequencies (roughly > 10 GHz), the parasitic effects associated with the packaging components (e.g., bonding wires and soldering pads) are no longer negligible. Then, the device characteristics measured by the S-parameters are inevitably, and in some cases intentionally, distorted by the effects of the packaging components. As a result, the frequency responses of packaged high-frequency transistors often do not

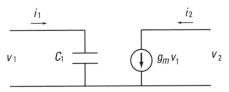

**Figure 5.2** Simplified small signal equivalent circuit for FETs.

follow the inverse proportionality in the usable frequency range, rendering $f_T$ almost meaningless in predicting the device performance at frequencies other than $f_T$ itself (where $h_{21} = 1$), as illustrated in an example shown in Figure 5.3.

We first note that $h_{21}$, which is defined in terms of the input and output currents, is not a parameter directly measurable in RF systems. In practice, $h_{21}$ is calculated from the measured S-parameters using the conversion formula for $S$ to $h$ parameters [5, 6]:

$$h_{21} = \frac{-2S_{21}}{(1 - S_{11})(1 + S_{22}) + S_{12}S_{21}} \tag{5.4}$$

In addition to $f_T$, another parameter is commonly used for frequency characterization in the RF business, namely, the maximum oscillation frequency (or simply maximum frequency), $f_{MAX}$, which is defined as the unity gain frequency of either unilateral gain or *MAG* (see Section 2.3). Here we use the latter definition. Although $f_{MAX}$ appears more relevant to RF designs since it is defined in RF gains, it also fails to predict the entire frequency response for a packaged device when the frequency in question is sufficiently high.

Figure 5.3(a) shows plots of $|h_{21}|^2$ and *MAG* along with $|S_{21}|^2$ versus frequency for a GaAs FET designed for applications around 20 GHz. For comparison, the same plots are shown in Figure 5.3(b) for a traditional Si BJT with $f_T = 10$ GHz. The frequency responses in Figure 5.3(b) follow a slope of −20 dB/decade, reasonably well over more than a decade, indicating the validity of (5.3). In contrast, from the plot in Figure 5.3(a), $h_{21}$ starts to deviate at around 4 GHz from the −20-dB/decade relationship and is completely irrelevant at higher frequencies (as is $f_T$,) in predicting the gain performance over the frequency. The huge peak in $h_{21}$ around 20 GHz is due to the two terms in the denominator in (5.4) nearly canceling each other in that frequency range.

This example explains why manufacturers of high-frequency transistors, particularly FETs, do not always provide an $f_T$ specification in their datasheets. The design of a modern high-frequency transistor is often optimized for a specific high-volume application, which is usually reflected in the frequency chosen for the specifications of the datasheet. The best practice for the designer to determine the usable frequency range of a transistor is to use the S-parameters in combination with the frequency specified on the datasheet.

### 5.1.3.2 Maximum Voltage and Maximum RF Power Ratings

All datasheets for RF devices have a section on maximum ratings. Unlike most other specifications on the datasheet, a maximum rating generally cannot be

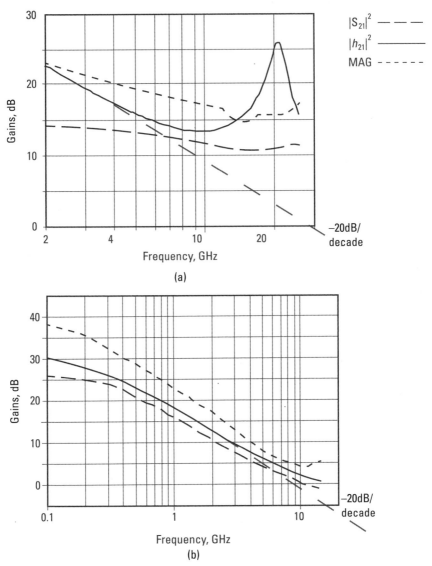

**Figure 5.3** Gains versus frequency: (a) a 20-GHz GaAs FET and (b) a Si BJT.

independently verified by the user in the sense that there is usually no immediate observable effect upon exceeding the specification. The determination of a maximum rating is usually associated with the long-term reliability data for the specific device. The majority of specifications in this category are self-explanatory. This section focuses on two maximum ratings that are often somewhat confusing or simply missing from datasheets, namely, maximum ratings for voltage and maximum RF input power.

The RF performance of an amplifier, particularly the power-related performance specifications, can often be improved by increasing the bias voltage. The question is then how high the bias voltage can be set for a given transistor. The maximum voltage rating is obviously most relevant in regard to this design consideration, but perhaps this specification is also the least understood in terms of how to apply it in actual circuit designs. The main difficulty is the lack of a clear connection between the upper limit on the DC bias voltage that can be set by the user and the maximum voltage rating. The latter is generally understood as the total voltage under all conditions including DC and RF voltages as well as any possible transient voltage spikes, which can, at best, only be estimated for an actual circuit condition. In practice, the situation in this regard varies greatly among different devices and manufacturers. The RF operation power of a specific device is usually the key factor in the consideration. This discussion is intended as a general overview on the topic rather than a quantitative guideline.

For low- and medium-power RF transistors (1W or less), the maximum rating for the voltage across two specific transistor terminals is usually related to the corresponding breakdown voltage. A detailed description of the physics and circuit behaviors of various breakdown processes in transistors can be found in a number of textbooks on device physics and the circuit theory of transistors, such as [1, 4]. Some manufacturers, especially BJT manufactures, list the breakdown voltages on the datasheets and use the minimum breakdown voltages for the maximum voltage ratings. In other cases, maximum voltage ratings are determined based on measured breakdown voltages with a certain amount of back-off.

For BJTs, the breakdown voltages are usually measured with an open circuit at the third terminal. For example, $BV_{CEO}$ is the breakdown voltage of the collector-emitter voltage with the base unconnected (open). However, in real application circuits, the base is always terminated with a load, and the breakdown voltage under that condition can be considerably higher than $BV_{CEO}$. To amend this situation, manufacturers sometimes also provide $BV_{CES}$, which is the same breakdown voltage with the base shorted or even a $BV_{CER}$ plot that shows the breakdown voltage versus base resistance $R_{EB}$. Figure 5.4 shows such a plot for two NPN transistors. (Their selection has no specific significance other than to demonstrate the typical range of the $BV_{CER}$ vs. $R_{EB}$ relationship.) Since $R_{EB}$ in a practical circuit is usually less than 1 k$\Omega$, from the plots we can see that the actual breakdown voltage in a circuit is typically two to three times higher than the datasheet specification $BV_{CEO}$. For this reason, setting an operation voltage at the maximum voltage rating, which is the minimum of $BV_{CEO}$, is usually considered acceptable for small signal BJTs.

For FETs intended for RF amplifiers, the source is always ground. Therefore, the breakdown voltages are usually characterized between the drain-source terminals and the gate-source terminals. However, breakdown voltages are not always specified by FET manufacturers. The user should use maximum voltage ratings in the design process. Unlike BJTs, the external DC load at the gate generally has no effect on the drain-source breakdown voltage. Therefore, the maximum voltage rating for an FET must be below the actual breakdown voltage regardless of the gate resistance.

Another layer of complication is the effect of RF voltage swing. Breakdown voltages are normally measured in the DC condition. The mechanism for breakdown associated with RF power can be considerably different depending on the frequency. Furthermore, the RF voltage at the device is not a directly measurable quantity, although it can be estimated using the RF power and the load impedance at the transistor output. In general, an accurate account of the effect of an RF swing is unrealistic for practitioners. A rule sometimes cited in the literature regarding the bias point is based on the operation principle of class A amplifiers (Chapter 6 and [1, 8]), which states that the amplitude of RF swing equals the DC voltage at the onset of nonlinearity. Then, the DC voltage should be no higher than the midpoint of the maximum rating in order to keep the total voltage below the maximum rating. While this rule is reasonable and useful in practice, if the input RF power is expected to reach

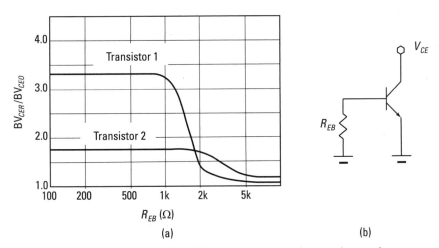

**Figure 5.4** (a) Collector-emitter breakdown voltage versus base resistance for two transistors and (b) measurement circuit.

the compression point (see Chapter 6), it is definitely too restrictive for small signal operations. As a practical matter, the bias condition used in datasheet specifications is usually a good starting point. If a benefit for increasing the bias voltage is clearly identified, the user may be able to increase the bias voltage by a certain amount without any compromise in long-term reliability. Generally, the risk of a bias point being too close to the maximum rating drops with the RF power level. Ultimately it is an engineering decision by the designer.

For ultra high-power RF amplifiers (several tens of watts or higher), the situation is more complex. Besides steady-state breakdown voltages, thermal and transient characteristics are also critical considerations for DC bias voltages. The concern of transient characteristics is due to possible high-voltage spikes induced by a sudden change in RF power, whether it is intentional as in the RF-pulsed amplifiers, or unintentional as in the case of a high-RF power reflection from the amplifier output caused by a sudden disconnection of the output load. Generally, the user of ultra high-power transistors has little discretion in selecting the bias voltage point and should strictly follow the manufacturer's recommendation regardless of the maximum voltage rating on the datasheet. In other words, the maximum voltage rating, which can be as high as three times higher than the recommended DC bias voltage, is of little practical use in ultra high-power amplifier designs.

Maximum input RF power is another specification that is frequently questioned by the user. This maximum rating is usually provided for RF amplifier ICs but is often absent in RF transistor datasheets. The absence of this maximum rating is primarily due to it usually being affected by the matching networks used in the circuit as well as by the operation frequency. In other words, a rating on the maximum input RF power is more of a specification for a circuit than one for a transistor. For a specific circuit, the input RF power is in fact limited by one of the maximum specifications of the transistor, including junction currents specified at DC (e.g., maximum gate current and maximum drain current for FETs) and power dissipation. The generation of DC currents by RF power is due to the rectification effect. Chapter 7 explains this effect in more detail. The increase in power dissipation is simply due to the increased output RF power. Since all these specifications are DC parameters (note that the power dissipation inside the transistor is the DC power minus RF output power) and can be relatively easily measured, the maximum input RF power for a circuit can be determined by monitoring these parameters while gradually increasing the RF power until one of these parameters reaches its maximum rating. For FETs, the maximum gate current is often the limiting parameter.

#### 5.1.3.3 Gain Specifications

There are various gain specifications on manufacturers' datasheets. The names used for the gain definition vary and are not always clearly defined. In practice, the symbols are more consistent. There are generally three types of gains provided: first, maximum stable gain (MSG) (if the stability factor $K < 1$) at the specified frequency or maximum available gain $MAG$ (for $K > 1$); second, $|S_{21}|^2$. For BJTs, it is often seen as $|S_{21e}|^2$ with "e" for the common emitter configuration; and third, for a low-noise device, $G_a$ (associated gain) or $G_{ma}$ (the maximum available gain) is often used, which is the available gain when $\Gamma_{opt}$ is used for $\Gamma_S$ in (2.14). For power devices, a more relevant gain specification is that at specified output power.

## 5.2 Stability Considerations

In RF amplifier designs, perhaps nothing frustrates an engineer more than unintended oscillations. These oscillations often occur at certain discrete frequencies and their harmonics. They can also appear on a spectrum analyzer as a broadband background noise with excessively high amplitude plus some random spikes. In some cases, the oscillation is also sensitive to the termination condition of the amplifier. While the detail varies, these oscillations are usually caused by the instability of the circuit. (One possible exception is the cavity effect discussed at the end of this section.) This section uses the stability factors $K$, as well as $\mu$ or $\mu'$ parameters introduced in Chapter 2, which are effective indicators for the stability of an amplifier. At the same time, this section points out the wide misconception that unconditional stability across frequency must be obtained at all cost and under all circumstances. In fact, an RF amplifier with conditional stability can work perfectly fine in certain applications. This is because the condition $K < 1$ is a necessary but not sufficient condition for oscillation. Section 5.2.1 examines the conditions under which an oscillation actually takes place. Then Section 5.2.2 describes how to use the stability parameters in the circuit design and evaluation. Section 5.2.3 discusses mechanisms for several common types of unintended oscillations.

### 5.2.1 Conditions for Oscillation

This section uses the block diagram of an amplifier circuit in Figure 5.5(a) to facilitate our analysis. Figure 5.5(a) features several circuit elements and parameters that are important to the circuit stability, and we consider each in detail. We first utilize $\Gamma_{out}$ and $\Gamma_L$ (see Figure 5.5) to describe the conditions

# RF Amplifier Designs in Practice

for oscillation. (The same analysis can be performed using $\Gamma_{in}$ and $\Gamma_S$, as explained in Chapter 2.) According to (2.24), for a system to start oscillation, the following condition is required:

$$\Gamma_{out}(\omega) \cdot \Gamma_L(\omega) = 1 \qquad (5.5)$$

Here both $\Gamma_{out}$ and $\Gamma_L$ are expressed explicitly as a function of frequency to suggest that an oscillation only occurs at certain frequencies where (5.5) holds.

Let $Z_L (= R_L + jX_L)$ and $Z_{out} (= R_{out} + jX_{out})$ be the corresponding impedances for $\Gamma_L$ and $\Gamma_{out}$; then from (1.13), we have

$$\Gamma_L = \frac{Z_L - Z_0}{Z_L + Z_0} \qquad (5.6)$$

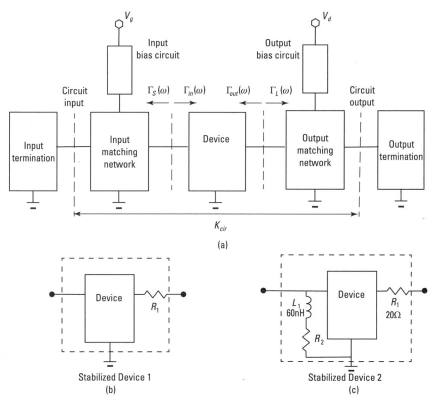

**Figure 5.5** (a) Block diagram of an RF amplifier for stability analysis, (b) stabilized device configuration 1, and (c) stabilized device configuration 2.

and

$$\Gamma_{out} = \frac{Z_{out} - Z_0}{Z_{out} + Z_0} \tag{5.7}$$

Substituting (5.6) and (5.7) into (5.5), the real and the imaginary parts of the resulted equation are

$$R_L(\omega) + R_{out}(\omega) = 0 \tag{5.8}$$

$$X_L(\omega) + X_{out}(\omega) = 0 \tag{5.9}$$

For a passive load $|\Gamma_L| < 1$, (5.5) is possible only if $|\Gamma_{out}| > 1$. In fact, $|\Gamma_{out}| < 1$ was the starting point for the unconditional stability condition (2.33) in our analysis in Chapter 2. Recall that $K < 1$ only implies that there is an unstable region inside the unit circle in the $\Gamma_S$ plane. For a system to oscillate, two additional conditions on $\Gamma_{out}$ and $\Gamma_L$ need to be met at a certain frequency or over a frequency range: First, $\Gamma_S(\omega)$ is actually in the unstable region, that is, $|\Gamma_{out}(\Gamma_S)| > 1$; second, in the frequency range where $|\Gamma_{out}| > 1$ a load impedance $Z_L(\omega)$ exists that satisfies (5.8) and (5.9). Therefore, an amplifier with $K < 1$ can still be stable (without oscillation) as long as the above conditions are avoided.

The condition $|\Gamma_{out}| > 1$ is considered a strong indication of instability. In practice, if a measurement of $S_{11}$ or $S_{22}$ indicates a value greater than unity, it is customary to refer to the system as having negative resistance. The reason for this term is clear from (5.8), since $R_{out}(\omega)$ must be negative if the load is passive ($R_L < 1$).

In oscillator designs, the condition on the imaginary parts in (5.9) is equally important as that on the real part in (5.8), but it is less often discussed in the context of unintended oscillations caused by instability. The essence of that condition is that between $X_{out}$ and $X_L$, one of them must be inductive (positive reactance) and the other capacitive (negative). Note that both $X_{out}$ and $X_L$ are an equivalent reactance of a one-port network and generally cannot be identified with any particular circuit element, especially for $X_{out}$. Typically, when an amplifier circuit oscillates, the inductive part is from $X_L$ due to the inductance in the bias circuit. The required inductance that satisfies (5.9) becomes unrealistically high at very low frequencies, which is why unintended oscillations of an RF amplifier circuit rarely occur below 10 MHz regardless of the $K$ value.

This analysis of unintended oscillations follows an oscillator design technique commonly known as the negative-resistance method. The framework of this technique was first outlined by Kurokawa in 1969 [9], and other treatments using this technique can also be found in the literature [6, 10, 11]. For an oscillator to function properly, the oscillation has to be stable both in frequency and amplitude. The steady state requirement imposes additional conditions on the amplitude dependence of $Z_{out}$ and $X_L$. This dependence is ignored in our analysis because for an RF amplifier circuit any spurious oscillation is absolutely not acceptable, whether it is stable or not.

### 5.2.2 Stability Analysis Using Stability Parameters

The above analysis, which is based on (5.5), provides a more precise condition (in comparison with $K < 1$) for determining whether or not a circuit oscillates. However, it is difficult to use (5.5) in a practical design. The main difficulty is that in order to compute $\Gamma_{out}$ and $\Gamma_L$, full knowledge of all circuit elements [see Figure 5.5(a)] is required prior to stability analysis. In fact, in some cases, the terminations are simply unknown, making calculations of $\Gamma_{out}$ and $\Gamma_L$ impractical. A more realistic approach commonly employed in practice is to utilize one of the stability parameters, $K$, $\mu$ and $\mu''$, to evaluate the stability characteristics of the device across frequency. Upon initial examination, if it is determined that stability improvement is required, the first step of this approach is to stabilize the device. Two stabilized devices are shown in Figure 5.5 (b, c). After the device is sufficiently stabilized, matching and bias circuits can be designed. In discussing this technique, we use $K$ in our analysis, for it seems more customary in practice. Note that despite the lack of a well-defined relationship with the stability margin, the numerical value of $K$ can still be used for practical devices as an indicator of stability improvement.

First, let's consider the effect of the lossless component (versus resistance) on stability. The $K$ invariance mentioned in Chapter 2 implies that adding a lossless component cannot change a circuit from being conditionally stable to unconditionally stable. This is equivalent to the statement that a network (or impedance) transformation using a lossless network cannot bring a $\Gamma$ point from outside the unit circle to inside. The reader can verify this by a Smith chart operation using any of the nonresistive elements, $C_p$, $C_s$, $L_p$, $L_s$, $TL_S$, $TL_{p\_short}$, $TL_{p\_open}$ considered in Section 3.5.2. Analytically, if we realize that $|\Gamma| > 1$ is equivalent to a negative resistance, then the proof for the cases of $L$ and $C$ is obvious since any transformation using a reactive component does not change the real part of the impedance. For transmission lines, it is still straightforward to show, using (1.20), that none of the three transmission line transformations can change the real part of an impedance from negative to positive. It is important to note that the above argument is not to say that lossless networks have no effect on the stability margin. In fact, the $\mu$ and $\mu'$ parameters generally do change when lossless networks are added to the device, as explained in Section 2.2, but numerically they will never cross the unity line.

Thus, we conclude that to make a transition from conditional stability to unconditional stability, a resistance must be used. There are two considerations in adding the resistance, namely, location (at the input or output) and configuration (series or shunt). Generally, for LNAs, the stabilizing resistance should be placed at the device output to avoid severe impact on noise figure (see Chapter 8), while for power amplifiers (PAs), the resistance should be at the device input to avoid degradation in power and efficiency. Regarding the configuration of the resistance, given the complexity of $K$ as a function of $S_{11}$ and $S_{22}$, we do not attempt to provide a rigorous proof analytically here. Instead, we intuitively point out that if an added resistance moves $S_{11}$ or $S_{22}$ toward the center of the Smith chart (reducing its magnitude), it is more effective in improving the stability. This can be seen by realizing that, for a practical device, the center is undoubtedly in the stable region, whereas the unstable region (when $K < 1$) is always in the border areas of the chart. (See Fig. 2.3 for examples.)

The following example uses a low-noise GaAs FET designed for applications around 4 GHz to demonstrate the stabilization process outlined here. Figure 5.6(a) shows the $K$ factor and $F_{min}$ (see Section 4.4) versus frequency of the original device. (Other curves in the plots are explained in the next two paragraphs.) Also shown in Figure 5.6(b) are the $S_{11}$ and $S_{22}$ plots on the Smith chart. From the plot we can see $K < 1$ at the operation frequency 4 GHz and drops to below 0.1 at low frequencies (< 500 MHz), indicating a strong instability.

Since this is an LNA design, the stabilizing resistance should be added at the output. Based on the location of $S_{22}$ at 4 GHz, either a series or a shunt resistance can improve $K$ at the in-band frequency. We use the series configuration as sketched in Figure 5.5(b) for it is more convenient in circuit implementation (to avoid a DC shunt). The value of $R_1$ in Figure 5.5(b) can be determined by an EDA tuning process. The $F_{min}$ and $K$ plots for two $R_1$ values (10 and 2Ω) are shown in Figure 5.6(a). The curves show, as expected, that as $R_1$ is increased, the stability is improved at the operation frequency the expense of noise figure. The figure also shows that $R_1$ has little impact on $K$ at frequencies around 500 MHz and below. This is because the coupling

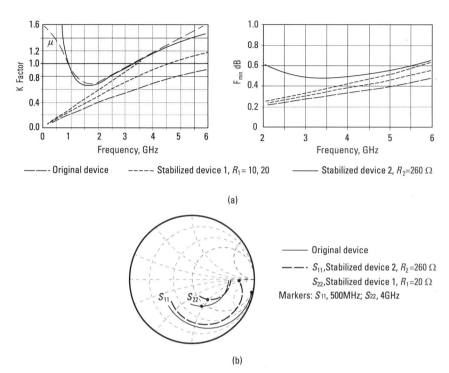

**Figure 5.6** Stabilization process characterized by (a) $K$ and $F_{min}$ and (b) $S_{11}$ and $S_{22}$.

between the device input and output becomes weaker (smaller $S_{12}$) as frequency decreases.

In this example, we select $R_1 = 20\Omega$ for it makes the device unconditionally stable at the operation frequency ($K \cong 1.2$ at 4 GHz). The same simulation shows that the *MAG* at 4 GHz is about 14 dB, which is reasonable. Thus, with $R_1 = 20\Omega$, we have achieved unconditional stability with a modest degradation in the in-band noise figure (about 0.1 dB). To improve the stability at the low-frequency end, where unintended oscillation often occur, it is necessary to add a resistance at the device input. From the location of $S_{11}$ at 500 MHz, we can see that a shunt resistance is required because a series one simply moves $S_{11}$ toward open. Obviously, this shunt resistance would severely degrade the noise figure. In this case, there is enough frequency separation between 4 GHz and 500 MHz that a series inductance can be used, as shown in Figure 5.5(c), to minimize the impact on the in-band noise figure while still allowing the desired stability improvement at low frequencies. A clever way to implement this is with an inductor whose self resonance frequency (SRF) (where the impedance is essentially infinite; see Section 9.1.2) happens to be close to the operation frequency so that the impact of the resistance on the in-band noise figure can be virtually eliminated. Since the main purpose of this exercise is to illustrate the basic concept, we do not attempt to model the inductor more accurately. Instead, we assume a reasonable value of 60 nH for the inductance and run a tuning process in simulation to determine the resistance value for $R_2$ with a goal of a good balance between the in-band noise figure and the stability at low frequencies. Based on the simulation results, we choose $R_2 = 265\Omega$. The curves labeled "Stabilized device 2" in Figure 5.6(a) are the results. The $\mu$ parameter is also plotted along with the $K$ factor as a reference. The device stability characterized by such a $K$ or $\mu$ curve (larger than unity below 1 GHz and above 3.5 GHz) almost guarantees an oscillation-free circuit regardless of the implementation of matching and bias circuits. The price for this stabilization is a 0.2-dB increase in noise figure. The same simulation also generates a new set of the S- and noise parameters (not shown). With these new parameters we can proceed with the matching network design for the RF performance using the technique to be outlined in the next section. A side benefit of resistance used for stabilization is easier balance between noise figure and return losses, a topic discussed extensively in Section 5.2.3. As a result, all things considered, the degradation in noise figure due to the device stabilization can be acceptable in many applications.

Generally, the circuit element of $R_2$ and $L_1$ in Figure 5.5(c) can be handily incorporated in the gate bias circuit for FETs—note that a DC supply is an AC ground—in most cases, since the gate current is negligibly small. For BJTs, the chosen $R_2$ may not be appropriate for the required DC base current. In that case, three factors, namely, the DC bias point, the in-band noise figure (for LNA design), and the stability at low frequencies, should be considered together in the design of the base resistance network.

For PA circuits, a similar technique can be employed by placing a series small resistance at the device input. If the resistance cuts too much in-band gain, a parallel capacitance (with the resistance) may be used to partially bypass the resistance at higher frequencies. At the PA output, no resistance should be added to the DC path. A more elaborate method is required if the bias circuit is used as part of the stabilization scheme. Section 5.2.3 discusses one such method, using ferrite beads.

In summary, the technique of adding some resistance for stability improvement is convenient in terms of design process but generally requires some compromise in RF performance. It is often employed in low- and medium-power amplifier designs. It is rarely used in high-power circuits because of the unavoidable power loss in the added resistance. The example we outline simply illustrates the basics of this technique. The process can be modified in a number of ways. The task for each individual design is to make the circuit sufficiently stable while maintaining the RF performance according to the application requirements. In terms of circuit implementation, it is not necessary (although it is convenient for simulation) to place the stabilizing resistances next to the device. They actually can be placed anywhere in the circuit, as long as they are effective. Finally, it is always a good practice to check the stability of the entire circuit by examining $K_{\text{cir}}$ across frequency [see Figure 5.5(a)], which can be calculated from an S-parameter measurement at the circuit input and output points. If the amplifier is intended as a stand-alone circuit, due to the unpredictable nature of the source and the load to be connected to the circuit, it is highly desirable to achieve the condition $K_{\text{cir}} > 1$ over the entire frequency range with perhaps the exception at very low frequencies for the reason explained above. On the other hand, if an amplifier is used in a system where it interfaces with a fixed source or a fixed load, then it is possible to limit the circuit in the stable regions even if $K < 1$. Such examples are an LNA in a receiver (fixed load) or a PA in a transmitter (fixed source).

### 5.2.3 Some Common Unintended Oscillations and Solutions

Here we consider solutions to several types of unintended oscillations commonly experienced in practice by engineers.

#### 5.2.3.1 Oscillation Associated with Bias Networks

Recall that it is not always desirable to make a device unconditionally stable across frequency by adding a resistance because of the unavoidable detrimental impacts on the RF performance. This is particularly true for RF PAs due to the high-power (DC or/and RF) dissipation in the resistance. On the other hand, if there are no resistive elements in the network for $\Gamma_L$ [see Figure 5.5(a)], the DC bias network is virtually a short circuit or a very low impedance at low frequencies because it must be connected to a DC source (AC short). As noted, this condition may cause stability problems. As a matter of fact, in practical RF amplifier designs, it is a high possibility that a low-frequency unintended oscillation (typically from tens of megahertz to several hundreds of megahertz) can be suppressed by improving the DC bias circuits. In general, a bias circuit for a PA amplifier should serve three basic functions at a reasonable cost: (1) sufficient RF isolation; (2) DC voltage and current supply; and (3) adequate resistive load at low frequencies for stability. These functions inevitably impose conflicting requirements on the circuit configuration and component selection. We now discuss each function individually and then introduce a circuit that provides a good balance for all three functions.

*RF isolation.* Chapter 3 introduced RF choke (usually an inductor) as a means of isolating an RF network from the DC source. As depicted in Figure 5.5(a), the bias circuits can be considered in a shunt connection to the RF matching network. Then, for the inductance to be effective as an RF choke, its impedance $\omega L$ should be sufficiently high compared with the network impedance at the point where the inductor is connected. For PA amplifiers, the impedance at the device output (drain or collector), $Z_{\text{Load}}$, is usually significantly smaller than the termination load 50Ω (see Chapter 6). Under such a condition, the closer the choke inductance is to the device the lower the required value for $L$. A numerical example is provided in the circuit shown in Figure 5.9. The RF choke scheme is generally convenient in bias circuit design for it allows tuning for RF performance without any concern of the DC bias circuit. On the other hand, as discussed in Section 3.5.3, in some cases, it can be advantageous (fewer components) to have the inductance as part of matching network. In that case, the tolerance of the inductance must be considered.

On paper, there is no upper limit on the value of $L$ in terms of RF isolation. However, from the viewpoint of DC specifications, $L$ should be kept as low as possible for the reason of the DC performance described in the next paragraph. Another practical constraint is the SRF of the inductor. When the operation frequency is considerably higher than the SRF, the inductor no longer behaves as an RF choke. For this reason, at high frequencies (around 10 GHz or higher), a quarter-wavelength transmission line is normally used, instead.

*DC characteristics.* The inductor used in a bias circuit needs to fulfill two requirements related to DC specifications: having sufficiently high current rating for the specific application and minimizing the voltage drop across the inductor. The former is mostly related to the wire size, and the latter is simply the product of current and the ESR of the inductor (see Chapter 9). The ESR value is, in turn, determined by the wire size and length (proportional to the inductance). For RF PA applications where the DC current can be in amperes or higher, the required inductors are usually bulky and costly. Any unnecessary increase in inductor size and inductance value is generally undesirable. Both should be kept just sufficient, but not much more, for the intended functions according to the DC circuit requirements.

*Impedance at low frequencies.* Recall that unintended oscillations often occur at a frequency significantly lower than the operation frequency. Thus, the low-frequency characteristics of the bias circuit are also of importance for stability reasons. Unlike in the case of RF isolation and DC specifications, there is not clear-cut guidance in terms of the effects of the inductance value on the stability. Nevertheless, it needs to be part of consideration in the bias circuit design. The following example illustrates this point.

We consider a 1-GHz PA design using a 25-W RF LDMOS. As explained in Chapter 6, for a given transistor and a specific frequency, optimal power performance is achieved with a set of impedances specified at the device input and output. They are often denoted as $Z_{Source}$ and $Z_{Load}$, (or in some variations), respectively in the manufacturer's datasheet. In this case, $Z_{Source} = 0.9 - j\, 1.75\,\Omega$ and $Z_{Load} = 3.5 + j\, 0.5\,\Omega$ at 1 GHz.

Here we center our discussion on stability. We first examine the overall stability characteristics of the device by plotting $\mu$ and $\mu'$ as shown in Figure 5.7(a). The choice of $\mu$ and $\mu'$ here as opposed to $K$ is to illustrate the effects of the input and output networks separately, as discussed in Chapter 2. From the plots we foresee potential instabilities in the frequency region below 200 MHz owing to the small values of the parameters, particularly $\mu$. In this case, adding a resistance directly to the device is not an option because the device has to be matched to $Z_{Source}$ and $Z_{Load}$ for the in-band performance. Now we will design the matching networks and then show how to improve the circuit stability through the bias circuit.

Matching networks for the required transformations can be devised on the Smith chart. Figure 5.8 shows the transformation for $Z_{Load}$ as an example. The corresponding schematic is shown in Figure 5.9(a). We focus on the effect of the drain bias circuit on the amplifier stability and assume that the gate is biased with a high resistance such that the gate bias circuit can be ignored in our analysis. In some practical situations, the gate biasing scheme also has a significant impact on the circuit stability. Should that be the case, the analysis technique and circuit solution outlined here can be applied to the gate side as well.

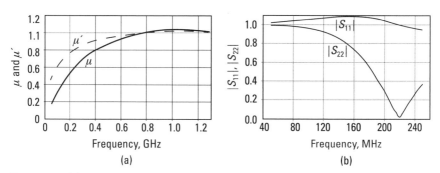

**Figure 5.7** (a) $\mu$ and $\mu'$ of the device only and (b) $S_{11}$ and $S_{22}$ at the input and output of the circuit shown in Figure 5.9(a).

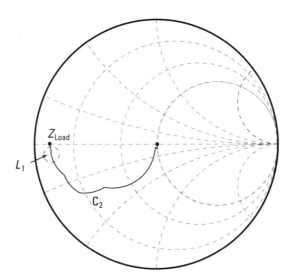

**Figure 5.8** Matching circuit design on the Smith chart for $Z_{Load}$.

The 10-nH shunt inductance $L_1$ in the output matching network is intended as an RF choke for the drain biasing. It can be seen in the Smith chart in Figure 5.8 that this inductance has virtually no effect on the matching network, if placed at the location as shown. This can be confirmed numerically by realizing that the impedance of 10 nH at 1 GHz is 63Ω, which is more than an order of magnitude larger than the network impedance at that point. In contrast, if the same inductance were placed at the 50-Ω side of the network, it would drastically change the impedance. This observation reaffirms our statement that the RF choke should be located as close as possible to the drain (or collector) for high-power devices.

We now directly evaluate the circuit stability by examining $|S_{11}|$ and $|S_{22}|$ at the circuit input and output points [RF_in and RF_out in Figure 5.9(a)]. The results are shown in Figure 5.7(b) for a frequency range up to 250 MHz. We immediately see that $|S_{11}| > 1$ around 200 MHz, indicating a severe instability. The fact that it is $|S_{11}|$ that exceeds unity is consistent with the observation that $\mu$ is worse than $\mu'$ as shown in Figure 5.7(a) (recall that when $\mu < 1$, a smaller $\mu$ implies a larger area in the $\Gamma_L$ plane that yields $|\Gamma_{in}| > 1$.) This is not merely a theoretical concern. In reality, this kind of circuit (designed only based on the in-band matching requirements) often exhibits oscillations even if the amplifier is terminated at 50Ω. The oscillation frequencies are typically in the range plotted in Figure 5.7(b). The root cause of this type of oscillation is the inductance for DC biasing. To find a solution, we

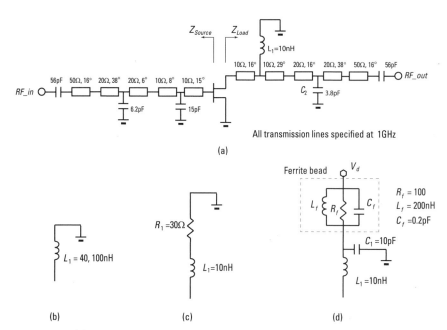

**Figure 5.9** (a) Schematic of the PA circuit based on RF performance, (b) the tuning of bias inductance $L_1$, (c) the addition of a series resistance $R_1$ to $L_1$ and (d) the employment of a ferrite bead.

experiment with different configurations and component values in place of the original $L_1 = 10\text{nH}$ as shown in Figure 5.9(b, c, d). In Figure 5.9(b), the inductance value is simply increased. The responses of $|S_{11}|$ for two values are shown in Figure 5.10, which indicates that the region where $|S_{11}| > 1$ is pushed toward the low frequency ends as the inductance value increases. Figure 5.10 also shows that a small resistance in series with the inductance [Figure 5.9(c)] can drastically improve the stability at lower frequencies.

However, neither solution (increasing inductance or adding a resistance) is acceptable from the DC consideration due to the reasons already discussed here. What we need in this case is a device that has very low resistance at DC but significantly high resistance at the frequency range where the stability is a concern. Ferrite bead is such a device. Chapter 9 provides a detailed discussion of ferrite beads. As described in Chapter 9, various equivalent circuit models with varying degrees of complexity can be found from ferrite bead vendors. For our purposes, we use a simple RLC parallel circuit to represent a ferrite as depicted in Figure 5.9(d). The resonance frequencies of practical ferrite beads typically range from a few tens of megahertz to 100–200 MHz. Thus, a ferrite bead allows DC to go through with minimal loss while imposing high impedances at the frequency range where we have stability concerns. The DC loss of a ferrite bead is specified by DCR (DC resistance), which is not included in our equivalent circuit because it is not relevant to our discussion here. The component values for the ferrite in Figure 5.9(d) are typical for this type of devices, and the simulation result for the bias circuit with the ferrite bead included is shown in Figure 5.10. We can see that the stability is drastically improved.

In this scheme, the selection of the bypass capacitance $C_1$ in Figure 5.9(d) is also of importance: it should provide a low impedance point at the operation frequency so that the RF matching network is isolated from the ferrite bead at this frequency; on the other hand, its capacitance should be small enough not to shunt the ferrite bead at low frequencies.

#### 5.2.3.2 Oscillation Due to Poor Ground Via Design

Another type of instability frequently encountered in practice is unintended oscillations in high frequency (usually > 10 GHz) amplifier designs. We will specifically consider an FET in a common-source circuit since this is the configuration that is employed by almost all RF amplifiers in this frequency range. It is generally known among RF engineers that the PCB layout of ground vias is critically important in this case, because any inductive element between the transistor source and ground can significantly increase the gain and reduce the stability of the circuit at high frequencies. The mechanism of this effect can be understood if we consider the source inductance as a feedback element from the output to the input of the amplifier. The concept and analysis of

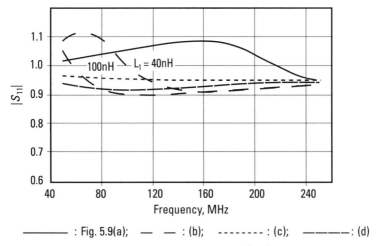

**Figure 5.10** Effects of the bias circuits in Figure 5.9 on $|S_{11}|$.

# RF Amplifier Designs in Practice 165

this feedback mechanism can be found in many textbooks on analog circuit designs [1, 8]. Now we will use a real-world case to demonstrate this effect without any further analytic treatment.

Figure 5.11(a) shows a PCB layout used for the S-parameter measurement by the manufacturer of a GaAs FET designed for LNA and oscillator applications in a frequency range of approximately 10–20 GHz. The device

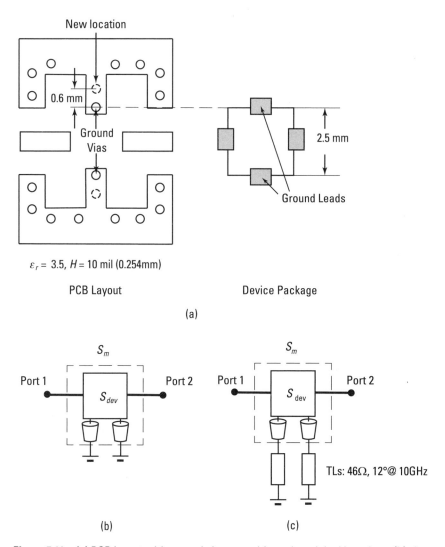

**Figure 5.11** (a) PCB layout with ground vias moved from the original locations, (b) vias included in the measured S-parameters, and (c) simulation setup for the effect of extra source traces.

has two source leads, both of which should be tied to ground for achieving low inductance ground paths. (Generally, in a PCB layout for RF applications, any piece of metal that is unintentionally left floating should be avoided for any RF path including grounding.) In the original layout, the ground vias are underneath the leads, which avoids any unnecessary PCB trace between the ground and the source. Some PCB assembly houses prefer not to have any vias on a soldering pad. To accommodate this requirement, in Figure 5.11, the ground vias are moved outside the soldering pads in a place labeled as "New location." After this change, the circuit became unstable, exhibiting oscillations.

To see how this happened, we note that the change in ground via location is equivalent to adding a short trace between the transistor sources and ground vias, represented by a transmission line with the parameters estimated using the dimensions of the trace. The effect of this extra transmission line is examined in a simulation using the S-parameters provided by the manufacturer. The simulation results of $|S_{21}|$ and $K$ for the configurations depicted in Figures 5.11(b, c) are compared in Figure 5.12. As shown in Figure 5.12, a seemingly trivial change in the ground via locations reduces the $K$ factor considerably at the high-frequency end of the plot, taking the device from unconditionally stable to potentially unstable. In fact, in high-frequency amplifier designs, it is not uncommon that the poor grounding in a layout design is the root cause of an unintended oscillation. This is a case when the RF performance must overweigh the convenience of PCB assembly in board layout designs.

Please note regarding the setups for simulation that the S-parameter of a device provided by the manufacturer is usually the measurement data on a specific PCB without de-embedding the effect of ground vias; that is,

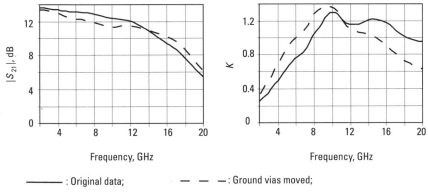

**Figure 5.12** Effects of extra traces on $|S_{21}|$ and $K$.

$S_m$ shown in Figure 5.11(b) instead of the true device S-parameter denoted as $S_{dev}$ in Figure 5.11(b). (The effects of transmission lines and connectors are still de-embedded; see Chapter 10 for details.) In theory, if the ground vias used in a specific application are different in dimensions from those by the manufacturer, some discrepancy is expected between the simulation and actual performance on the board. For practical purposes, however, this difference is small enough to be ignored in most cases. For the reader who wants to study the effect of vias more closely, most EDA tools include vias in their component libraries. A via in the simulation is treated as an inductor with the equivalent inductance value given by a formula initially proposed by [12] and can be found in texts such as [13] and online sources. From Figure 5.11(c) we can see that both ground leads should be tied to ground, because otherwise the equivalent source inductance would be doubled.

Readers may also notice that the arrangement of the ground vias and the traces in the simulation setup in Figure 5.11(c) has the order reversed compared with the actual condition. It can be shown that by using the impedance transformation formula of a transmission line introduced in Chapter 1 (1.20) that when $Z_c \gg \omega L$ where $Z_c$ is the trace characteristic impedance and $L$ is the equivalent inductance of the via, the two configurations are almost identical. We leave the proof to the interested reader by pointing out that the values of $Z_L$ in (1.20) are 0 and $\omega L$, respectively, for the two cases. Therefore, the simulation result is a meaningful representation of the reality.

Finally, the same plot also indicates that the $K$ factor is actually improved at low frequencies with this extra source inductance (Note this source inductance affects both the input and output of the network, hence is different from a lossless network placed at either input or output.) In fact, in the early days of RF transistors when the transition frequency $f_T$ was still low (10 GHz or below), a source (or emitter) inductance was sometimes intentionally introduced for better matching in LNA designs [6]. However, with considerable improvements in noise performance and substantially higher $f_T$, even for transistors intended for sub-gigahertz applications, this technique becomes less desirable, given its benefit and the potential risk of unintended oscillations. Another related technique is the source self-bias [2] (also known as single-supply bias [6]) used for depletion-mode FETs to avoid negative supply. In this scheme a source resistance is inserted between the device source and the ground to set up a negative gate-source voltage. Although a bypass capacitance is always used to provide the RF ground, the parasitic inductance associated with the extra circuit elements is generally difficult to eliminate completely at high frequencies. Again, if the chosen FET has significant gains above 10 GHz, caution should be exercised in implementing this bias scheme.

### 5.2.3.3 Cavity Effect

According to the electromagnetic field theory, the radiation power of a simple dipole antenna is proportional to the square of the frequency [14]. While none of the components in an amplifier circuit are intended as an antenna, any component that carries electrical currents radiates somewhat, which is especially true at high frequencies ($\gtrsim 10$ GHz) due to the square-law dependence of the radiation power on frequency. For any individual component, the amount of radiation is usually negligibly small to have any noticeable effect. However, in an amplifier circuit, any radiation from a component at the amplifier output can be coupled to the input, forming a signal loop. If the amplifier gain is high enough to exceed the radiation coupling, an oscillation can happen due to the positive loop gain. To prevent this kind of output-to-input coupling, multiple-stage RF amplifiers are usually housed in a compartmentalized metal structure. At high frequencies, the dimension of each compartment can be comparable to the wavelength (e.g., at 10 GHz, $\lambda = 3$ cm in the air), making the metal compartment a potential resonant cavity in the sense that it supports certain EM field patterns (at the standing-wave frequencies) with little attenuation. Because of this cavity effect, a single-stage amplifier with a modest gain that is perfectly stable in an open space can start to oscillate once placed in a metal enclosure.

In fact, the effect of a metal enclosure on the circuit stability can be interpreted in a more general term than the resonant cavity, which requires the existence of standing wave patterns. As stated in the beginning of this book, RF circuit theory essentially deals with EM propagating waves confined in transmission lines and circuit components. In that sense, the EM radiation from a circuit element is part of the circuit characteristics, and the presence of a metal wall certainly alters the radiation pattern, which in turn affects how the circuit behaves. This argument explains a common observation that the performance (most noticeably the gain) of a high-frequency amplifier is changed when the circuit is measured in a metal enclosure compared to in an open space. An unintended oscillation can be considered the most severe consequence of this mechanism.

Regardless of the details of the mechanism, cavity-related oscillation can usually be identified with some changes on the cavity structure such as removing the cover or placing some microwave absorbing material (e.g., Eccosorb®) inside the enclosure. However, this type of oscillation is generally difficult to tackle in a systematic approach because of the unknown nature of the radiation pattern involved. Practical troubleshooting mostly involves experimenting with different components and board layouts (not a convenient task). In a worst-case scenario, the engineer may have to try a transistor from a different manufacturer.

## 5.3 Low-Noise RF Amplifier Design with the Constant-Circle Method

Chapter 2 proves that when a device is unconditionally stable ($K > 1$), the simultaneous conjugate matching yields a perfect match at both input and output and a MAG. The required source and load reflection coefficients for this condition, $\Gamma_{SC}$ and $\Gamma_{LC}$, are given by (2.54) and (2.55). In addition, Section 2.3 shows that for the case of conditional stability ($K < 1$), the targets $\Gamma_S$ and $\Gamma_L$ can be determined using the constant-circle technique based on the required input and output specifications.

When noise performance is considered, it is no longer immediately clear what the desired $\Gamma_S$ and $\Gamma_L$ are that produce optimal performance even for the case of $K > 1$. This is because the input matching that yields a minimum noise figure ($\Gamma_S = \Gamma_{opt}$, as discussed in Chapter 4) is usually not the same as $\Gamma_{SC}$. In some references, it is suggested that $\Gamma_{opt}$ should always be chosen for the source impedance to achieve the minimum noise figure. In actual applications, however, there usually is a minimum requirement on the input return loss for the reasons explained in Section 5.1. This requirement very often cannot be achieved with $\Gamma_S$ being at $\Gamma_{opt}$. In that case, the noise figure has to be compromised in order to achieve the required input return loss. Adding to the complications, for the bilateral case ($S_{12} \neq 0$), the selection of $\Gamma_L$ also affects the input matching condition. Hence, the design task of a LNA is essentially to select a pair of $\Gamma_S$ and $\Gamma_L$ such that the three performance parameters, input return loss $RL_{in}$, output return loss $RL_{out}$, and noise figure $F$ meet the target specs simultaneously. Mathematically, $RL_{in}$ and $RL_{out}$ as a function of $\Gamma_S$ and $\Gamma_L$ (through $M_S$ and $M_L$) are given in (2.20) and (2.18), and $F$ is given in (4.29); therefore, it is conceivable that an algorithm can be developed that allows optimization of $RL_{in}$, $RL_{out}$, and $F$ over the space of $\Gamma_S$ and $\Gamma_L$ (four-dimensional since both are complex numbers). In practice, however, LNA designs are not as difficult as the mathematical solution suggests. Most vendors provide some forms of design examples for their devices. A designer can use the vendor's example as a starting point. Usually, through trial and error tunings on the bench or with a simulator, an adequate result can be achieved. Nevertheless, some designers may find the trial-and-error method less satisfactory and prefer a more systematic design approach. The constant-circle method introduced in Chapter 2 turns out to be especially effective in this case. Let's now outline that process.

Noise figure $F$ as a function of $\Gamma_S$ is given by (4.29). Using the same technique for constant gain and mismatch circles in Chapter 2, we can show that the contour for a constant $F = F_0$ ($F_0$ must be $> F_{min}$) on the $\Gamma_S$ plane is a circle whose center and radius are given by [5, 15]:

$$\Gamma_{F0} = \frac{\Gamma_{opt}}{1 + N_0} \quad (5.10)$$

and

$$R_{F0} = \frac{\sqrt{N_0^2 + N_0\left(1 - |\Gamma_{opt}|^2\right)}}{1 + N_0} \quad (5.11)$$

where

$$N_0 = \frac{(F_0 - F_{min})|1 + \Gamma_{opt}|^2}{4r_n} \quad (5.12)$$

Since any point inside the circle described by (5.10) and (5.11) corresponds to a noise figure less than $F_0$, a larger $R_{F0}$ implies less sensitivity of the noise figure to the mismatch between $\Gamma_S$ and $\Gamma_{opt}$. Equation (5.11) indicates that $R_{F0}$ is a monotonically increasing function of $N_0$. $N_0$ in turn is scaled with the factor of $(F_0 - F_{min})/r_n$ as suggested by (5.12). This is the analytic basis for the statement made in Section 4.4 that a smaller $r_n$ yields a larger $R_{F0}$ for a given $F_0$.

We use the following set of noise and S-parameters of a real device (an earlier generation SiGe bipolar transistor with a moderate noise figure) to illustrate the design process [16].

$f = 5.8\text{GHz}$;
$S_{11} = 0.325 \angle 123°$;
$S_{21} = 3.23 \angle 8.9°$;
$S_{12} = 0.08 \angle 7.0°$;
$S_{22} = 0.531 \angle -103°$;
$F_{min} = 1.33\text{dB}$;

$\Gamma_{opt} = 0.3\angle -178°$;

$r_n = 0.09$.

In this example, $K > 1$. However, the design process can be easily extended to the case of $K < 1$ in the same manner as illustrated in Chapter 2. The reader should attain familiarity with the constant-circle method before proceeding with the analysis in Sections 5.3.1 and 5.3.2.

### 5.3.1 Estimate of $RL_{in}$ When $F = F_{min}$ and $RL_{out}$ Is Perfect

Similar to the process in Section 2.3, we start the design by plotting several gain and noise constant circles as shown in Figure 5.13. In addition to the two gain circles $G_A$ and $G_P$, two constant $F$ circles are added. Three special points $\Gamma_{opt}$, $\Gamma_{SC}$, and $\Gamma_{LC}$ are also labeled. Again, this family of plots provides a quick, rough idea of where $\Gamma_S$ and $\Gamma_L$ will be. For example, if conjugate matching is implemented, the noise figure would be approximately 2.3 dB, almost 1 dB higher than $F_{min}$. Also, we can see that for $F = 1.4$ dB (any point on the 1.4-dB circle), point $C$ should be selected for best possible return loss

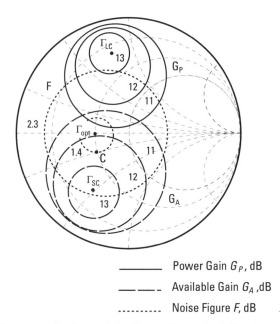

———— Power Gain $G_P$, dB

— — — Available Gain $G_A$, dB

·········· Noise Figure $F$, dB

**Figure 5.13** Constant noise figure plots along with constant $G_A$ and constant $G_P$ plots.

**Table 5.1**
Determination of $RL_{in}$ Using the Constant-Circle Method

| | Operation | Notes |
|---|---|---|
| 1 | Noise circles, locate $\Gamma_{opt}$ for $\Gamma_S$ | Not shown |
| 2 | $G_A$ circles | |
| 3 | Pick the $G_A$ (=12 dB) circle going through $\Gamma_{opt}$ | Select the point on $G_A$ at $\Gamma_{opt}$ |
| 4 | $\Gamma^*_{out}$ circle mapped from the chosen $G_A$ circle, locate point $A$ corresponding to $\Gamma_{opt}(=\Gamma_S)$ | |
| 5 | $G_P$ circles, pick the one (=13 dB) going through point $A$, locate point $B$ for $\Gamma_L$ | Points $A$ and $B$ are at the same location but on different circles |
| 6 | $\Gamma^*_{in}$ circle mapped from the chosen $G_P$ circle, locate point $C$ corresponding to point $B$ | |
| 7 | Mismatch circles for $\Gamma^*_{in}$ at point $C$, the one going through $\Gamma_{opt}$ determines the mismatch and return loss | Conclusion: $\Gamma_{opt}$ and point $B$ are chosen for $\Gamma_S$ and $\Gamma_L$, $RL_{in}$ = 6 dB ($M_S$= −1dB) |

since it is closest to $\Gamma_{SC}$. It is difficult to make such determinations based only on analytical expressions.

To achieve the target specifications, we need to have $\Gamma_S = \Gamma_{opt}$ and $M_L = 0$ dB. Using the process shown in Figure 5.14 and summarized in Table 5.1, we conclude that the input return loss is about 6 dB ($M_S = 1$ dB).

### 5.3.2 Trade-Offs Among $F$, $RL_{in}$, and $RL_{out}$

We now seek a design that has a moderate return loss of 10 dB at both input and output and a best noise figure under these conditions. It usually takes several iterations to reach the target specifications, but the process is straightforward. Figure 5.15 illustrates the final result, providing a brief note for each step in its legend. In Figure 5.15, point $A$ is chosen for the $\Gamma_S$ that yields a noise figure better than 1.4 dB (inside the $F = 1.4$ dB circle) and $RL_{in}$ of about 10 dB (slightly outside the $M_S = -0.4$ dB circle [see step (7) in Table 5.1], and point $C$ for $\Gamma_L$, which corresponds to an $RL_{out}$ better than 10 dB (inside the $M_L = -0.4$ dB circle).

The numerical values of point A and point C are: $\Gamma_A = 0.28\angle-170°$ and $\Gamma_C = 0.435\angle 126°$. Figure 5.16 shows the matching networks, devised on the Smith chart. In Figure 5.6, a capacitance of 2.2 pF is chosen for the DC

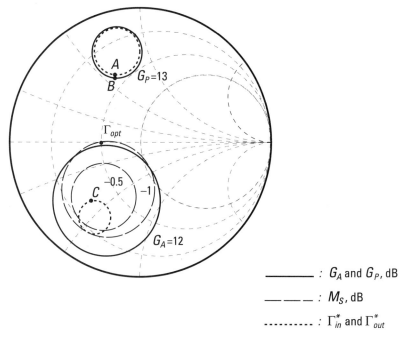

**Figure 5.14** Estimate $RL_{in}$ using constant-circle method.

blocking capacitors for its SRF (see Chapter 9) close to 5.8 GHz. Thus, the impedance transformations start effectively at the first shunt capacitances, 0.3 pF and 0.4 pF for the input and the output matching, respectively. For DC bias, a quarter-wavelength transmission line is used for RF choke at the drain, and at the base the inductor not only provides the in-band RF isolation but also reduces the gain at low frequencies. Measurement data on an actual circuit (at the board connectors) are compared with the design specifications as shown in Table 5.2 [16].

The board loss before the transistor is about 0.1dB at 5.8 GHz. After taking this loss into account, the measured noise figure is in good agreement with the design target, and so are other parameters listed in Table 5.2.

A further note on the subject of noise figure versus input return loss: As mentioned in Chapter 4, most vendors use $F_{min}$ as the noise figure specification on their data sheets. It is a reasonable practice since there is no obviously better alternative for the specification. In real designs, however, the noise figure may have to be significantly degraded from $F_{min}$ to achieve the required input return loss. For this reason, $F_{min}$ alone is not always sufficient for predicting

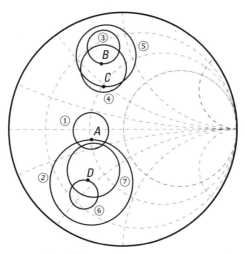

① F=1.4dB
② $G_A$=12.3dB, "A" chosen for $\Gamma_S$
③ $\Gamma_{out}^*$ for $G_A$ =12.3dB, "B" mapped from "A"
④ $M_L$ = −0.4dB for "B"
⑤ $G_P$= 12.5dB, "C" chosen for $\Gamma_L$
⑥ $\Gamma_{in}^*$ for $G_P$ =12.5dB, "D" mapped from "C"
⑦ $M_S$= −0.4dB for "D"

**Figure 5.15** Design for balance among $F$, $RL_{in}$, and $RL_{out}$.

the noise figure of an LNA circuit. Some preliminary evaluation, similar to that performed in this chapter, is recommended before the final selection of a device. For a specific transistor, generally, the noise figure and the return loss can be reasonably balanced if the operation frequency is within the frequency range recommended by the manufacturer or around the typical frequency on the datasheet. When the intended application frequency is significantly outside the recommended range, particularly on the lower side, the designer needs to pay special attention to the trade-off between the noise figure and

**Table 5.2**
Design Specs from Simulation and Measured Results

|  | NF (dB) | $RL_{in}$ (dB) | $RL_{out}$ (dB) | Gain (dB) |
|---|---|---|---|---|
| Simulation | 1.4 | 10 | 10 | 11.9 |
| Measured | 1.5 | 9.8 | 11.8 | 11.0 |

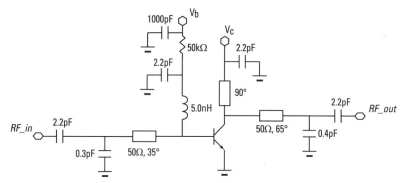

**Figure 5.16** Circuit realization based on the results obtained in Figure 5.15.

return loss. Specifically, the designer should consider the values of both $F_{min}$ and $\Gamma_{opt}$ in the initial device screening. A device can be unsuitable at certain frequencies due to the difficulty in matching for the return loss despite its superior $F_{min}$ values.

An alternative approach to deal with the balance between the noise figure and return loss is to use a balanced circuit configuration, which is discussed in Section 5.4.

## 5.4 Configurations for RF Amplifiers

This section discusses several circuit configurations for RF amplifiers. Each finds its application in practice because of its unique functions or features.

### 5.4.1 Balanced Amplifiers

A balanced amplifier, shown in Figure 5.17, consists of two identical 90-degree hybrid couplers (see Chapter 9 for a detailed description of this component) and two identical amplifier circuits. Figure 5.17 represents the amplifier circuit by a transistor plus input and output matching networks. This balanced configuration should not be confused with a differential amplifier whose output signal is sometimes said to be in balanced mode. (A circuit for differential to single-ended conversion is covered in Section 5.4.3. As shown in Figure 5.17, an input RF signal at port I1 splits into two paths through the input hybrid. After amplification, each signal is taken to the output hybrid. Since the two signals have theoretically the same phase shift from the input (port I1) to the output (port O4), they are combined in phase at the output of the balanced amplifier.

Some key features and design considerations of the balanced amplifiers are summarized as follows (refer to Chapter 9 for the properties of the hybrids discussed here):

- *No benefit in gain.* Following the signal path from the input to the output, we can see that the effect of the two hybrids in terms of signal power level is to divide at the input first and then combine at the output; therefore the gain of a balanced amplifier is the same as that of the individual amplifiers between the two hybrids.
- *Matching networks.* According to property 4 (in other words, that the impedance looking into ports I2, I3 and O2, O3 is 50Ω), the same matching networks designed for the transistor in a 50-Ω environment can be retained.
- *Good input and output return losses.* From property 2 (perfect match at input), both return losses at the input (port I1) and the output (port O4) are perfect over the specified frequency range of the couplers. In principle, this feature gives the designer the freedom to design matching networks for noise and power without a compromise in return loss specifications.

However, in practical LNA designs, the lowest insertion loss of a commercial hybrid coupler is still in the range of a fraction of decibels. In comparison, the trade-off in the noise figure for a good return loss is about 0.1 or 0.2 dB for a modern low-noise transistor, as explained in Section 5.3. Therefore, for narrowband applications, a balanced configuration offers little or no benefit in terms of good balance between the noise figure and input return loss. Furthermore, a balanced amplifier requires a considerably higher component

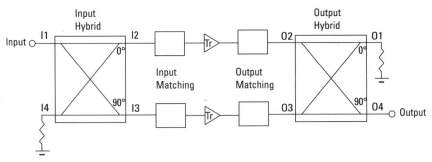

**Figure 5.17** Balanced amplifier.

budget and a larger PCB space as well as double the DC power consumption, compared with a single-transistor circuit. For these reasons, the balanced configuration is rarely used in commercial wireless systems.

On the other hand, balanced amplifiers are widely used in stand-alone RF amplifiers designed for testing and instrumentation and for some special industrial applications. Good return loss over a relatively broad frequency band is one of the justifications for this circuit configuration. It also increases the circuit reliability because the unit is still functional in the case that one of the amplifiers fails. In addition, note that Chapter 6 explains that the 1-dB compression point (and IP3 as well) is increased by 3 dB in a balanced configuration in comparison with these of the individual amplifiers.

For high-power amplifiers, the circuits at ports O2 and O3 can be considerably mismatched (see Chapter 6 for power matching), and if the output (O4) is open or poorly terminated, all or a significant amount of the output power is reflected back to the hybrid and dissipates at port O1, according to property 3. Therefore, the power rating of the resistance at port O1 should be chosen to be comparable to the power capability of the amplifier.

### 5.4.2 Broadband Amplifiers with Negative Feedback

In addition to return loss, the power gain of an amplifier is another circuit parameter whose frequency response is often an important design consideration. Specifically, a flat gain response (i.e., independent of frequency) is required in certain applications. As discussed in Section 5.1.3.1 (see Figure 5.3), the power gain of a transistor-based amplifier generally rolls off with frequency. One of the methods to achieve the flat gain uses the negative feedback technique. The concept of this technique is well explained in a number of textbooks on analog electronics (e.g., [1, 2]). Analyses there show that if the basic amplifier is assumed to be unilateral, which is generally true at analog frequencies, and to have a high enough gain, the gain of an amplifier with negative feedback is only dependent on the feedback element. Therefore, the gain is flat if the feedback is through a frequency-independent component such as a resistance. For an RF circuit, the assumption of the basic amplifier being unilateral is no longer valid due to various capacitance couplings, making any quantitative analysis much more complex. For practical designs, it is more convenient to run a simulation using the device's S-parameters to evaluate the circuit performance rather than relying on analytical results. Figure 5.18(a) shows a simulation setup that allows component tuning. In Figure 5.18(a), the feedback path consists of an R, an L, and a C. The main component is the R, which determines the amount of feedback. The capacitance is for DC blocking only and should be large enough to be short-circuited at the operation frequency range. The inductance is used for creating an upward tilt at higher frequencies.

For a transistor to be suitable for a negative feedback amplifier, its gain needs to be sufficient at the upper limit of the required operation frequency range but not excessively high to avoid any potential instability problems associated with the feedback at higher frequencies. In a simulation environment, the suitability of a given transistor can be quickly evaluated by a tuning process, as illustrated in the simulation results in Figure 5.18(b).

In this specific case, using the component values given below Figure 5.18(b), a flat gain of about 13 dB from near DC to almost 3 GHz is obtained. The function of the inductance in adjusting the gain curve is more prominent in the actual circuit than the simulation results indicate because the loss

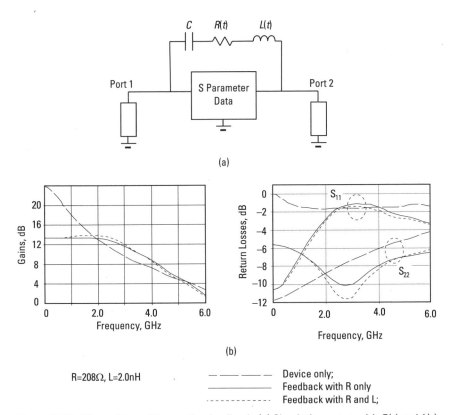

**Figure 5.18** RF amplifier with negative feedback: (a) Simulation setup with $R(t)$ and $L(t)$ as tunable parameters, and (b) gain and return losses of the amplifier.

at high frequencies is not included in the simulation. The poor input return loss over the entire operation frequency may not be acceptable. A small series resistance in combination with a larger shunt resistance usually can improve that situation. Finally, some bench tuning will be required to optimize the actual circuit performance.

Practically, this type of negative feedback configuration can be implemented from megahertz up to 3–4 GHz, depending on the characteristics of the transistor to be used. Beyond that frequency range, control of the phase of the feedback circuit becomes difficult.

In a final remark on broadband RF amplifiers, we mention the distributed amplifier [10] that can offer gains over ultrabroad frequency ranges up to the tens of gigahertz. Although, in principle, this circuit configuration can be implemented with a PCB-based circuit, practical distributed amplifiers are almost all MMIC devices.

### 5.4.3 LC Balun for Differential and Single-Ended Circuits

RF amplifiers in wireless transceiver ICs are often chosen to be in differential (also called balanced) mode, as opposed to single-ended (unbalanced) mode, for its high common mode rejection characteristics [1, 8]. On the other hand, most surface-mount technology (SMT) components are for single-ended signals. To bridge the difference, a differential to single-ended conversion (for transmitters), or vice versa (for receivers), is required at some point in the transceiver signal chains (between the IC and the antennas). In some cases, such conversions are implemented inside the ICs so that the ICs' input and output ports for the RF transmitter and receiver are in single-ended mode. The entire board design is then for single-ended signals. Alternatively, other transceiver ICs keep the differential mode up to the ICs' Tx and Rx ports. In such a situation, the board designer has to design a mode-conversion circuit on the PCB that is commonly referred to as a balun (balanced to unbalanced). Similar to matching networks for which resistive elements are usually excluded to avoid unnecessary power losses, most baluns used in practical designs are made of LC circuits or transmission lines. At frequencies roughly below 6 GHz, an LC balun is usually chosen for applications because it is more compact in size, and at higher frequencies transmission-line–based baluns are often advantageous in terms of insertion loss and cost. An extensive treatment of transmission-lines based baluns can be found in [17].

For popular wireless bands, commercial integrated LC baluns are readily available. On the other hand, discrete LC baluns still find applications in practice because of their flexibility in configuration and specification selections.

Figure 5.19(a) shows the simplest form of such a balun. This type of balun was employed as early as the 1930s as an antenna interface circuit. For a pair of differential and single-ended impedances, $R_d$ and $R_s$, $L$ and $C$ are determined by:

$$L = \frac{\sqrt{R_d R_s}}{\omega} \qquad (5.13)$$

$$C = \frac{1}{\omega \sqrt{R_d R_s}} \qquad (5.14)$$

Note that $L$, $C$, and $\omega$ follow the relationship for resonance frequency, namely, $LC = 1/\omega^2$. For a typical WiFi 5-GHz band application, the impedances

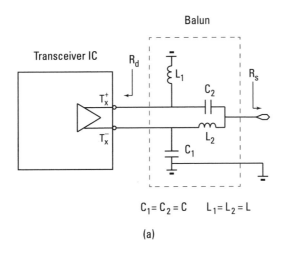

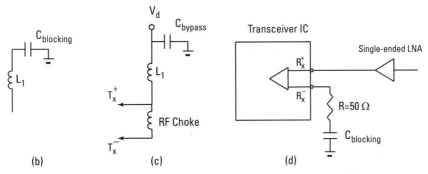

**Figure 5.19** (a) Lumped LC balun, (b) use of DC blocking capacitance, (c) incorporation of bias circuit, and (d) termination of unused port.

usually are $R_s = 50\Omega$ and $R_d = 100\Omega$, and then at the midpoint of the band, $f = 5.5$ GHz, we have $L = 2$ nH and $C = 0.4$ pF; both values are reasonably practical. It takes some equation manipulation to prove the formulas in (5.13) and (5.14). Interested readers can try to work out the proof or refer to the literature (e.g., [18]).

In the balun circuit in Figure 5.19(a), one of the inductors, $L_1$, is tied to ground. If DC voltages are present on the RF ports, a DC block capacitor is needed to prevent the DC shunt as shown Figure 5.19(b). Figure 5.19(c) shows a circuit implementation where the IC RF amplifiers are DC-biased by an external supply through the RF ports. $L_1$ along with an RF choke is used for the DC supply. In some practical cases where the total RF gain is not critical, such as the case for the receiving chain when an external LNA is used, the designer can simply use one of the RF ports while terminating the other as illustrated in Figure 5.19(d). This configuration simplifies the design at a price of a 3-dB loss in gain. (Note that comparing with the differential signal, the voltage swing in this configuration is reduced by half but the load impedance is also halved.) Generally, the datasheet is a good reference on how to implement the differential to single-ended conversion for a specific transceiver IC.

# References

[1] Gray, P., et al., *Analysis and Design of Analog Integrated Circuits*, Fourth Edition, John Wiley and Sons, 2001.

[2] Millman, J., and A. Grabe, *Microelectronics*, Second Edition, McGraw-Hill, 1987.

[3] Schwierz, F., and J. J. Liou, *Modern Microwave Transistors: Theory, Design, and Performance*, John Wiley and Sons, 2003.

[4] Sze, S. M., and K. K. Ng, *Physics of Semiconductor Devices*, Third Edition, John Wiley and Sons, 2007.

[5] Gonzalez, G., *Microwave Transistor Amplifiers Analysis and Design*, Second Edition, Prentice Hall, 1997.

[6] Vendelin, G. D., A. M. Pavio, and U. L. Rohde, *Microwave Circuit Design Using Linear and Nonlinear Techniques*, John Wiley and Sons, 1990.

[7] Gilmore, R., and L. Besser, *Practical RF Circuit Design for Modern Wireless Systems, Volume II*, Norwood, MA: Artech House, 2003.

[8] Sedra, A. S., and K. C. Smith, *Microelectronic Circuits*, Sixth Edition, Oxford University Press, 2009 (and later edition).

[9] Kurokawa, K., "Some Basic Characteristics of Broadband Negative Resistance Oscillator Circuits," *The Bell System Technical Journal*, Vol. 48, 1937, 1969.

[10] Pozar, D. M., *Microwave Engineering*, Second Edition, John Wiley and Sons, 1998.

[11] Rhea, R. W., *Oscillator Design and Computer Simulation*, Second Edition, Noble Publishing Corporation, 1995.

[12] Goldfarb, M. E., and R. A. Pucel, "Modeling Via Hole Grounds in Microstrip," *IEEE Microwave Guided Wave Letters*, Vol. 1, No. 6, 1991.

[13] Bahl, I., *Lumped Elements for RF and Microwave Circuits*, Norwood, MA: Artech House, 2003.

[14] Balanis, C. A., *Antenna Theory*, Third Edition, John Wiley and Sons, 2005.

[15] Collin, R. E., *Foundations for Microwave Engineering*, Second Edition, McGraw-Hill, 1992.

[16] Dong, M., and B. Urborg, "Constant-Circle Approach for Low Noise Microwave Amplifier Design," *Microwave Product Digest*, Sept. 2003.

[17] Mongia, R., I. Bahl, and P. Bhartia, *RF and Microwave Coupled-Line Circuits*, Norwood, MA: Artech House, 1999.

[18] Reynaert, P., and M. Steyaert, *RF Power Amplifiers for Mobile Communications*, Springer-Verlag, 2006.

# 6

# RF Power Amplifiers

This chapter is concerned with the power performance of an RF amplifier. The RF power amplifier is perhaps the most widely researched subject in the RF engineering field. This is the result of the unique importance of power amplifiers in practical applications, as well as the engineering challenge of meeting the performance requirements for high power, high linearity and high efficiency, all at high frequencies. This chapter covers only a few very basic concepts in power amplifier design, from the application perspective. Readers may find more information on the subject in specialized texts (e.g., [1–4]). For practical design, industry white papers and manufacturers' application notes are also useful sources of information on specific devices in target applications.

Chapters 2 and 4 on linear and low noise RF circuit designs describe how to devise the input and output matching impedances that yield the desired performance, whether it is gain (and return losses) or noise figure. In comparison, RF power amplifier design is about the maximization of output power. For many modern communication systems, power performance also includes linearity and efficiency, parameters that most often have a negative correlation. Sections 6.1 and 6.2 examine the conditions for power optimization. Section 6.3 describes two popular nonlinearity specifications of the 1-dB compression point and the third-order intercept point. Chapter 7 considers high-efficiency RF power amplifiers.

**Table 6.1**
Technologies for RF Power Devices

| Device | Upper Frequency Range | Comments |
|---|---|---|
| Si BJT | 4 GHz | Still offers the best long-term reliability |
| | | May be gradually phased out of the market |
| Si MOSFET | ≤GHz | A cost-effective solution for the frequency range |
| Si LDMOS | 4 GHz | Offers competitive performance/cost solutions but is facing competition from GaN |
| GaN | Tens of gigahertz | Currently growing in market share |

For the design of ultra high power ($\gtrsim$ 20–30W) amplifiers, there are a number of critical factors to consider, including component selection, DC bias circuit design, thermal properties, the electrical characteristics at extreme conditions (e.g., nearly complete output power reflection), and, of course, the impedances for matching. While there are general principles on each of these considerations, many practical techniques for handling them are device- and application-specific and are best learned in practice.

In today's markets, most medium-power (several watts to 10W) and almost all high-power (> 10W) RF power amplifiers are based on discrete transistors. In some cases they may come as a module with internal matching circuits for a target application. Table 6.1 lists the major device technologies currently on the market for RF power applications. The GaAs technology still finds its applications in various areas, but for high-power RF amplifiers, it is mostly replaced by GaN devices for new designs. The listed frequency range for each category is for reference only. The technologies in this field are constantly evolving. It is always a good practice to survey the market before selecting a device for a specific application. The discussions here and in Chapter 7 are at a conceptual level, with LDMOS being used for illustration purposes. The results are, in general, applicable to any RF power amplifier.

## 6.1 DC Characteristics of RF Transistors

In selecting a transistor for RF power amplifier applications, there are two basic device parameters that are critical to the amplifier's performance: power capability (how much power is deliverable to the load) and usable frequency range. To the first-order approximation, the RF power capability of a transistor is determined by its DC capacity in voltage and current (Note that it is S-parameters that are used in small-signal RF amplifier design). For the frequency range, the determining parameters are the drain-gate and drain-source capacitances. (The frequency range goes down as the capacitances increase.) Since both power capability and capacitance are scaled with the device dimensions, high power and high frequency are generally two conflicting requirements in an RF transistor design. The discussion here focuses on the load condition that optimizes power performance for a given device. We start with the DC characteristics of an RF transistor.

The basic amplifier function of an FET can be explained by a simple equivalent circuit of a gate-voltage-controlled current source, as illustrated in Figure 6.1. In this highly simplified model, the drain current $I_d$ is a function of the gate voltage $V_g$ and the drain voltage $V_d$. The plots in Figure 6.2 represent the transfer function $I_d$-$V_g$ and $I_d$-$V_d$ curves of an actual 10-W LDMOS device. The concepts of the equivalent circuit and these characteristic curves are covered in introductory textbooks in electronics (e.g., [5, 6]). We make use of them here in the context of power applications. For brevity, we use $V_d$, $V_g$, and $I_d$ without "s" in the indices, because the source of an FET in RF power amplifier applications is always the common port and tied to ground. In actual circuits, the grounding, which is usually through vias, is critical not only for the stability reasons discussed in Section 5.2 but also for reasons of thermal management.

In reference to the transfer function shown in Figure 6.2, $I_d$ steadily increases with $V_g$ after the threshold voltage $V_{th}$ (also known as pinch-off voltage) has been reached, and the rate of increase starts to slow down around 3.5A. $I_d$ is expected to eventually reach a saturation point. It is almost a standard practice in the literature to use the piecewise approximation for an actual transfer function, which is illustrated in Figure 6.3. This approximation allows an analysis that would provide insight into some characteristics of the device/

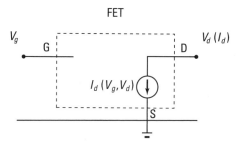

**Figure 6.1** DC equivalent circuit of a FET.

circuit while keeping the complexity level manageable. From Figure 6.3, the approximation consists of two steps: (1) linearization of the curve after $V_g > V_{th}$; and (2) a hard transition point (at $I_{max}$) from the linear to the saturation regions. The first step is usually a reasonable representation of the actual device behavior. However, caution should be exercised if the circuit designer attempts to apply the results of the second step to any real device or circuit for two reasons: First, as indicated in Figure 6.2, the maximum current rating, denoted as $I_{d\_rating}$ in this book, of this device is 3A, which would be below the transition point represented by $I_{max}$ in Figure 6.3. This is actually a common situation among practical devices because the maximum current rating is usually determined by the manufacturer based on the reliability data such as the current capabilities of the metal traces and wires used in the device. The practical implication of $I_{d\_rating} < I_{max}$ is that any effect associated with the hard limit on the drain current is not significantly relevant to the actual circuit because it cannot reliably operate at the $I_{max}$ level. Second, even if the maximum current rating is above the current at the transition point, the curve in Figure 6.2 shows that $I_d$ is still an increasing function with a slower rate rather than a flat curve. This detail makes the determination of $I_{max}$ somewhat arbitrary. Despite the limitations of this approximation, we use it in our analysis for its simplicity.

The $V_d$ dependence of $I_d$ is characterized by the $I_d$-$V_d$ curves in Figure 6.2. We notice that for a specific $V_g$, an $I_d$-$V_d$ curve can split into two regions separated by the knee voltage (labeled as $V_k$ in Figure 6.2). Below $V_k$ is the ohmic region. It is so called, because in this region $I_d$ is approximately proportional to $V_d$. The resistance in this region is known as the drain-source on-resistance, denoted as $R_{on}$. In the active region, $I_d$ is almost flat, indicating the independence of $V_d$. For this reason, $I_d$ will be considered being controlled only by $V_g$ in our analysis in Section 6.2. Furthermore, the active region is

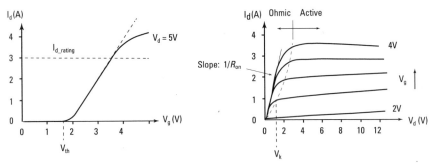

**Figure 6.2** Transfer function $I_d$-$V_g$ and $I_d$-$V_d$ curves of a 10-W LDMOS.

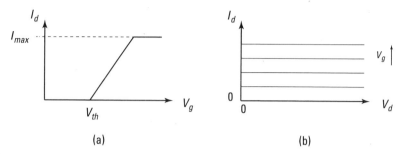

**Figure 6.3** Idealized FET characteristics: (a) Transfer function with the piecewise approximation and (b) $I_d$-$V_d$ curves with $V_k = 0$.

where the amplifier operates, that is, when $V_d < V_k$ the circuit stops working as an amplifier. To simplify our analysis in this chapter, we assume that $V_k = 0$. Based on this assumption, we impose a condition on the drain voltage: $V_d > 0$. Using these two assumptions, the idealized $I_d$-$V_d$ curves are sketched in Figure 6.3(b).

## 6.2 Load-Line and Matching for Power

We now turn to the question of how to design a power matching that yields the targeted power performance. Unlike a linear RF amplifier for which the gain is equally affected by the input and output matching impedances (see the transducer gain in Section 2.1), the power performance of an amplifier is dominated by the output matching. The input matching has a secondary effect and is usually chosen for good return loss at a specific power level.

Consider the idealized amplifier circuit drawn in Figure 6.4, which is constructed from the ideal transistor model of Figure 6.1 with a few additional components. The output impedance, $Z_{out}$ is drawn for later discussions in this section and Section 7.5.1 but is not included in our analysis here. The drain and gate are DC-biased by the supply voltages $V_c$ and $V_{gq}$ respectively. The signal source $v_g$ provides the RF swing at the gate. Both the RF choke RFC and the DC-blocking capacitor $C_{blk}$ are assumed ideal; that is, the RFC is a perfect conductor at DC and an open circuit for any RF signals, and $C_{blk}$ is totally transparent to RF signals. (RFC and $C_{blk}$ in a practical design are covered in Section 9.1.) The key concept here is that a component is assumed to have characteristically different behaviors in different frequency ranges (e.g., RFC is a DC through and an RF open). This simplification technique turns out to be particularly convenient in the RF power amplifier analysis. Another example is the harmonic trap (introduced in Chapter 7), which is a load that behaves as a through at the fundamental frequency and a short circuit at all harmonics.

Then the drain voltage and current, $V_d$ and $I_d$, which have both the DC and RF components, should be analyzed separately in the DC and RF networks shown in Figure 6.4(b). Combining them, we have the following relationships

$$V_d = V_c + v_R \tag{6.1}$$

$$I_d = I_c - i_R \tag{6.2}$$

$$v_R = i_R R \tag{6.3}$$

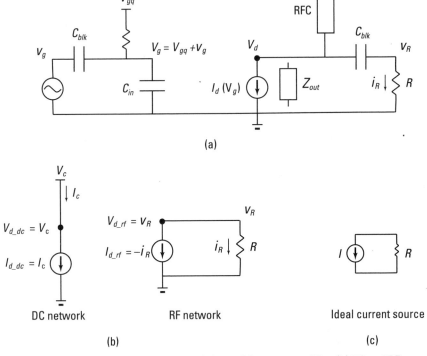

**Figure 6.4** (a) Idealized equivalent circuit for an RF power amplifier; (b) DC and RF networks for $V_d$ and $I_d$ analysis; and (c) ideal current source and a load $R$.

It is clear the DC drain voltage stays at $V_c$ regardless of the input signals because of the ideal RFC assumption. In contrast, there is no such restriction on the DC drain current $I_d$. In general, the DC drain current changes once the input RF power reaches a certain level. It is customary in the RF business to use the term of "quiescent" current to specify the bias current when no RF signal is applied. To focus on the topic of this section, we limit our discussion to a condition where the DC drain current is a constant $I_c$ and is not changed with the input signal. A general case when an additional DC current is induced by an RF input signal is analyzed in Chapter 7.

From (6.1) to (6.3), $I_d$ and $V_d$ are related by

$$I_d = I_c - \frac{V_d - V_c}{R} \tag{6.4}$$

Equation (6.4) implies that during an RF cycle the drain current and voltage of the device must move along a straight line with a negative slope of $1/R$ in the $I_d$-$V_d$ plane and that the DC bias point ($I_c$, $V_c$) must be on this line. Thus, this line, which is known as a load line in the literature, is fixed by the values of a set of parameters of $R$, $I_c$, and $V_c$.

The goal of our analysis is to understand the waveforms of $I_d$ and $V_d$ under an input signal $V_g$. These three variables are related by the device characteristics represented by the transfer function $I_d(V_g)$ and the $I_d$-$V_d$ curves, plus the load line equation (6.4). A general analytic solution for such a case can be complex, even under a greatly simplified condition, as illustrated in an example in Section 7.3. As an alternative, a graphic method proves to be convenient in facilitating our discussion. Figure 6.5(a) demonstrates the process of how a pair of $I_d$ and $V_d$ is determined for a given $V_g$ ($V_g \rightarrow I'_d \rightarrow V'_d$). For a sinusoidal input signal, using the same technique, we can quickly locate the two extrema for $I_d$ and $V_d$ and thus the entire wave form, as illustrated in Figure 6.5(b).

Now the question is: For a given supply voltage $V_c$ and the idealized transfer function that is featured with $I_{max}$ as shown in Figure 6.3, what is the optimal value of $R$, denoted as $R_{opt}$, that produces a maximum RF power deliverable to the load? We first note that for a true ideal current source (see Figure 6.4(c)) there is no limit on the power deliverable to the load $R$. The existence of $R_{opt}$ is due to the abovementioned constraints on $I_d$ and $V_d$. Once $R_{opt}$ is known, the RF design is reduced to a lossless matching network that transforms the load (usually 50Ω) to $R_{opt}$, a process that is conceptually identical to that in the linear amplifier design.

From Figure 6.4(b), the RF power delivered to the load is simply $(1/2)|v_R||i_R|$ ($|\cdot\cdot|$ means amplitude). Since $v_R$ and $i_R$ are equal to the RF

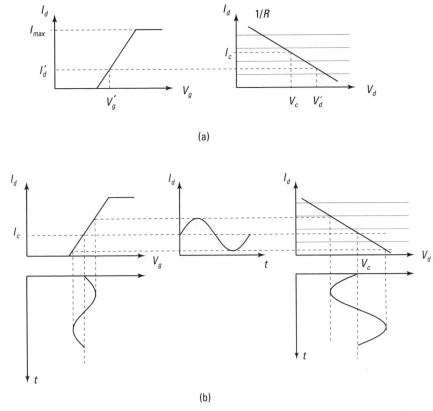

**Figure 6.5** (a) $I_d$ and $V_d$ values corresponding to a $V_g$ and (b) waveforms of $I_d$ and $V_d$ under an excitation of $V_g$.

components in $V_d$ and $I_d$ respectively [see (6.1) to (6.2)], the maximum power occurs when the RF swings of both $I_d$ and $V_d$ are maximized. To keep our analysis reasonably simple, we impose a distortion-free condition in the analysis; that is, the RF wave forms for all three variables, $V_g$, $I_d$ and $V_d$, are perfect sinusoidal functions. From the transfer function in Figure 6.3, we can see that to have a maximal RF swing in $I_d$ within the condition $0 < I_d < I_{max}$, the quiescent $I_{dq}$ must be selected at the midpoint of $I_{max}$. Figure 6.6(a) shows the corresponding waveforms for $V_g$ and $I_d$.

Next we need to decide the conditions for the maximum RF swing in $V_d$. Three situations are depicted in Figure 6.6(a–c). In Figure 6.6(a), $I_d$ reaches the maximum swing before $V_d$ does. In Figure 6.6(b), $V_d$ reaches the maximum swing before $I_d$. Figure 6.6(c) is the condition when the RF swings of both $I_d$ and $V_d$ reach the maximum at the same time. Thus, we determine graphically

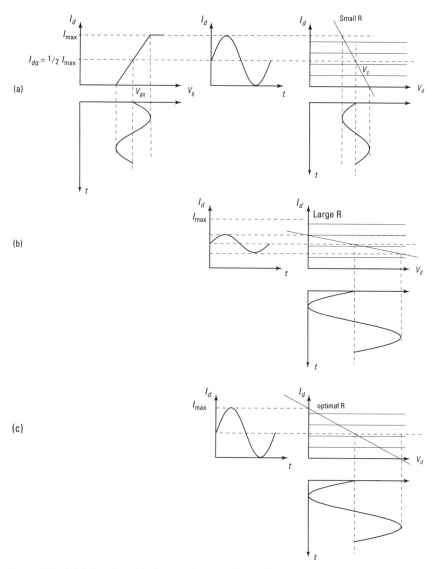

**Figure 6.6** (a) Selection of $I_{dq}$ for maximum swing and load line of small $R$; (b) load line of large $R$; and (c) load line of $R_{opt}$.

that the maximum deliverable RF power $P_{\text{rf\_max}}$ under the distortion-free condition occurs at

$$R_{\text{opt}} = \frac{2V_c}{I_{\max}} \qquad (6.5)$$

and

$$P_{rf\_max} = \frac{I_{max}V_c}{4} = \frac{V_c^2}{2R_{opt}} \qquad (6.6)$$

Equations (6.5) and (6.6) are among the most cited equations in the literature, as the basic operation principles in power RF amplifier designs. While these equations offer a loose estimate of the relationships among the bias conditions, the optimal load and the maximum output power (distortion free), they are generally not accurate enough to be used as a practical design guide in any quantitative manner. Some of the reasons for this imprecision are as follows.

First, the selection of quiescent current, combined with the condition of distortion-free, in our analysis, is in fact the so-called class A operation mode. As discussed in Chapter 7, most practical RF power amplifiers do not operate in the class A mode. Consequently, it is expected that the results in (6.5) and (6.6) may not predict the performance accurately in those cases.

Second, two conditions, $V_k = 0$ and $I_d < I_{max}$ are used in our analysis. The first one allows the maximum RF swing to be $V_c$. However, when $V_k$ is included, the effective maximum voltage swing will be less than $V_c$, which reduces the maximum RF power. $I_{max}$ in the second condition is from the idealization of the transfer function and is not necessarily related to any device characteristics, as discussed earlier. In addition, the RF current is not a measurable quantity in an RF circuit. Consequently, the moment when the RF swing reaches the maximum under the distortion-free condition [Figure 6.6(c)] can only be estimated from RF measurements. The ambiguity of $I_{max}$ in these aspects makes it difficult to quantitatively compare an actual circuit performance with the results in (6.5) and (6.6). Another unrealistic assumption in our equivalent circuit in Figure 6.4 is the omission of the source impedance [$Z_{out}$ in Figure 6.4(a)]. Once $Z_{out}$ is included, the optimal load for power matching is expected to be a complex impedance rather than a pure resistance.

Despite the limitation of (6.5) and (6.6) in quantitatively predicting the actual performance, the above analysis reveals the essence of the power matching: that the optimal load resistance (or impedance) should be chosen that maximizes the RF swings of the drain voltage and current simultaneously. Such an optimal load resistance is determined by the external bias voltage and the characteristics of the transistor's transfer function. Then we ask the following question.

Recall that for a linear amplifier design, assuming unconditional stability, the required load impedance for the maximum gain is the conjugate matching $\Gamma_{LC}$ (see Section 2.3), which is completely determined by the device's S parameters. The selection of $\Gamma_{LC}$ is termed linear gain matching in our discussion, as opposed to power matching (when $R_{opt}$ is chosen). There is no fundamental circuit principle that dictates the relationship between $R_{opt}$ and $\Gamma_{LC}$. However, all practical RF power transistors are designed such that $R_{opt}$ and $\Gamma_{LC}$ are reasonably close on the Smith chart. In other words, the gain at low input powers of an amplifier that is matched for power is usually within a few decibels or less of the linear gain (when the same device is matched for gain). Figure 6.7 illustrates this situation.

To see how this can happen, Figure 6.8(a) provides a much more realistic equivalent circuit for an FET. Actual circuits for device modeling usually have more circuit elements. With this kind of equivalent circuit, we can imagine that the S-parameters can be adjusted by changing a number of elements. Each element is usually related to one or several physical attributes of the device. This is how a device designer can tweak the device characteristics.

In practice, instead of providing the DC characteristic curves, many manufacturers of RF power transistors provide the data of the load and source impedances for the power matching (common notations include $Z_{load}$, $Z_L$, $Z_{opt}$ and $Z_{source}$, and $Z_s$) for a series of sets of frequency and bias conditions. In some cases, load-pull data, which consists of contours of constant output powers on the Smith chart of the load impedance [1, 9], is also provided, allowing the user to assess the sensitivity of the power performance to the matching condition. If the impedance data is not available, the designer can start with the linear matching, assuming the availability of the S-parameters, and then tune the circuit for the power performance.

Finally, note that the actual load impedance used for a power amplifier, whether provided by the manufacturer or determined by the user, often

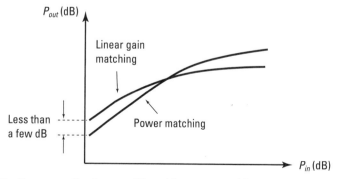

**Figure 6.7** $P_{out}$ versus $P_{in}$ of an amplifier with power matching and linear gain matching.

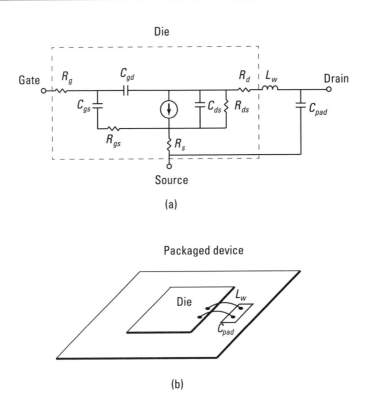

**Figure 6.8** (a) A more realistic equivalent circuit for FET and (b) bonding wires and soldering pad.

appears to be significantly larger than $R_{opt}$. One of the possible causes for this discrepancy is that the transistor is prematched internally using bonding wires (equivalent to an inductor) and a soldering pad (equivalent to a capacitor) as illustrated in Figure 6.8(b). This L-C matching network transforms the impedance downward in the relevant region on the Smith chart, making the actual impedance at the transistor die lower than that at the package lead. For high-power transistors, prematching inside the package can significantly ease the on-board matching design.

## 6.3 Nonlinearity Specifications

The concept of a linear system is widely used in many engineering fields. There are various more rigorous definitions for a linear system in the literature. For practical RF amplifier designs, a simple definition in terms of the amplifier gain is traditionally used. That is, if the gain is independent of the input power, the amplifier is said to operate in a linear mode; otherwise, nonlinearity occurs. In this section we first describe how to specify the nonlinearity of an amplifier using its gain characteristics. Then we explain that the gain alone is often insufficient to fully characterize the nonlinearity performance of an amplifier, especially in modern communication systems. As a result, additional nonlinearity specifications are used in practice.

We begin our analysis on nonlinearity by considering the input power dependence of the gain of an amplifier. Let $v_i$ and $v_o$ be the input and output signals of the amplifier; both are linear variables (either voltage or current), then the amplifier is linear if

$$v_o = av_i \tag{6.7}$$

where the coefficient $a$ is a constant and can be dimensionless or dimensional, depending on whether $v_o$ and $v_i$ are the same type of variable or not. For example, for a transimpedance amplifier, $v_i$ is actually the input current and $v_o$ the output voltage, and $a$ has the dimension of resistance. In addition, we simply take $v_i^2$ and $v_o^2$ for the input and output power, respectively. This is permissible, because the parameters to be derived are a power ratio and as such the actual load is not relevant. Clearly, a linear amplifier has a constant power gain as $a^2$. As the input signal increases, at a certain point, the gain starts to deviate from the constant value; that is, the amplifier is no longer a linear system. Mathematically, the nonlinear behavior of an amplifier can be analyzed by a Taylor expansion of $v_o$ as a function of $v_i$:

$$v_o(v_i) = a_1 v_i + a_2 v_i^2 + a_3 v_i^3 + \cdots \tag{6.8}$$

where the coefficients $a_i$'s are related to the $i$th order derivative of $v_o(v_i)$. In addition, note the following three comments on using this expression:

- First, this expansion usually is only valid over a certain range of $v_i$ due to the interactive nature of $a_i$'s and $v_i$. We will limit our analysis in a region where the amplifier gain just starts to compress (decrease) from the linear behavior. As such, the linear term $a_1 v_i$ is assumed to be still dominant, and the nonlinear effects are considered only up to the third order. We also note that $a_3$ must be negative so that $v_o$ increases at a slower rate than $v_i$ (gain compression). However, this is not necessarily always true, because over a wider input power range, $a_3$ could be positive, corresponding to a gain expansion.

- Second, the $a_2$ term is usually neglected in an analysis at the complexity level of this book, unless the second harmonic is considered. This can be somewhat justified by realizing that $a_2$ is the second derivative of the function $v_0(vi)$. The first derivative (the gain curve) of a practical device often exhibits a change in moving direction (from increasing to decreasing), indicating a region where $a_2 \cong 0$. Furthermore, this approximation is actually necessary to keep the results simple enough to be useful in practice.
- Third, all $a_i$'s in (6.8) are assumed to be independent of time. This assumption is only true if the system is memoryless (i.e., the output at any instance only depends on the input at that instance). Being memoryless is a nontrivial condition for practical devices. In fact, memory effects, which are usually due to charge trapping, can have significant impacts on the nonlinearity characteristics of certain devices and circuits and are widely studied in the literature (for example, [2,3,7]). A more detailed discussion on the subject is beyond the scope of this book.

In summary, these three simplifications allow us to derive some useful results using (6.8) with a manageable complexity. These results generally capture the basic nonlinearity behaviors of an RF system, but they are not invariably in quantitative agreement with measurements because these simplifications are not always fully justifiable for practical cases.

We return to (6.8). Consider an input signal $v_i = A\cos(\omega t)$; using (6.8), the output is

$$v_o = a_1 A \cos(\omega t) + a_2 A^2 \cos^2(\omega t) + a_3 A^3 \cos^3(\omega t) \qquad (6.9)$$

Using the trigonometry identities, we have

$$v_o = \frac{1}{2}a_2 A^2 + \left(a_1 A + \frac{3}{4}a_3 A^3\right)\cos(\omega t) + \frac{1}{2}a_2 A^2 \cos(2\omega t) + \frac{1}{4}a_3 A^3 \cos(3\omega t) \qquad (6.10)$$

We notice that the $\cos^2(\omega t)$ term in (6.9) generates new frequency components at 0 (DC) and $2\omega$, while the $\cos^3(\omega t)$ term generates two extra terms at $\omega$ and $3\omega$, respectively. These additional frequencies associated with the nonlinear terms are the reason that harmonics of a power amplifier must be properly characterized and controlled.

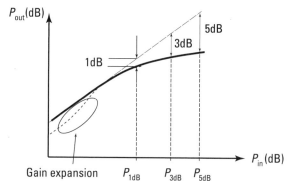

**Figure 6.9** Gain compression points on a $P_{out}$ versus $P_{in}$ plot.

Another conclusion we can draw from (6.10) is that the coefficient of the fundamental frequency, which is the gain, becomes

$$a_1 + \frac{3}{4}a_3 A^2 \qquad (6.11)$$

Recall that $a_3$ is negative. Equation (6.11) implies that as the input amplitude $A$ increases the gain is compressed. It is a standard practice in RF engineering to use gain compression points to specify the nonlinear behavior of an amplifier. The most popular one is the 1-dB compression point, which is the input power, denoted as $IP_{1dB}$, (or the output power, denoted as $OP_{1dB}$), where the gain drops by 1 dB from the linear value. In linear terms, the 1-dB compression point is when the gain drops to 79% of the original value. Thus, as noted, the compression point is a gain ratio. The gain compression points can be easily determined on an output power versus input power plot, as illustrated in Figure 6.9. Figure 6.9 also shows that after reaching the $IP_{1dB}$ point the gain continues to decrease as the input power increases. The rate of the gain's decreasing varies widely from circuit to circuit. For this reason, other gain compression points, such as the 3-dB and 5-dB points, are also used in practice.

In a practical circuit, the amplifier gain is not always constant before reaching the gain compression region. In some cases, particularly when the quiescent DC bias current is low, the circuit may exhibit a gain expansion in a certain region before the compression, as indicated by the dashed line in Figure 6.9. Clearly, (6.11) with a negative $a_3$ cannot account for this effect. Also, the varying gain before the compression makes the determination of the $P_{1dB}$ point somewhat arbitrary. The gain expansion usually occurs due to an increased DC bias current induced by the RF power. (See Chapter 7 for a discussion of this effect.)

Let $A_{1dB}$ be the amplitude of the input signal at the $P_{1dB}$ point, then by the definition, we have

$$20\log\frac{a_1 + \frac{3}{4}a_3 A_{1dB}^2}{a_1} = -1 \text{ dB}$$

This leads to

$$\text{IP}_{1dB} = A_{1dB}^2 = 0.145\frac{a_1}{|a_3|} \qquad (6.12)$$

Equation (6.12) itself has little practical use because neither $a_1$ nor $a_3$ is known. However, it allows an analysis that relates $P_{1dB}$ to other nonlinearity parameters as discussed later in this section. We can also use (6.12) to derive a formula for a cascaded case (see Chapter 8).

The $P_{1dB}$ point may be approximately considered a midpoint between the fully saturated state (the output power no longer increases with the input power) and the linear state of a PA. In practice, only in certain cases, an RF PA may actually operate at the $P_{1dB}$ point. However, because of its simplicity, the $P_{1dB}$ specification is widely used as a figure of merit for both power capability and linearity for RF amplifiers. To assess the actual performance on either side (saturation and linear), the PA has to be measured at the specific condition. This is especially true for a PA used in a communication system where the PA usually operates at a point that is significantly backed off from the 1-dB compression point. In that case, the actual PA performance has to be evaluated using a different nonlinearity specification pertinent to the specific application.

This leads to another widely used nonlinearity specification, the third-order intermodulation. This specification is defined with a two-tone measurement where the input signal for the measurement comprises two equal-amplitude signals at frequencies, $\omega_1$ and $\omega_2$ ($\omega_2 > \omega_1$), namely, $v_i = A\cos(\omega_1 t) + A\cos(\omega_2 t)$. Using (6.8), the output is

$$v_o = a_1 A\left(\cos\omega_1 t + \cos\omega_2 t\right) + a_2 A^2\left(\cos\omega_1 t + \cos\omega_2 t\right)^2 + a_3 A^3\left(\cos\omega_1 t + \cos\omega_2 t\right)^3$$
$$(6.13)$$

We have observed that the nonlinear terms generate new frequency components. In this case, in addition to the harmonics of $\omega_1$ and $\omega_2$, there exists a number of so-called intermodulation (IM) frequencies, such as $\omega_2 + \omega_1$ and $2\omega_1 - \omega_2$. For a reason explained in Chapter 8, we only consider the case where the separation of the two frequencies, $\Delta = \omega_2 - \omega_1$, is much smaller than $\omega_1$ and $\omega_2$ themselves (i.e., $\Delta \ll \omega_1, \omega_2$). Then there are two intermodulation frequencies that are close to $\omega_1$ and $\omega_2$, namely, $\omega_2 + \Delta$ and $\omega_1 - \Delta$, as depicted in Figure 6.10(a). When $\Delta \ll \omega_1, \omega_2$, these two IM frequencies are so close to the two signals that they cannot be filtered in practical circuits. These frequencies generated by a nonlinear effect are the origin for the so-called spectrum regrowth, which is a critical specification in broadband communication systems (see Chapter 8). It is easy to verify that both terms are generated by the cubic term and have the same power. Therefore they are referred to as third-order intermodulation power (also known as third-order

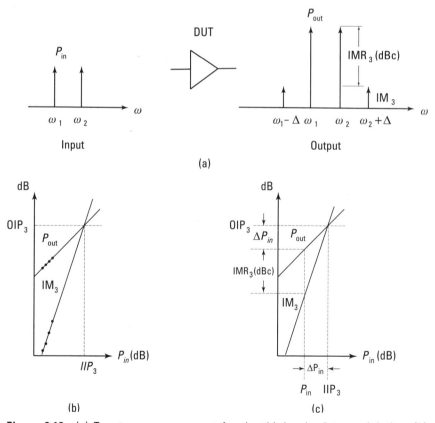

**Figure 6.10** (a) Two-tone measurement for the third-order intermodulation; (b) determination of $IP_3$; (dots represent measured points) and (c) relationship of $IMR_3$ and $IP_3$.

intermodulation distortion). It is denoted by $IM_3$ and is given by (again the algebra is left out):

$$IM_3 = \left(\frac{3}{4}a_3 A^3\right)^2 \qquad (6.14)$$

Equation (6.14) indicates that the $IM_3$ term increases with the input power in a third-order power. In this case, similar to (6.11), there is also an extra term, $(9/4)a_3 A^2$, in the linear gain. However, in practical applications, the intermodulation becomes significant, in terms of the system's nonlinearity specification, at an input power level long before the gain compression is measurable. In other words, we can use (6.14) for the intermodulation term while assuming that the gain is still in the linear region. This is the reason for the earlier statement that the gain alone is insufficient for nonlinearity specification. As a result, we expect that the $IM_3$ term increases three times faster than the output power in a log-log plot, as shown in Figure 6.10(b). Then at a certain point these two curves intercept. This point is called the third-order intercept point, denoted as $IP_3$. $IP_3$ can be either in input power, $IIP_3$ or output power, $OIP_3$. They are related by $OIP_3 = G \cdot IIP_3$, where $G$ is the linear gain.

Such a defined intercept point does not depend on specific measurement conditions (input power and frequency separation $\Delta$). Therefore it is convenient to use $IP_3$ as a figure of merit in comparing different circuits.

Unlike $P_{1dB}$, the $IP_3$ point cannot be directly measured, because it is far beyond the compression point. In addition, the measurement of $IM_3$ usually has high uncertainty because of relatively low power levels. A common practice in the measurement of $IP_3$ is to make a series measurement of $P_{out}$ and $IM_3$ in the linear region and then graphically determine the $IP_3$ point through a curve fitting and extrapolation, as shown in Figure 6.10(b).

Returning to (6.14), let $A_{IP3}$ be the input amplitude at $IP_3$; then by definition $A_{IP3}$ satisfies the following equation:

$$\frac{3}{4}|a_3| A_{IP3}^3 = a_1 A_{IP3} \qquad (6.15)$$

Note that (6.15) is also a ratio of two power terms (the ratio being 1). Thus we have

$$IIP_3 = A_{IP3}^2 = \frac{4a_1}{3|a_3|} \qquad (6.16)$$

In practice, IM$_3$ in an absolute power level has little meaning without reference to the fundamentals. Instead, the third-order intermodulation ratio, IMR$_3$, which is the ratio of IM$_3$ to the fundamental power, is usually used [see Figure 6.10(a)]. For an input power $P_{in} = A^2$, using (6.14) and (6.16) IMR$_3$ is

$$\text{IMR}_3 = \left(\frac{3a_3 A^2}{4a_1}\right)^2 = \left(\frac{A^2}{A_{IP3}^2}\right)^2$$

Then in decibel terms,

$$\text{IIP}_3(\text{dB}) = P_{in}(\text{dB}) - \frac{\text{IMR}_3(\text{dBc})}{2} \qquad (6.17)$$

Here IMR$_3$ (dBc) in (6.17) is negative. Equation (6.17) is an equation widely used in practice in the calculation of IMR$_3$ from IP$_3$ or vice versa at a specific input power. The same equation can also be derived using the graphic solution illustrated in Figure 6.10(c) (note that from the relationship of the two slopes we have $\Delta P_{in} + \text{IMR}_3 = 3\Delta P_{in}$).

Last, we can relate IIP$_3$ to IP$_{1dB}$ by comparing (6.12) and (6.16). The result is

$$\text{IIP}_3 = \text{IP}_{1dB} + 9.6 \text{ dB} \qquad (6.18)$$

The P$_{1dB}$ specification is usually supplied by the manufacturer or can be easily measured, whereas the IP$_3$ specification is not always available. In the RF business, an easy formula, IP$_3$ = P$_{1dB}$ + 10 dB, is commonly used to estimate IP$_3$ from P$_{1dB}$. Equation (6.18) is the analytical justification for this practice. In some cases, IP$_3$ = P$_{1dB}$ + 12 dB is also cited as a practical rule of thumb.

While most measurement results on actual amplifiers are in a reasonable agreement with this simple formula, some circuits exhibit a substantial deviation that can be as large as 5 dB in either direction. This is perhaps not surprising, because otherwise there is no need to measure IP$_3$. The often observed discrepancy between the analytic result in (6.18) and actual measurements stems from highly simplified assumptions in the derivation of (6.12) and (6.16).

For certain PA technologies, the IM$_3$ versus $P_{in}$ plot exhibits the so called sweet spot where IM$_3$ reaches a local minimum at a certain input power level and a bias condition [2, 3, 8]. Extensive efforts have been made by manufacturers to utilize this characteristic in the device design to minimize the intermodulation effect. The mechanism for the sweet spot is left to [2, 3, 8]. In practical amplifier design, for the optimization of intermodulation performance, besides device selection, it is important to ensure that the DC supply line is sufficiently bypassed at the frequency range comparable to the signal bandwidth (or $\Delta$ in the two-tone measurement) [1, 9].

# References

[1] Sechi, F., and M. Bujatti, *Solid-State Microwave High-Power Amplifiers*, Norwood, MA: Artech House, 2009.

[2] Colantonio, P., F. Giannini, and E. Limiti, *High Efficiency RF and Microwave Solid State Power Amplifiers*, John Wiley and Sons, 2009.

[3] Cripps, S. C., *Advanced Techniques in RF Amplifier Design*, Norwood, MA: Artech House, 2002.

[4] Albulet, M., *RF Power Amplifiers*, Noble Publishing, 2001.

[5] Sedra, A. S., and K. C. Smith, *Microelectronic Circuits*, Sixth Editio), Oxford University Press, 2009 (and later edition).

[6] Gray, P., et al., *Analysis and Design of Analog Integrated Circuits*, Fourth Edition, John Wiley and Sons, 2001.

[7] Vuolevi, J. H. K., T. Rahkonen, and J. P. A. Manninen, "Measurement Technique for Characterizing Memory Effects in RF Power Amplifiers," *IEEE Trans. Microwave Theory & Tech.*, Vol. 49, No. 8, 2001.

[8] Pedro, J. C., and N. B. Carvalho, *Intermodulation Distortion in Microwave and Wireless Circuits*, Norwood, MA: Artech House, 2003.

[9] Cripps, S. C., *RF Power Amplifiers for Wireless Communications*, Second Edition, Norwood, MA: Artech House, 2006.

# 7
# Efficiency of RF Power Amplifiers

Traditionally, high-efficiency RF amplifiers are achieved through techniques of waveform manipulation. Sections 7.1–7.4 discuss some of those techniques that are commonly seen in the literature [1–4]. This discussion, which focuses on the relevance to the actual circuit performance rather than analytic sophistication, shows clearly that the desired waveforms for the efficiency enhancement are usually generated when the RF transistors are driven near or into the compression region. Therefore, these techniques are intrinsically in conflict with the high-linearity requirements of modern wireless communication systems, a subject discussed Chapter 8. Various linearization techniques [3, 5, 6], which allow an amplifier to operate near the compression point while maintaining the required linearity, are either presently implemented or are actively being investigated. Among them, the Doherty amplifier is of particular importance in today's RF industry. Section 7.5 briefly discusses the Doherty amplifier.

## 7.1 RF Cooling and Efficiencies of RF Power Amplifiers

For readers who have experience in power management of electronic systems, the calculation of the system power dissipation is usually straightforward:

(supply voltage) × (supply current). Where an RF power amplifier is concerned, the situation is generally more complex owing to the so-called RF cooling effect. This is because part of the DC supply power is converted to the RF power delivered to the load (or radiated in the case of a wireless system). In other words, the system power dissipation is reduced (or "cooled") by taking the RF power out of the system. The RF power relative to the DC supply power is the essence of the efficiency of an RF power amplifier. Technically, there are two efficiency parameters commonly used in the RF business.

Efficiency, denoted as $\eta$ and also known as drain efficiency for FETs and collector efficiency for BJTs, is defined as

$$\eta(\%) = \frac{\text{RF output power}}{\text{DC supply power}} \times 100 \tag{7.1}$$

Equation (7.1) is simple but not completely fair in calculating DC-to-RF conversion efficiency because the RF input power, which is necessary to generate the RF output power, is not included. Therefore, power-added efficiency (PAE) is more often used in practice. It is calculated as

$$\text{PAE} = \frac{\text{RF output power} - \text{RF input power}}{\text{DC supply power}} \times 100 \tag{7.2}$$

Obviously, PAE is always less than $\eta$. For an amplifier with a sufficient gain (e.g., > 15 dB), the difference may be ignored in practice.

To see how the RF cooling occurs in an actual circuit, see Figure 7.1(a), which reexamines the circuits of Figure 6.4(a), and rewrite (6.1) to (6.3) as (note the notation of $I_c$ is replaced with $I_{dc}$ for the DC current)

$$V_d = V_c - i_d R \tag{7.3}$$

$$I_d = I_{dc} + i_d \tag{7.4}$$

Section 6.2 explains these equations in detail.

The average power dissipated inside the transistor is given by

$$\begin{aligned} P_{\text{diss}} &= \frac{1}{T}\int_0^T V_d I_d \, dt \\ &= V_c I_{dc} + \frac{V_c}{T}\int_0^T i_d \, dt - \frac{I_{dc} R}{T}\int_0^T i_d \, dt - \frac{R}{T}\int_0^T i_d^2 \, dt \end{aligned} \tag{7.5}$$

Here $T$ is the period of the RF signal, and (7.3) and (7.4) are used. The second and third terms in (7.5) are zero, and the fourth term is the same, numerically, as the power delivered to the load, since $|i_d| = |i_R|$. The negative RF power implies outgoing power and hence the cooling effect. By examining Figure 7.1, we realize that this negative sign is the result of the drain current and voltage waveforms being out-of-phase. In fact, the basic idea of achieving high efficiency in the circuit architectures discussed in Sections 7.2–7.5 is to manipulate the RF current and voltage waveforms such that the power dissipation inside the transistor is minimized with respect to the RF output power.

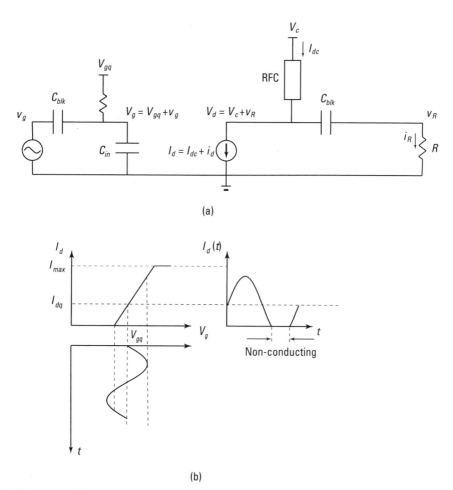

**Figure 7.1** (a) Equivalent circuit for DC and RF variables and (b) drain current waveform with clipped bottom.

## 7.2 Conduction Angle and Class A, AB, B, and C Power Amplifiers

We first complete the calculation of (7.5) for the case of maximum power under the distortion-free condition. From (6.6), the RF output power is $(1/2)V_c I_{dc}$ (note $I_{dc} = (1/2)I_{max}$). Hence, half of the DC supply power goes to RF power. Then the efficiency in this case is 50%. In fact, this is the result for the class A amplifier, which is the topic of Section 7.2.1.

### 7.2.1 Conduction Angles

The concept of the conduction angle of an amplifier is best explained by examining how the waveform of $I_d$ in Figure 6.6(a) will change if we lower the quiescent point of $I_{dq}$ (by reducing $V_{gq}$) while keeping the amplitude of the input signal unchanged. Figure 7.1(b) shows the result.

We notice that a portion of the waveform $I_d(t)$ in Figure 7.1(b) is "clipped" during the "nonconducting" state ($I_d = 0$). The amount of the time that the transistor is conducting within one period is described by the conduction angle, which is either in degrees or in radians. It is specified this way, because the conduction angle is automatically normalized to 360° or $2\pi$, whereas the conduction time would require an additional specification of the wave's period. For equation manipulation, the cosine function turns out to be more convenient than the sine function because of its symmetry with respect to $t = 0$. In the following analysis, we use the waveform shown in Figure 7.2, where $2\theta_c$ is the conduction angle, $I_p$ is the amplitude of the sinusoidal wave, and $\theta = \omega t$. From Figure 7.2, we can see that $I_d(\theta)$ is described by

$$I_d(\theta) = \begin{cases} I_{dq} + I_p \cos\theta & |\theta| < \theta_c \\ 0 & \theta_c < |\theta| < \pi \end{cases} \quad (7.6)$$

The conduction angle is essentially an operating condition (quiescent bias point and RF input level) of a power amplifier. However, traditionally, instead of being given directly in degrees or radians, the conduction angle is labeled in a series of classes defined as follows:

- Class A: $2\theta_c = 2\pi$;
- Class AB: $\pi < 2\theta_c < 2\pi$;
- Class B: $2\theta_c = \pi$;
- Class C: $2\theta_c < \pi$.

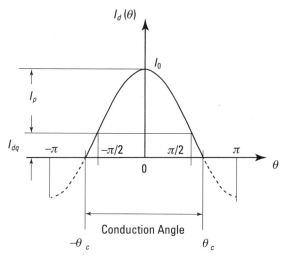

**Figure 7.2** Conduction angle.

In this convention, class A and B amplifiers operate at a specific conduction angle while class AB and C operate at a range of conduction angles.

Now we return to (7.6). This function is no longer a pure sinusoidal function. We expect harmonic components to be generated. Since $I_d(\theta)$ is an even function, we can use the Fourier cosine series

$$f(\theta) = \frac{a_0}{2} + \sum_{n=1}^{\infty} a_n \cos n\theta$$

where

$$a_i = \frac{2}{\pi} \int_0^{\pi} f(\theta) \cos n\theta \, d\theta, \quad i = 0, 1, \cdots$$

to calculate its harmonics. We will only work out the DC and fundamental components ($a_0$ and $a_1$) here

$$I_{dc} = \frac{1}{\pi} \int_0^{\theta_c} \left( I_{dq} + I_p \cos\theta \right) d\theta = \frac{1}{\pi} \left( I_{dq} \theta_c + I_p \sin\theta_c \right) \quad (7.7)$$

$$I_1 = \frac{2}{\pi} \int_0^{\theta_c} \left( I_{dq} + I_p \cos\theta \right) \cos\theta \, d\theta = \frac{2}{\pi} \left[ I_{dq} \sin\theta_c + \frac{I_p}{2} \left( \theta_c + \sin\theta_c \cos\theta_c \right) \right] \quad (7.8)$$

The reader may work out the results for the higher terms or find them in the literature (e.g., [3, 4]). Note the DC component given in (7.7) is no longer the quiescent value $I_{dq}$. The waveform of (7.6) is determined by $I_{dq}$ and $I_p$. The conduction angle can be calculated by letting $I_d(\theta_c) = 0$ in (7.6):

$$\cos\theta_c = -\frac{I_{dq}}{I_p} \quad (7.9)$$

We notice that if $I_p$ is kept constant while $I_{dq}$ is being lowered, the highest point of $I_d(\theta)$ ($I_0$ in Figure 7.2) is lowered as well. The standard condition used in the literature for the conduction angle analysis is to keep $I_0$ constant while varying $\theta_c$. That is, $I_p$ is adjusted upon a change in $I_{dq}$ such that

$$I_0 = I_{dq} + I_p \quad (7.10)$$

is a constant.

Before proceeding with our analysis, let's clarify again the three "maximum currents" used in this chapter and Chapter 6:

- $I_{max}$: The maximum linear drain current in the piecewise transfer function (see Figure 6.3); it is a device characteristic, albeit a highly idealized one.
- $I_{d\_rating}$: The maximum current rating on the datasheet; it may be higher or lower than $I_{max}$.
- $I_0$: The maximum current used in the conduction angle analysis [defined in (7.10)]; it must be in the linear region; that is, $I_0 \leq I_{max}$, and it can be controlled by the circuit designer.

We are chiefly interested in the device/circuit performance characteristics in this section; hence, we are not concerned with $I_{d\_rating}$ in our analysis, since it is mainly determined by the reliability data, as explained in Chapter 6. In practical designs, $I_{d\_rating}$ must be carefully considered for the long-term reliability.

In most texts on the subject, $I_0 = I_{max}$ is assumed, because it is the condition for the maximum linear output power. However, Section 7.3 shows that there are several advantages in terms of efficiency by selecting a lower $I_0$. Section 7.2.3 also explains how $I_0$ can be determined by the load line selection in an actual circuit.

Under the condition of (7.10) ($I_0$ being fixed), we can study the effect on the waveform $I_d(\theta)$ due to $\theta_c$ alone. To this end, we rewrite $I_p$ and $I_{dq}$, using (7.9) and (7.10), as

$$I_p = \frac{I_0}{1-\cos\theta_c} \qquad (7.11)$$

and

$$I_{dq} = -\frac{I_0 \cos\theta_c}{1-\cos\theta_c} \qquad (7.12)$$

Substitution of (7.11) and (7.12) into (7.7) and (7.8) leads to $I_{dc}$ and $I_1$ expressed in $\cos\theta_c$ and $I_0$ as

$$I_{dc} = \frac{I_0}{2\pi}\frac{2\sin\theta_c - 2\theta_c \cos\theta_c}{1-\cos\theta_c} \qquad (7.13)$$

$$I_1 = \frac{I_0}{2\pi}\frac{2\theta_c - \sin 2\theta_c}{1-\cos\theta_c} \qquad (7.14)$$

$I_{dc}$ and $I_1$ as a function of the conduction angle, $2\theta_c$, are plotted in Figure 7.3, where the functions are normalized to $I_0/2$. As explained in Section 7.1,

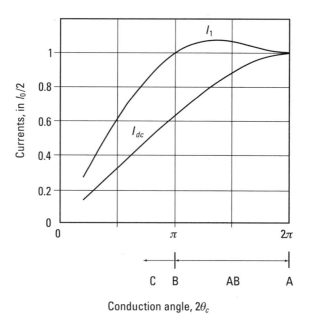

**Figure 7.3** Conductance-angle dependence of the DC and fundamental components of the drain current.

$I_1$ is related to the RF power (at the fundamental frequency) delivered to the load, while $I_{dc}$ is the current from the DC supply. Therefore, these plots are the basis for our analysis on the amplifier efficiency, outlined in Section 7.2.2.

### 7.2.2 Efficiency of Class A, AB, B, and C Amplifiers

To analyze the power distribution between the load and the transistor, given the drain current waveform expressed in (7.6), we need to consider the waveform of the drain voltage, which in turn is determined by the load condition. If we still use the simple circuit configuration depicted in Figure 6.4, the RF power dissipated in the load would include all the harmonic components generated in $I_d$. This would be an unfavorable condition for several reasons. First, in practice, powers at harmonics are not only wasted but also highly undesirable in any wireless systems for the reason of regulatory compliance. Second, higher-order harmonics will be affected by the output capacitance (included in $Z_{out}$ in Figure 6.4); hence the simple circuit shown in Figure 6.4 may not be an realistic representation in terms of harmonic performance. Third, adding harmonic components, even if only for the first few orders, makes analysis much more complex. For these reasons, the standard approach in the literature is to add a harmonic trap in parallel with the resistance load, as shown in Figure 7.4(a). As mentioned in Section 6.2, the harmonic trap has two distinct circuit characteristics: It has no impact on the fundamental signal but completely shorts circuit at all harmonics. The circuit realization of this harmonic trap can be a high-$Q$ parallel LC resonant circuit.

With the harmonic trap, the RF voltage on the load has only a fundamental component $|v_1|$ and its amplitude is given by

$$|v_R| = |v_1| = I_1 R \quad (7.15)$$

It is critically important to note the difference between (7.15) and (6.3) from which we introduced the concept of the load line in Figure 6.5. At the presence of the harmonic trap, only the fundamental component in $I_d$ contributes to $v_R$. As a result, the load line is no longer a straight line defined by the slope of $1/R$ and the quiescent point of $(I_{dq}, V_c)$. To derive the load line for this case, we simply need to remove $\cos\theta$ in the expressions of $I_d(\cos\theta)$ and $V_d(\cos\theta)$. For this end, we explicitly write $V_d$ as

$$V_d = V + v_1 = V_c - I_1 R \cos\theta \quad (7.16)$$

*Efficiency of RF Power Amplifiers* 211

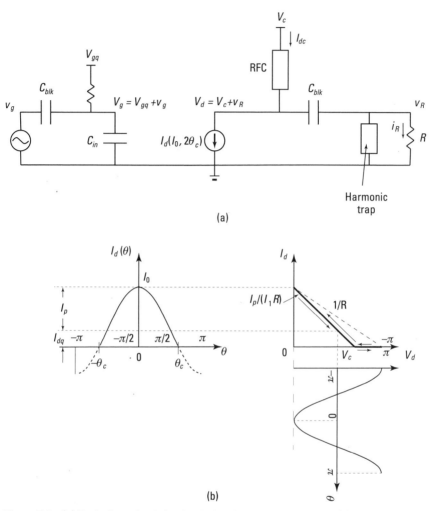

**Figure 7.4** (a) Equivalent circuit for the drain efficiency analysis and (b) waveforms of $I_d$ and $V_d$ for a class AB amplifier. Arrows are directions of movement in the $I_d$-$V_d$ plane for one RF cycle.

$I_d$ only has nonzero values within the conduction angle. Then, using (7.6) and (7.16), we have

$$I_d = I_{dq} - \frac{I_p}{I_1 R}(V_d - V_c) \text{ for } |\theta| < \theta_c \qquad (7.17)$$

The modified load line still goes through the quiescent point as expected, but its slope is increased by a factor of $I_p/I_1$, as sketched in Figure 7.4(b).

Also, the slope in (7.17) is no longer predetermined since it depends on the drain current waveform ($I_p$ and $I_1$). Therefore, (7.17) has to be interpreted as a steady-state solution rather than an instantaneous one. Note that the portion of the trajectory where $I_d = 0$ is consistent with the waveform of $I_d(\theta)$ shown in Figure 7.4.

The maximum RF swing of $V_d$ is still set by $V_c$, as shown in Figure 7.4(b); that is,

$$\left|v_1\right|_{max} = V_c = I_1 R_\theta \tag{7.18}$$

$R_\theta$ is the optimal load resistance, the same concept as $R_{opt}$ derived in (6.5). For a given pair of $I_0$ and $\theta_c$ (hence $I_1$ is known),

$$R_\theta = \frac{V_c}{I_1} \tag{7.19}$$

yields the maximum power delivered to the load. Under this condition, the DC and the RF powers are given by

$$P_{dc} = V_c I_{dc} \tag{7.20}$$

$$P_{RF} = \frac{1}{2}\left|v_1\right|_{max} I_1 = \frac{1}{2} V_c I_1 \tag{7.21}$$

Then the efficiency is

$$\eta = \frac{P_{RF}}{P_{dc}} = \frac{I_1}{2 I_{dc}} \tag{7.22}$$

Clearly, $\eta$ is also a function of $\theta_c$. For Class A and B amplifiers, $\eta$ and $R_\theta$ are calculated from (7.13) and (7.14) as follows:

- Class A, $I_0 = I_{dc} = (1/2)I_0$, $\eta = 50\%$ and $R_\theta = 2V_c/I_0 = R_{opt}$ [when $I_0 = I_{max}$; see (6.5)].
- Class B, $I_1 = (1/2)I_0$, $I_{dc} = (1/\pi)I_0$, $\eta = \pi/4 = 78.5\%$, and $R_\theta = 2V_c/I_0$

We notice that in the above analysis of the drain voltage and current that as the DC and RF loads are different (as discussed in Section 6.2) so are the loads at fundamental and harmonic frequencies (the harmonic trap). Section 7.4 discusses more variations of the load frequency characteristics.

Figure 7.5(a) shows plots for $\eta(\theta_c)$ and $R_\theta(\theta_c)$, where $R_\theta$ is normalized to $2V_c/I_0$. To remind readers of the two conditions under which these results are obtained, Figure 7.5(b) also presents the waveforms of $I_d$ for the four classes and the waveform for $V_d$ at the maximum RF swing. The improvement in $\eta(\theta_c)$ as the conduction angle reduces can be understood if we realize that the conduction-angle reduction is achieved by the combination of a decrease in $I_{dq}$ (DC) and an increase in $I_p$ (RF).

Section 6.3 explains how harmonics are generated by a nonlinear effect. In this case, as evidenced in the Fourier expansion, harmonic components are also generated due to the truncated waveform. However, the transfer function (7.6) is still a linear function in the sense that all the harmonics are linearly dependent on the input variable, $I_p$; that is, there are no $I_p^n$ terms. Therefore, the amplifier classes discussed in this section are considered linear amplifiers. As discussed in Section 6.3, the more problematic nonlinear effect for communication systems is the intermodulation distortion, which can only be produced from a higher-order term in the transfer function.

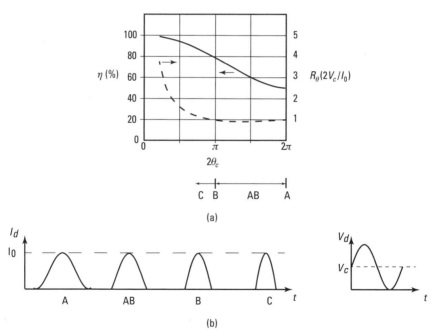

**Figure 7.5** (a) Conduction-angle dependence of efficiency $\eta$ and load resistance $R_\theta$ (normalized to $2V_c/I_0$) and (b) waveforms used in calculations of $\eta$ and $R_\theta$

The two efficiencies, 50% for class A and 78.5% for class B, along with the efficiency plot in Figure 7.5(a), are widely cited in the RF community. However, the application of these results to a real-world amplifier is not straightforward, because other than $I_{dq}$ and $V_c$, the parameters $I_0$ (or $I_p$) and $R_\theta$ are not directly known. Section 7.2.3 examines these parameters more closely in the context of actual circuit performance.

### 7.2.3  $I_0$ and Load Line Selection

We first consider how $I_0$ is set by the load line selection. Recall that when $I_0 = I_{max}$ the required load resistance is given by $R_{opt} = 2V_c/I_{max}$ [see (6.5)]. Now if a load resistance, $R > R_{opt}$ is chosen, the drain voltage $V_d$ will reach the maximum swing before $I_d$ reaches $I_{max}$. Beyond this point, the drain current can no longer increase freely with the input signal $v_g$. This is how the linear region of the drain current, which is characterized by $I_0$, is controlled by the load resistance. Quantitatively, we note from Figure 7.3 that $I_1(\theta_c)$ is fairly constant in the range of conduction angles from $\pi$ to $2\pi$. Then we can have an approximation: $I_1(\theta_c) \approx I_0/2$ for this conduction-angle range. For a given load resistance $R > R_{opt}$, the maximum voltage swing occurs at $I_1 R = V_c$. Thus, $I_0$ is approximately determined by

$$I_0(R) \approx \frac{2V_c}{R} \tag{7.23}$$

when the amplifier operates between the class A and B modes. If the quiescent current changes, the slope of the load line changes correspondingly, but $I_0$ remains roughly constant. Figure 7.6 illustrates this concept for two load resistances at two $I_{dq}$ conditions.

The load resistance of an actual amplifier is fixed for a target output power. Then the efficiencies at lower input powers are unavoidably lower due to the reduced $I_1$ as well as the reduced drain voltage swing. An ideal solution

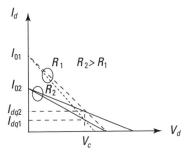

**Figure 7.6**  Load lines at two $I_{dq}$ conditions.

for achieving high efficiencies over a range of input power is a tunable load that varies with the input power so that the drain voltage swing is kept at $V_c$. In fact, a tunable load scheme is at the core of the Doherty amplifier. (See the discussion in Section 7.5.)

### 7.2.4 Class A to Class B Modes in Actual RF Amplifiers

In practice, the performance parameters of a power amplifier, such as output power, gain, and efficiency, are typically characterized as functions of input power. The plot in Figure 7.5(a) should not be interpreted as the input-power dependence of the efficiency even though the input power is indeed related to the conduction angle (through $I_p$).

To see how the analytic results in this section can be applied to the measurement data of an actual amplifier, we note that in our analysis, $I_p$ is assumed to be proportional to $|v_g|$. Then, the input power is $\propto I_p^2$. Similarly, in the linear region, the output power is $\propto I_1^2$. In our discussion, $I_p$ is limited to the range of $I_{dq} < I_p < I_0 - I_{dq}$ ($I_{dq} < I_0/2$). The first inequality is due to the fact that if $I_p < I_{dq}$, the condition is trivial, since $I_1 = I_p$ (with no clipping at the bottom). The second inequality is required for the system to be in the linear region. Within this range, we can calculate the output power and efficiency as a function of $I_p$ with $I_{dq}$ as a parameter.

We assume $V_c$ and $I_0$ [from the load resistance; see (7.23)] are specified. For a given $I_{dq}$, the half conduction angle $\theta_c$ can be calculated from (7.9) when $I_p > I_{dq}$. Then $I_{dc}$ and $I_1$, as a function of $I_p$, are determined from (7.7) and (7.8). The efficiency can still be calculated using (7.22) with $P_{RF}=I_1^2R/2$ and $R = 2V_c/I_0$ (see (7.23)). Now, $I_p$ is a variable in the range of $I_{dq}$ to $I_0 - I_{dq}$. Figure 7.7 plots the results for the output power ($\propto I_1^2$) and $\eta$ versus the input power ($\propto I_p^2$) for three $I_{dq}$ values. As in the previous analysis (Figures 7.3 and 7.5), all currents are normalized to $I_0/2$ and the load resistance to $2V_c/I_0$. The proportionality factors in the input and output powers are not important, since we are only interested in the relative relationship in the logarithm scale.

The range for the input power in Figure 7.7 reduces as $I_{dq}$ rises because of the restriction of $I_p > I_{dq}$. At the class A mode, it becomes one point ($I_p = I_{dq} = I_0/2$). In addition, we realize that the efficiency in Figure 7.5(a) is the value at the high end of each curve where $I_p = I_0 - I_{dq}$. [The exact values are slightly different, because a fixed load resistance ($2V_c/I_0$), rather than $R_\theta(\theta_c)$, is used in our calculation.]

Figure 7.7 summarizes the essence of the class A through class B in the context of a typical RF power amplifier characteristics: (1) The maximum linear output powers are approximately given by $I_0V_c/4$, regardless of the quiescent

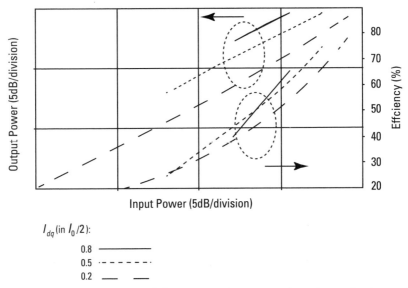

**Figure 7.7** Output power and efficiency versus input power from the conduction-angle analysis.

current $I_{dq}$, and (2) lowering $I_{dq}$ improves the efficiency at the maximum linear output power but increases the required input power to reach that point. For example, the required $I_p$ is doubled (for a given $I_0$) from the Class A mode to the Class B mode, which means that the gain is reduced by 6 dB. Therefore the higher efficiency is achieved at the expense of the gain. The following example illustrates how these results can be related to the performance data of an actual amplifier.

Figure 7.8 shows the performance plots of a 460-MHz power amplifier using a 10-W LDMOS transistor. Both the input and output matching networks are composed of multiple sections of shunt capacitors and series transmission lines, a typical configuration for power amplifiers in this frequency range (see, for example, Figure 5.9). It also happens to be a low-pass network, which, in combination with the device output capacitance (not shown), can function similarly to the harmonic trap assumed in our analysis. The drain voltage is 7.5V, and the circuit is tuned for the maximum saturation output power. The estimated load resistance is $R = 3\Omega$, based on the output matching network. Using (7.23) and (7.21), we have $I_0 = 2V_c/R = 5A$ and $P_{RF} = 9.4W$ ($I_1 \approx I_0/2$ is used). If a knee voltage of 1.5V is taken into account, replacing $V_c$ with $V_c - V_k$, these two figures become 4A and 6W, respectively. (See Section 7.3.2 for further discussion on the knee voltage). The maximum current rating of the transistor is 3A. As evidenced by the plots in Figure 7.8(b), the

# Efficiency of RF Power Amplifiers

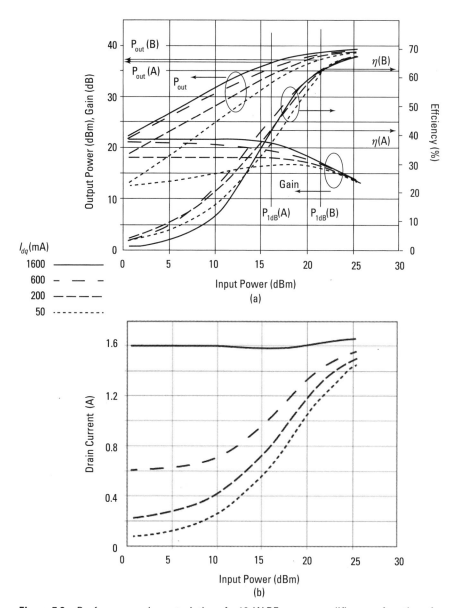

**Figure 7.8** Performance characteristics of a 10-W RF power amplifier as a function of input power at four quiescent bias conditions: (a) output power, gain, and efficiency and (b) current.

DC current converges to about 1.6A in high-saturation conditions, which indicates that the maximum drain current for linear operation is limited by the transfer function rather than the load line because a 3-$\Omega$ load line would allow the DC current to grow up to around 2A. This is actually consistent with the design goal of achieving maximum output power at the given supply voltage because a smaller [than the value in (7.23)] load resistance allows the drain current to increase further after the $I_0$ point. Section 7.3 provides a more detailed description of the process.

In our analysis, $I_{dq}$ = 1.6A is used as a representative for the class A operation. For the class B, $I_{dq}$ = 50 mA is used. Only these two conditions are analyzed here. We use the 1-dB compression point as the boundary of the linear region (that is, where $I_0$ is reached). The $P_{1dB}$ point for each condition, along with the corresponding output powers $P_{out}$(A) and $P_{out}$(B) and efficiencies, $\eta$(A) to $\eta$(B), is marked on the plot. The difference between the $P_{1dB}$ point for class A and B is about 5.5 dB, reasonably close to the theoretical value of 6 dB. The two output powers are quite close, slightly above 37 dBm (5W), which is also consistent with the analytic results, considering that the actual $I_0$ is about 3.2A instead of 4A. Finally, the plots show that the efficiencies at the $P_{1dB}$ points are about 42% and 63% for class A and B, respectively. If the knee voltage is included in the calculation, these efficiency numbers are in good agreement with the theory.

This analysis illustrates that the analytic result for an amplifier class considered in this section is just one data point on a general performance plot and that the determination of this point is somewhat arbitrary. The choice of the $P_{1dB}$ point as the maximum power in the linear region is mainly for convenience and is perhaps a little too high.

From these plots, we immediately see that higher efficiencies can be achieved by driving the amplifier into deep saturation. Furthermore, we notice that all the curves in the plots are converging at high input powers, indicating that the quiescent current has almost no effect on the ultimate efficiency achievable by the amplifier. Section 7.4 details how driving the transistor into saturation can further increase efficiency.

In terms of linear performance, the basic message of the conduction angle theory is that the designer can trade the gain of an amplifier for efficiency by lowering the conduction angle (i.e., lowering the quiescent bias current) without affecting the maximum achievable linear power. In practice, however, it is often experimentally determined that there exists an optimal quiescent current that produces a best combination of output power, linearity, and efficiency. While the conduction angle may still have some effect, this optimal bias point is usually associated with a specific characteristic of the transfer function of the chosen transistor.

In summary, our analysis here, which is based on the conduction angle technique, provides an insight into how the RF output power and the efficiency are determined by the bias voltage, the drain current characteristic, and the load-line selection. The results of this analysis, however, have limited value as quantitative guides in practical design. In practice, when an amplifier is said to operate in a certain class, it usually indicates how the quiescent current is selected. For example, a class B or C implies that the quiescent gate voltage is selected at or below the threshold voltage. In some cases, a class AB may indicate that the amplifier is specified for linear operation. Beyond that, the class label generally does not have a significant implication in terms of performance.

Traditionally, another factor that is important in selecting the quiescent current is the idle time (not transmitting RF powers), which the amplifier may experience in applications. Since any DC power during the idle time is totally wasted, a low-quiescent current is always desirable. However, in modern communication systems, the DC supply for the power amplifiers is almost always controlled by a microcontroller and is only turned on when the system is ready for transmission. Therefore, the idle time is no longer a significant design concern in these systems.

## 7.3 Operation in Overdriven Conditions

So far, we have limited our analysis to the linear operations, or mathematically, the RF amplitude $I_p$ is constrained by $I_p < I_0 - I_{qd}$. However, as shown in Figure 7.8, after the linear region (marked by the $P_{1dB}$ point in Figure 7.8), both the output power and efficiency continue to increase with the input power. This is particularly true for efficiencies of the class A and high class AB. In fact, it is common knowledge among practicing RF engineers that driving an amplifier into saturation almost always yields a higher efficiency. This section studies how an amplifier behaves after the drain current exceeds the linear limit $I_0$. Recall that Section 7.2 explains that $I_0$ is either limited by the device characteristic $I_{max}$ or designed by the load line selection. We consider the two cases separately in Sections 7.3.1 and 7.3.2.

### 7.3.1 $I_0 = I_{max}$

If the linear region of the drain current is not limited by the load line, it generally can continue to rise with the input signal $v_g$ after the linear region. In this case, the drain current is described by the transfer function, which is no longer a linear function of $v_g$. For simplicity, a hard limit is usually assumed (see Figure 6.3). Under this condition, the drain current waveform is clipped at the top as shown in Figure 7.9(a) and can be modified from (7.6) as

$$I_d(\theta) = \begin{cases} I_{max} & |\theta| < \theta_b \\ I_{dq} + I_p \cos\theta & \theta_b < |\theta| < \theta_c \\ 0 & \theta_c < |\theta| < \pi \end{cases}$$

We can see that, if there is no restriction from the load line, the flat portion at the top of $I_d(\theta)$ becomes wider ($\theta_b$ becomes larger) as the transistor is driven further into saturation. The extreme case is a square wave, which is the condition for the so-called switching-mode operation. Sections 7.4 consider some examples of this mode of operation. From the symmetry of a square wave, we can intuitively see, without mathematical proof, that the DC level converges to the midpoint of $I_{max}$ as the waveform becomes more flattened at both the bottom and top. Furthermore, it can be proved [2] that the efficiency for a drain current of the square waveform is 64%. Both are consistent, although not necessarily in a quantitative manner, with the measurement results plotted in Figure 7.8.

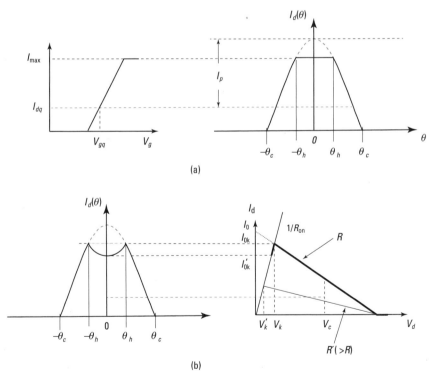

**Figure 7.9** Clipping of drain current due to the (a) hard limit in the transfer function and (b) the load line restriction.

In this analysis, the drain current is assumed to be completely free from any load-line restriction. In actuality, the drain current and voltage are interdependent, as explained in Section 7.3.2.

### 7.3.2  $I_0$ by the Load Line Selection

We now consider the case shown in Figure 7.9(b) where $I_0$ is limited by a specific load resistance $R(> 2V_c/I_{max})$. To make the case more realistic, Figure 7.9(b) includes the knee voltage effect. As shown in Figure 7.9(b), the inclusion of $V_k$ moves $I_0$ [determined by (7.23)] to a lower value $I_{0k}$, because once $V_d$ reaches $V_k$ the current must go down, as dictated by the $I_d$-$V_d$ curve. The exact amplitude of the drain voltage is difficult to determine in this case for two reasons: (1) $V_d$ can still move slightly toward $V_d = 0$ after the $V_k$ point as depicted by the $I_d$-$V_d$ trajectory [the bold line in Figure 7.6(b)], the drain voltage swing is essentially determined by $V_c - V_k$. Considering that the DC power is still $I_{dc}V_c$, the $V_k$ effect on the efficiency is a reduction by a factor of $(V_c - V_k)/V_c$. For a larger load resistance $R'(> R)$, $V_k$ is smaller. [See Figure 7.9(b).] Therefore, the factor $(V_c - V_k)/V_c$ explains why a higher efficiency is usually achieved with a larger load resistance. It also implies that increasing the bias voltage $V_c$ always yields higher efficiency, provided that the load resistance is adjusted accordingly.

Figure 7.9(b) shows a complete waveform of $I_d$. We notice that $I_d$ continuously goes down after reaching $I_{0k}$ until the lowest point labeled as $I'_{0k}$ in Figure 7.9(b). The exact solution for the waveform is obtainable but not particularly useful for practical design. Interested readers may find the solution in [4, 9].

In addition to the advantage of a smaller $V_k$, a larger load resistance also forces the drain current into saturation early, a process that typically generates a significant amount of harmonics. Section 7.4 explains how the harmonics of the drain current may be used to further enhance the drain efficiency.

## 7.4  Harmonic Flattening and Class F and Inverse Class F Amplifiers

### 7.4.1  Second- and Third-Harmonic Flattening

The entire analysis in Section 7.2 is based on the assumption of the harmonic trap. Under this assumption the drain voltage is [repeating (7.15) and 7.16)]:

$$V_d = V_c - V_1 \cos\theta \qquad (7.24)$$

where $V_1 = I_1 R$.

In this condition, the constraint of $V_d > 0$, discussed in Section 6.1, limits $V_1$ to be $\leq V_c$ (or even less if $V_k$ is considered). The idea of harmonic flattening (also known as harmonic tuning) is to add a harmonic component in (7.24) such that $V_1$ can be increased beyond $V_c$ without violation of the condition $V_d > 0$. We consider two specific cases, second and third harmonics, in this section. To simplify our analysis, we rewrite (7.24) as

$$v_d = 1 - v_1 \cos\theta \tag{7.25}$$

where voltages are normalized to $V_c$. Note that the lower limit for the fundamental term is $-1$ to maintain $v_d > 0$

### 7.4.1.1 Second Harmonic Flattening

We start with the second harmonic case by examining the function:

$$y = -\cos\theta + x\cos(2\theta + \phi_2) \tag{7.26}$$

$\phi_2$ presents a general case for the phase between the second harmonic and the fundamental terms. We only consider the case of $\phi_2 = 0$, because it is the condition that yields the maximum desired effect. The reader can verify this statement by plotting the function in (7.26) or prove it analytically. Under this condition, the second harmonic term in (7.26) flattens the bottom portion of $y$ as shown in Figure 7.10(a) for two $x$ values. As a result, the fundamental term can further grow before reaching the original lower limit, $-1$. This is the idea behind harmonic flattening.

It is clear from Figure 7.10(a) that $x = 0.45$ is more favorable than $x = 0.15$ for the second harmonic flattening. To determine the optimal $x$ value, we first locate the extrema of the $y$ function by letting the derivative $y' = 0$; that is,

$$y' = \sin\theta - 2x\sin 2\theta = \sin\theta(1 - 4x\cos\theta) = 0$$

One of the solutions is obvious: $\sin\theta_1 = 0$, and for our case, $\theta_1 = \pi$. The other two solutions are $\cos\theta_{2,3} = 1/(4x)(>0)$. $\theta_2$ and $\theta_3$ must be in the first and fourth quadrants, respectively. Furthermore, the optimal $x$ must be greater than $1/4$. Then $y$ reaches the lowest value at these two points and it is given by

$$y(\theta_{2,3}) = -\frac{1}{4x} + x\left(\frac{2}{(4x)^2} - 1\right) = -x - \frac{1}{8x} \tag{7.27}$$

Here the trigonometric identity is used. We can determine the maximum distance of $y(\theta_{2,3})$ from $-1$ by letting $dy(\theta_{2,3})/dx = 0$. The result is

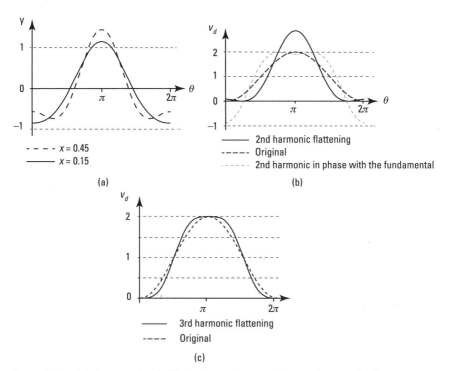

**Figure 7.10** (a) The waveform of (7.26) for $x = 0.45$ and 0.15, and $\phi_2 = 0$; (b) the waveform for maximum second harmonic flattening; and (c) the waveform for third harmonic flattening.

$$x = \frac{1}{\sqrt{8}} = 0.35 \qquad (7.28)$$

Then at this condition, using (7.27) we have

$$y\left(\theta_{2,3}, x = 0.35\right) = -0.707 \qquad (7.29)$$

Equation (7.29) implies that the fundamental can be increased by a factor of 1.4 (1/0.707). Using this result to replace $v_d$ in (7.25), the normalized drain waveform under the condition of the maximum second harmonic flattening is

$$v_d = 1 - 1.4(\cos\theta - 0.35\cos 2\theta) \qquad (7.30)$$

Its waveform is shown in Figure 7.10(b).
In the above analysis, the second harmonic term in (7.26) is explicitly assumed to be out of phase with the fundamental ($\phi_2 = 0$). The reader can

verify that there is a range of $\phi_2$ that produces the second harmonic flattening effect, but $\phi_2 = 0$ is the most favorable condition. This point has significance in actual circuit conditions, as outlined in Section 7.4.2.

### 7.4.1.2 Third Harmonic Flattening

Similar to (7.26), we consider a function including a third harmonic term:

$$y = -\cos\theta + x\cos 3\theta \qquad (7.31)$$

The same technique used for the second harmonic flattening can be applied to (7.31). However, the analysis turns out to be more complex in this case because of the $\cos 3\theta$ term. So we seek, instead, a maximum flat condition where the second derivative at the extrema is 0. Again by the symmetry of the function, we realize that $\theta = 0$ is one of the lowest points (the other one is at $2\pi$). Then the optimal value of $x$ in (7.31) is determined from the equation:

$$y''(\theta)\big|_{\theta=0} = 0$$

The result is

$$x = \frac{1}{9} \qquad (7.32)$$

Then the lowest point of $y$ is

$$y\left(\theta = 0, x = \frac{1}{9}\right) = -1 + \frac{1}{9} = -\frac{8}{9} \qquad (7.33)$$

The corresponding normalized drain waveform

$$v_d = 1 - \frac{9}{8}\left(\cos\theta - \frac{1}{9}\cos 3\theta\right) \qquad (7.34)$$

is plotted in Figure 7.10(c). The optimal value for $x$ is about 0.17 [2], somewhat larger than that in (7.32).

In the processes discussed above, the DC component is unchanged. Therefore, the increases in the fundamental components due to the harmonic flattening seem to suggest a respective improvement in efficiency of 41% and 11% for the second and third harmonic cases. Then we immediately see a problem with the class B amplifier because the efficiency would be 78.5% × 1.41 > 100% if the second harmonic flattening were employed. Section 7.4.2 examines how this paradox arises.

## 7.4.2 Class F and Inverse Class F Amplifiers

We consider the inverse class F operation first because it directly answers the question we just raised.

Recall that the fundamental component of the drain voltage originates from the fundamental current according to $V_1 = I_1 R$ [see (7.16)]. Naturally, the harmonic voltages would be expected to have the same origin. However, in a class AB amplifier, the second harmonic is in phase with the fundamental. (Readers can work it out using the formula provided in Section 7.1 or find it in the literature [2–4]). In contrast, the second harmonic in (7.30) is out of phase with the fundamental ($\phi_2 = 0$). If we change the sign of the $\cos 2\theta$ term in (7.30), the waveform is shown in Figure 7.10(b) (the "in phase" curve). As seen, a significant portion of the waveform is below $v_d = 0$. Therefore, this waveform cannot be supported by the transistor. This observation leads to the idea of the inverse class F amplifier

Figure 7.11(a) shows the basic circuit topology of an inverse class F amplifier, where the drain current waveform is assumed to be a square wave rather than a truncated sine wave as in the class B amplifier. The physical reality for this assumption is a situation when the transistor is heavily driven into saturation, which is briefly discussed in Section 7.3. From the symmetry, an ideal square wave does not have even harmonics. If a parallel resonant circuit at the second harmonic is connected at the drain, as shown in Figure 7.11, the load impedance at the second harmonic is infinite. We can conceptualize a process where some distortions from the perfect symmetry of the square wave generate a small second harmonic component with a favorable phase condition for the harmonic flattening. Then the infinite impedance can support a finite voltage out of this small current. This is the origin for the formula, $0 \cdot \infty =$ finite number, often seen in the discussion of this subject. In fact the essence of this operation is a self-enforcement process: (1) only the second harmonic with the favorable phase conditions allows the voltage to grow and (2) a large voltage swing causes more distortions in the current waveform (see Section 7.3). As a result, this process leads to a self-consistent solution between $I_d$ and $V_d$, which includes the harmonic flattening effects, as long as the load impedance at the second harmonic is sufficiently high (not necessarily a resonant circuit). For the efficiency calculation, the Fourier coefficients for the DC and the fundamental components of a square wave are [2]: $I_{dc} = 1/2$ and $I_1 = -2/\pi$, which results in a drain efficiency of $2/\pi = 64\%$. With the improvement factor due to the second harmonic flattening, the maximum possible drain efficiency is $64\% \times 1.4 = 90\%$. Therefore the condition $\eta > 100\%$ does not exist in reality. In fact, the inverse class F mode is often defined in the literature as having a truncated sine wave, which includes all even harmonic terms, for the drain voltage. It can be proved that (see, for example, [1, 2]) the efficiency under that idealized condition is 100%.

Since the loading condition for an inverse class F (only with second harmonic flattening) is an infinite load at the second harmonic and a short for all odd harmonics, the circuit implementation for the inverse class F amplifier is conceptually straightforward as shown in Figure 7.11(a).

To summarize, the process leading to the inverse Class F operation involves the following: (1) removing the in-phase second harmonic by driving the drain current waveform into a square wave; (2) assuming an infinitesimal of the out-of-phase second harmonic in the drain current; and (3) applying an infinite second harmonic load to support a finite drain voltage at the second

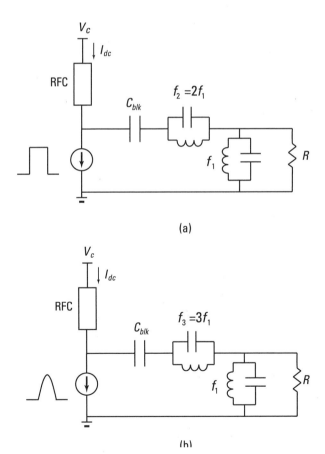

**Figure 7.11** Conceptual circuit realizations: (a) inverse class F with second harmonic flattening and (b) class F with third harmonic flattening

harmonic with the right phase. In an actual circuit, none of the above conditions needs to be perfect for achieving some efficiency improvement through the second harmonic flattening.

The drain current and voltage waveforms for a class F amplifier are reversed from those of the inverse class F; that is, the drain current is a truncated sine wave (class B) that only has even harmonics. For the same reasoning used above, the absence of the odd harmonics in the drain current allows a third harmonic flattening by an infinite load impedance at the third harmonic. Then the improved efficiency (from that of the class B amplifier) is 78.5% × 1.1 = 86%. Again, adding more odd harmonics in the drain voltage waveform eventually makes it a square wave, which is the condition used for the definition of the class F amplifier. The drain efficiency under this idealized condition is also 100% [1, 2]. Figure 7.11(b) shows the conceptual circuit realization for the class F amplifier.

Countless amplifier designs using the concept of the class F or inverse class F have been published; see, for example, [1–3] and the references therein. Those circuits usually employ a number of L-C or/and transmission-line resonant circuits to realize the required impedance conditions. Due to the nature of resonant circuits, the performance of such circuits is often susceptible to the component variation. Therefore, if such a circuit is intended as a product for mass production, care should be taken to ensure that the design is robust enough to allow potential component variations in production.

In practical design, if the drain efficiency is the primary design target, a higher load impedance (than the value for the maximum output power) usually turns out to be advantageous. Besides the benefit of a lower $V_k$, which is discussed in Section 7.3, such a condition allows the transistor to be driven into the saturation state at a lower input power, creating some harmonic contents, which can be potentially utilized for the harmonic flattening. At a sub-gigahertz range, an efficiency of 80% or higher can be achieved using this technique without more sophisticated resonant circuits [9,10]. The trade-offs for the high efficiency using this technique are the output power and gain.

Reference [11] is an excellent review article with a historical perspective on the class F and inverse class F RF amplifiers by two prominent contributors in the field.

## 7.5 Class E and Doherty Amplifiers

We complete this chapter with a brief discussion of class E and Doherty amplifiers. Both have had profound impacts on the RF industry.

### 7.5.1 Class E Amplifier

Similar to the inverse class F, a class E amplifier is also driven into heavy saturation. The drain current of the class E amplifier, however, is modeled by an on-and-off switch rather than an ideal square wave. Another important feature in the class E amplifier is the inclusion of a shunt capacitance $C$ at the transistor's drain terminal. Its minimum value is the transistor output capacitance, $C_{out}$. Figure 7.12(a) shows a complete class E circuit. The series resonant circuit ($L_0$ and $C_0$) is tuned at the fundamental frequency and is assumed to be ideal; that is, it has no effect on the fundamental frequency and blocks all harmonics. The extra inductance $L$ is for the extra phase shift required by the operation principle.

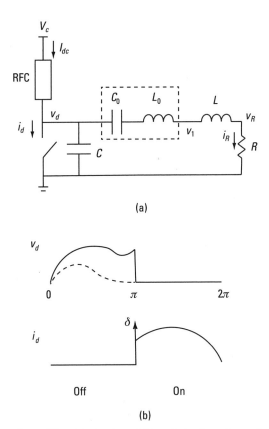

**Figure 7.12** Class E amplifier: (a) circuit and (b) optimization of waveforms for efficiency.

In this circuit model, the drain current and voltage are alternatively turned on and off as shown in Figure 7.12(b) for one cycle ($0-2\pi$). Since there is no overlap between $i_d$ and $v_d$, the integration used in the calculation for the "cooling" effect (7.5) seems to suggest a zero-power dissipation inside the transistor. This is not generally true because of the discontinuity of $v_d$ at the boundary of the on-and-off regions. The presence of $C$ provides an easy physical explanation: At the moment of the off-to-on transition, if the voltage $v_d$ (which is also the voltage across the capacitance) is nonzero, the energy stored in the capacitance, $CV^2/2$, must be dissipated through the drain channel. Mathematically, a $\delta$ function of $i_d$ at the boundary is used to account for this effect. From this analysis, the design equation for the class E operation is derived based on two criteria: $v_d = 0$ and $v'_d = 0$ at the boundary. The second condition ensures the continuity of $v_d$, as illustrated by the dash curve in Figure 7.12.

The key to the derivation of the design equations for $R$, $L$, and $C$ is that $v_1$ in Figure 7.12 is the fundamental component of $v_d$. Then the Fourier expansion of $v_d$ establishes additional conditions for the unknown variables. The detailed algebra is a little tedious and can be found in many texts such as [2–4]. Accordingly, we list the results here without derivation:

$$R = 0.58 \frac{V_c^2}{P_{out}}$$

$$C = \frac{0.18}{\omega R}$$

$$L = 1.15 \frac{R}{\omega}$$

$L$ is required for additional freedom in the adjustment of the phase of $v_d$. In practice $L$ can be absorbed into $L_0$ with a corresponding adjustment on $C_0$. Finally, there is a frequency range for the validity of the above results. Since the minimum $C$ is the transistor output capacitance $C_{out}$, the maximum frequency is given by

$$f_{max} = 0.051 \frac{I_{dc}}{C_{out} V_c}$$

At tens of megahertz frequencies, a class E amplifier design using the above design equations usually can yield a high efficiency with some minor bench tuning. At higher frequencies, other nonideal factors must be taken into account. Reference [12] provides a set of review articles on the class E amplifiers.

## 7.5.2 Doherty Amplifier

As discussed in Chapter 8, the signal of a modern wireless communication system generally has an exceedingly high peak-to-average power ratio (PAPR), which poses a challenge to the RF power amplifier design: The load has to be appropriate to support the output power at a high input power, and ideally the load can be increased to allow a good efficiency at a lower input power. Section 7.2 mentions the concept of "tunable load" (also known as "load modulation" [2]) for maintaining a high efficiency over a varying input power. This concept can actually be implemented in a circuit configuration known as the Doherty amplifier. Doherty amplifiers are widely employed in the today's markets, particularly in the base station industry. This section briefly describes the basics of a Doherty amplifier. Interested readers can find a great deal of information on the subject from texts [1–3] and from publications by the device manufacturers.

Figure 7.13(a) shows the basic configuration of a Doherty amplifier. It is composed of two amplifiers labeled as "carrier" (also known as "main") and "peaking" (or "auxiliary") along with an input power divider and two $\lambda/4$ transmission lines. The actual value of the load $Z_L$ is not important to our discussion. Each amplifier includes an internal matching circuit that transforms the load impedances $R_c$ and $R_p$ to the required load at the transistor drain. Unlike in a balanced amplifier discussed in Section 5.4, the two amplifiers are biased differently. The carrier amplifier is biased in the class B or low class AB to allow operation at low input powers while the peaking amplifier is biased in the class C and only operates when the input power exceeds a certain point.

To see how the load to the carrier amplifier is tuned, we consider the simplified case shown in Figure 7.13(b), where two RF currents $i_1$ and $i_2$ join in phase at the load $Z_L$. We point out that the load resistance is simply the ratio of voltage to current. Then the load to the source 1 is

$$R_{c1} = \frac{v}{i_1}$$

However, $v$ is

$$v = Z_L(i_1 + i_2)$$

Thus,

$$R_{c1} = \frac{i_1 + i_2}{i_1} Z_L \qquad (7.35)$$

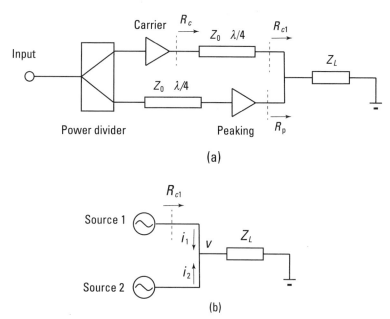

**Figure 7.13** (a) Basic circuit topology of Doherty amplifier and (b) load tuning by current.

This is how the load $R_{c1}$ is "tuned" by $i_2$.

Now we can understand how the Doherty amplifier operates. At low input powers, the peaking amplifier is off, and thus, $i_2 = 0$. The load to the carrier amplifier, $R_c$, is chosen to be high enough to ensure a good efficiency under the low input power conditions. Note that $R_c$ is related to $R_{c1}$ by

$$R_c = \frac{Z_0^2}{R_{c1}} \quad (7.36)$$

(See Section 1.3.3.) Once the input power reaches a certain point, the peaking amplifier starts to operate; that is, $i_2$ starts to rise, which results in an increased $R_{c1}$ according to (7.35). Thus from (7.36), $R_c$ is lowered as the input power increases.

Actual circuit implementation is much more complex than what is described in our example. Nevertheless, this analysis captures the two key ingredients of the Doherty amplifier: (1) load tuning to achieve a high efficiency over a wide input power range and (2) realization of load tuning through a peaking amplifier.

# References

[1] Grebennikov, A., N. O. Sokal, and M. Franco, *Switchmode RF and Microwave Power Amplifiers*, Second Edition, Academic Press, 2012.

[2] P. Colantonio, F. Giannini, and E. Limiti, *High Efficiency RF and Microwave Solid State Power Amplifiers*, John Wiley and Sons, 2009.

[3] Cripps, S. C., *RF Power Amplifiers for Wireless Communications*, Second Edition, Norwood, MA: Artech House, 2006.

[4] Albulet, M., *RF Power Amplifiers*, Noble Publishing, 2001.

[5] Ghannouchi, F. M., Q. Hammi, and M. Helaoui, *Behavioral Modeling and Predistortion of Wideband Wireless Transmitters*, John Wiley and Sons, 2015.

[6] Wang, Z., *Envelope Tracking Power Amplifiers for Wireless Communications*, Norwood, MA: Artech House, 2014.

[7] Sedra, A. S., and K. C. Smith, *Microelectronic Circuits*, Sixth Edition, Oxford University Press, 2009. (and later edition).

[8] Gray, P., et al., *Analysis and Design of Analog Integrated Circuits*, Fourth Edition, John Wiley and Sons, 2001.

[9] Dong, M., "Design Study of a High Efficiency LDMOS RF Amplifier," *High Frequency Electronics*, July 2011.

[10] Dong, M., "A High Efficiency Class J RF Power Amplifier," *Microwave Journal*, Vol. 58, No. 6, June 2015.

[11] Grebennikov, A., and F. H. Raab, "History of Class-F and Inverse Class-F Techniques: Developments in High-Efficiency Power Amplification from the 1910s to the 1980s," *IEEE Microwave Magazine*, Vol. 19, No. 7, November/December 2018.

[12] "Nathan Sokal and the Class-E Amplifier" *IEEE Microwave Magazine*, Vol. 19, No. 5, July/August 2018.

# 8

# RF Designs in Wireless Communications Systems

## 8.1 Introduction

Up to this point, this book has focused on individual RF circuits, especially RF amplifiers. While RF amplifiers are still widely used as stand-alone components in today's market, the majority of them are employed in various wireless communication systems. In these applications, RF amplifiers often need to meet certain application-specific requirements, in addition to the typical ones, such as gain and power. Sections 8.4.3 and 8.4.4 discuss two such application-specific requirements, error vector magnitude (EVM) and the effects associated with intermodulation. Furthermore, in wireless product development, RF engineers' responsibilities usually include the circuit implementation necessary to meet the required systems architecture as well as RF characterizations at the system level. Accordingly, RF engineers benefit from a basic understanding of some of the concepts in communication systems that are pertinent to RF performance. This chapter aims to present these concepts without the burden of analytical rigorousness. The chapter's focus is explaining why the issues under consideration are important from the standpoint of system performance; related circuit designs are not covered in any detail.

The most basic function of any wireless communication system is to transmit and receive through the air—with "air" used here as a general term including other unbounded transmission media—information-bearing RF signals. Almost all modern wireless systems are bidirectional, which requires that wireless devices fulfill both the transmitting and receiving functions. For this reason, they are often referred to as transceivers. In practice, there are enormous differences among wireless transceivers (e.g., consider a mobile phone and the device on a cellular tower known as a base station). For the purposes of our discussion, however, the simple conceptual block diagram shown in Figure 8.1 is sufficient to represent the typical architecture of wireless transceivers.

Figure 8.1 is constructed to emphasize the major RF-related building blocks. There are three of them: a transceiver IC, an RF front-end circuit, and an antenna. Preceding seven chapters have extensively discussed various basic concepts and design issues of the amplifier circuits in the front-end block, and antenna design is usually a separate field from RF circuit designs. This chapter describes the function of the transceiver IC—that is, how information is prepared for transmission and detected from an RF signal. From this, we can understand how the RF characteristics of a communication system affect the data transmission and hence the reason for certain requirements on RF performance. Sections 8.4 and 8.5 detail the specifications for transmitters and receivers.

The modulation and demodulation processes in the transceiver IC block are at the core of any wireless communication schemes: (1) In the modulation process, an RF signal of single frequency $f_c$ [usually denoted as the local oscillator (LO) in the context of modulation/demodulation] is modulated by

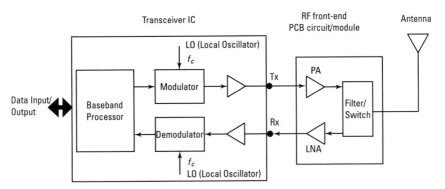

**Figure 8.1** Block diagram of a wireless transceiver.

a data-bearing baseband signal, and the modulated RF signal, after further amplification and filtering, is transmitted through the air; (2) on the receiving side the demodulation process reverses the sequence and regenerates the original baseband signal. Since the information is carried by an RF signal propagating through the air in this scheme, the wireless industry often refers to the RF signal as a carrier.

In terms of performance specifications, a communication system must at least be sufficiently reliable; that is, the error rate in the data transmission has to be within an acceptable level. The second critical design goal is high bandwidth efficiency (also known as spectral efficiency), which is the amount of data transmitted per unit frequency. This is especially critical for wireless systems, because the RF spectrum, as well as the RF power, for specific applications are strictly controlled by regulatory agencies. One example is the Federal Communications Commission (FCC) regulations for the 2.4-GHz and 5-GHz ISM bands for Wi-Fi applications. In fact, most increases in the transmission data speed of modern wireless systems are realized through improvements in bandwidth efficiency, although occasionally new bands are allocated to certain applications. In general, the requirements of low error rate and high bandwidth efficiency are in conflict with one other. Communication theory is basically a study of how to achieve better performance in these two areas and an optimal balance between them under the specific conditions of the application. Readers can consult [1–4] for a complete treatment of this subject.

This chapter presents its material in a considerably different form than that in Chapters 1–7, conducting the discussion mostly in an intuitive manner. For example, in most cases, the chapter omits mathematical proofs. Also, the chapter uses some common terms and concepts in communication theory without repeatedly citing references, assuming that interested readers can find more precise definitions and explanations in [1–4].

## 8.2 Baseband, Modulation, and Passband

### 8.2.1 Baseband and M-ary Scheme

Modern communication systems are almost exclusively in digital form. The basic function of such a system is to transmit a sequence of digital data from one device to another through a channel that can be either a set of wires/cables or the air, in the case of a wireless system.

We start with the binary sequence shown in Figure 8.2(a). A major portion of communication theory is on how to optimize this sequence for better system performance. However, this topic is not relevant to our discussion here. Instead, we are concerned with how to use this binary sequence to modulate (or control) an electrical signal at the transmitter side and demodulate (or recover) on the receiver side. Figure 8.2(b) shows a popular scheme where the binary states, 1 and 0, are represented by a positive voltage $A$ and a negative voltage $-A$, respectively, in a pulse sequence. This type of modulation scheme is called pulse-coded modulation (PCM) in communication theory.

The modulated pulse sequence is usually referred to as a baseband signal, and Figure 8.2(b) is clearly a time-domain representation. In wireless communications, this baseband signal is used to further modulate an RF carrier at a specific frequency. A simple modulation scheme is to multiply the RF signal with the baseband signal. Let $\omega_c$ be the angular frequency of the RF signal; the modulated signal is

$$S(t) = A_i(t)\cos\omega_c t \tag{8.1}$$

where $A_i(t)$ is the amplitude of the $i$th bit and can be either $A$ or $-A$ depending on the bit state. The waveform of $S(t)$ for the first four bits in Figure 8.2(b) is sketched in Figure 8.2(c). The physical device for the circuit realization of

(a)

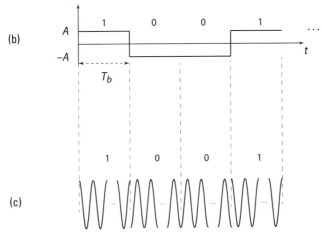

**Figure 8.2** (a) A binary sequence, (b) PCM, and (c) the waveform of a modulated carrier.

the multiplication of two signals is called a mixer. In most modern wireless systems, mixers are incorporated in the transceiver IC. We leave the detailed discussion of this device to the literature [5, 6].

Equation (8.1) is, in fact, a mathematical representation of the so-called amplitude modulation where the amplitude of a sinusoidal wave is modulated by a controlling signal. The same process can be viewed from a different angle if we use the identity $-\cos\omega_c t = \cos(\omega_c t - 180°)$ to rewrite (8.1) as

$$S(t) = A\cos(\omega t - \theta_i(t)) \quad (8.2)$$

Here $\theta_i$ can be either 0°, corresponding to $A_i = A$, or 180° corresponding to $A_i = -A$. This seemingly trivial step of mapping allows us to readily visualize $S(t)$ in a phasor diagram [7]: The binary state 1 is at $\theta_1 = 0°$ while the state 0 is at $\theta_2 = 180°$ as illustrated in Figure 8.3(a). It is customary to label the real and imaginary axes of this diagram as $I$ (in-phase) and $Q$ (quadrature), respectively, in communication theory. This $I/Q$ plane where any point represents a state turns out to be an extremely useful tool in system analysis as illustrated in the M-ary modulation scheme discussed next.

We first consider the bandwidth limitation. In the modulation scheme outlined here, for a given pulse duration $T_d$, the bit rate is $1/T_d$, since each pulse carries one bit. Obviously, to increase the data rate, $T_d$ must be shortened in this scheme. This is when we need to consider the bandwidth in the frequency domain. While a precise relationship between the bandwidth (denoted as B) and the pulse duration depends on several factors and is outside the scope of this book, we can qualitatively see that they are reciprocally proportional to each other; that is,

$$B \propto \frac{1}{T_d}$$

The point here is that the higher the bit rate, the wider, other things being equal, the bandwidth will be. That is why "bandwidth" and "data speed" are often used interchangeably in casual conversation. In wireless systems, the bandwidth is strictly limited by either the communication protocol or the spectrum regulation. Consequently, the pulse duration cannot be unrestrictedly reduced. However, if we can increase the number of bits within the same period, we can achieve a higher data rate without increasing the bandwidth. This is the core idea behind the M-ary scheme.

In this scheme, we first arrange $n$ consecutive bits into a group known as a symbol, which is illustrated in Figure 8.3(b) for $n = 4$. By transmitting $n$

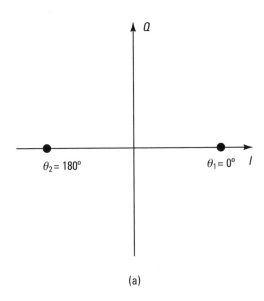

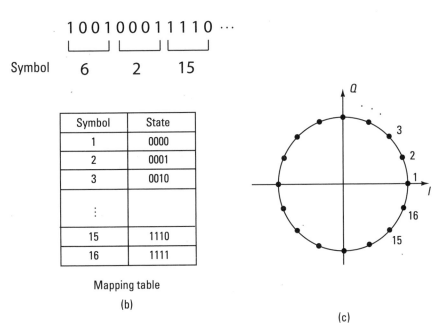

**Figure 8.3** (a) A phasor representation for binary states; (b) a symbol of 4 bits and mapping table; and (c) an *I-Q* space representation for 16 symbols.

bits within the same time duration we can increase the data rate by a factor of $n$ without any changes in bandwidth. We realize that there are $2^n$ different states in this case, as opposed to two states in the binary scheme. In general, a modulation scheme that allows $M (= 2^n)$ distinguished states is referred to as M-ary modulation. Figure 8.3(c) shows an example in which $\theta_i$ in (8.2) is taken for $M = 16$ values to represent 16 symbols.

Without getting into the detail of the demodulation technique at the receiver, we can imagine that as $n$ increases, detection becomes more challenging because of the reduced spacing between two neighboring points in the I-Q diagram. A better way to represent digital states on the I-Q diagram is to utilize the amplitude as well. Figure 8.4(a) illustrates this concept. Mathematically, if we let $A_i$ be the amplitude of the $i$th state, then the modulated signal can now be written as

$$S(t) = A_i \cos(\omega t - \theta_i(t)) \quad (8.3)$$

In Figure 8.4(b) the 16 states are rearranged such that the distance between any two neighboring states is constant, and hence they are evenly distributed. This diagram is called a constellation diagram. This type of modulation scheme represented on a I-Q diagram is called *M-ary* quadrature amplitude modulation (QAM). For the case shown in Figure 8.4(b), it is 16-*QAM*.

Using the trigonometry identity, (8.3) can also be expressed in the I-Q coordinates:

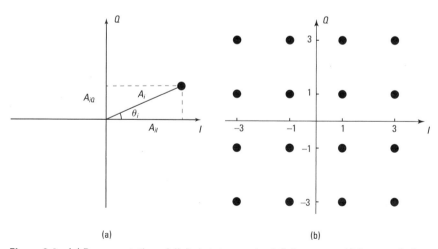

**Figure 8.4** (a) Representation of digital states on the I-Q diagram and (b) constellation diagram of 16-*QAM*.

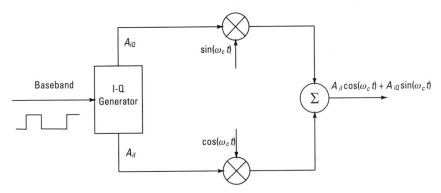

**Figure 8.5** Circuit implementation for *QAM* signal generation.

$$S(t) = A_{iI}\cos\omega_c t + A_{iQ}\sin\omega_c t \tag{8.4}$$

where $A_{iI} = A_i\cos\theta_i$ and $A_{iQ} = A_i\sin\theta_i$ are the *I* and *Q* components of the *i*th state in the *I-Q* diagram.

While (8.3) and (8.4) are mathematically identical, the latter suggests how a *QAM* signal is generated in a circuit, as shown in Figure 8.5. The actual circuit also includes various filters that are omitted in Figure 8.5.

### 8.2.2 Passband of Modulated RF Carriers

In the frequency domain, the spectrum of a modulated RF carrier is referred to as the passband. We now examine how the passband is related to the original modulating baseband signal. We first note that the spectrum of a pulse sequence such as the one shown in Figure 8.2(b) has a symmetry with respect to $f = 0$ as sketched in Figure 8.6(a).

The detail of the spectrum depends on the time-domain waveform and is not important to our discussion. The negative-frequency portion of the spectrum is simply a flip of the positive-frequency portion and therefore has no physical meaning in the sense that a spectrum measurement only shows the

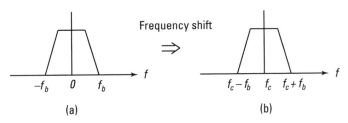

**Figure 8.6** Spectra of the (a) baseband, and (b) passband.

positive-frequency spectrum. However, when such a baseband signal is used to modulate a carrier [described mathematically by (8.1) where $A_i(t)$ represents the baseband waveform], the resulting passband spectrum is a duplicate of the entire baseband with the center frequency at $f_c$, as shown in Figure 8.6(b). This phenomenon is termed in communication theory as double-sideband (DSB) modulation. Since, in this case, half the passband carries no new information, the DSB technique is bandwidth-inefficient. In comparison, the *QAM* modulation scheme described in (8.4) uses two carriers of the same frequency but different in phase by 90° and two distinguished baseband signals ($A_{iI}$ and $A_{iQ}$). Since both modulated carriers occupy the same frequency band, the *QAM* scheme eliminates the bandwidth inefficiency problem. The price is the added circuit complexity due to the second modulation and the combining circuits. A more precise mathematical treatment on the subject can be found in [2]. The key concept here is that the passband spectrum of a *QAM* signal is, in principle, identical to that of the baseband with a frequency shift from $f = 0$ to $f_c$. Thus, in terms of spectrum characteristics, the passband signal is determined by the baseband signal, and the basic functions of the RF circuits are frequency up-conversion (at the transmitter) and down-conversion (at the receiver) with minimal added distortions.

We complete this section with a note on terminology in the frequency spectrum. Two terms, "band" and "channel," are commonly used in communication systems. These terms are not precisely defined in practice, but, if their context is properly understood, their usage generally does not cause any confusion. Take, for example, the 2.4-GHz ISM band. In the United States, this band ranges from 2.4 to 2.4835 GHz. Within this band, the spectrum is further divided into channels with the first channel starting at 2.412 GHz and the following channels having an incremental frequency of 5 MHz (see Figure 8.7). For typical Wi-Fi applications, the first 13 channels are allocated. Then

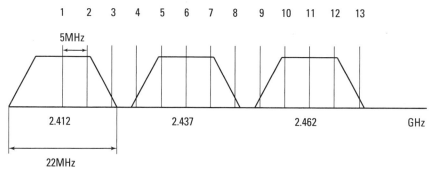

**Figure 8.7** The 2.4 GHz ISM band for Wi-Fi applications.

there are three nonoverlap channels centered at channel 1, 6, and 11 when a 20-MHz Wi-Fi signal (whose actual bandwidth is 22 MHz) is employed, as shown in Figure 8.7. In this description, passband is related to "channel," while "band" refers to the frequency range allocated for a certain application.

## 8.3 Cascaded Noise Figure and Linearity Specifications

In most applications of wireless systems, the RF portion of the circuits consists of several cascaded stages for amplification and filtering. We consider, in this section, several performance specifications for a cascaded configuration.

### 8.3.1 Cascaded Noise Figure

Figure 8.8 shows a configuration where the noise figure and gain of each individual stage are known, and our task is to determine the cascaded noise figure of the system, denoted as $F_{sys}$. It turns out that the definition of noise figure in the SNR [see (4.33)] is handy for this purpose. We only need to work out the result for the first two stages. The generalization is straightforward.

As explained in Section 4.5, the input noise power in computation of noise figure has to be the thermal noise, which is only true for the first-stage. For this reason, $(S/N)_{out}$ of the first stage cannot be used as $(S/N)_{in}$ for the second-stage noise figure calculation. Instead, the cascaded noise and signal powers at the second-stage output should be calculated separately. The results then can be used in the cascaded noise figure calculation. To simplify our

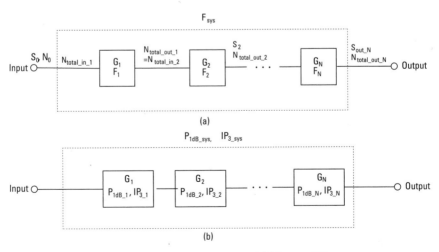

**Figure 8.8** Cascaded systems: (a) noise figure, and (b) $P_{1dB}$ and $IP_3$.

analysis, we will ignore any mismatch effect between stages. Then the signal at the system output is simply the source signal amplified by the system gain. We only need to consider the noise power in the following analysis.

We start with the total input noise power at the input of the first stage, which is denoted as $N_{\text{total\_in\_1}}$. It has two contributions: (1) the thermal noise of the source $N_0$ and (2) the noise power due to the first-stage device $N_{\text{in\_1}} = (F_1 - 1)N_0$ (see (4.15). Thus,

$$N_{\text{total\_in\_1}} = N_0 + N_{\text{in\_1}} = F_1 N_0.$$

By the same reasoning, the total noise power at the second-stage input is

$$N_{\text{total\_in\_2}} = G_1 N_{\text{total\_in\_1}} + N_{\text{in\_2}} = F_1 N_0 G_1 + (F_2 - 1)N_0$$

Then, we have the noise power at the second-stage output as

$$N_{\text{total\_out\_2}} = N_0 G_2 \left[ F_1 G_1 + (F_2 - 1) \right]$$

Since $S_{\text{out\_2}}/S_0 = G_1 G_2$, where $S_0$ and $S_{\text{out\_2}}$ are the respective signals at the source and at the system output, we obtain

$$F_{\text{sys}} = \frac{S_0/N_0}{S_{\text{out\_2}}/N_{\text{total\_out\_2}}}$$

$$= F_1 + \frac{(F_2 - 1)}{G_1} \qquad (8.5)$$

The same result can also be derived using a two-stage equivalent circuit. Interested readers can find a proof in [1].

From (8.5) we can deduce the general expression for an N-stage case as

$$F_{\text{sys}} = F_1 + \frac{F_2 - 1}{G_1} + \cdots + \frac{F_N - 1}{G_1 G_2 \cdots G_{N-1}} \qquad (8.6)$$

An important implication of (8.6) is that $F_{\text{sys}}$ is dominated by $F_1$, provided that $G_1$ is sufficiently large, which is the reason that the first-stage noise figure and gain are most critical in a cascaded LNA design. Also, as previously discussed, any insertion loss at the input of an LNA degrades the noise figure of the LNA by the same amount as the insertion loss. Now it becomes clear why that is the case, if we note that the noise figure of an attenuator is its attenuation, as explained in Section 4.5.

## 8.3.2 Cascaded P$_{1dB}$ and IP$_3$

### 8.3.2.1 P$_{1dB}$

Chapter 6 derived that if the output signal of an amplifier is expressed in the first three terms of a power series of the input signal as

$$v_o = a_1 v_i + a_2 v_i^2 + a_3 v_i^3$$

the input and output powers at the 1-dB compression point are

$$\text{IP}_{1dB} = -0.145 \frac{a_1}{a_3} \qquad (8.7)$$

and

$$\text{OP}_{1dB} = 0.79 \cdot G \cdot \text{IP}_{1dB} \quad \text{or} \quad \text{OP}_{1dB}(\text{dB}) = \text{IP}_{1dB}(\text{dB}) + G(\text{dB}) - \text{dB}$$

Here G is the linear gain of the amplifier. Note that $a_3 < 0$. Now we derive the formula for P$_{1dB}$ of a cascaded system. Similar to the case of the noise figure, we only need to consider a two-stage case. The power series for the first and second stages can be written as

$$v_{o\_1} = a_{1\_1} v_{i\_1} + a_{2\_1} v_{i\_1}^2 + a_{3\_1} v_{i\_1}^3$$
$$v_{o\_2} = a_{1\_2} v_{i\_2} + a_{2\_2} v_{i\_2}^2 + a_{3\_2} v_{i\_2}^3$$

Since $v_{i\_2} = v_{o\_1}$ (the second stage input is the same as the first stage output), $v_{o\_2}$ can be expressed in a power series of $v_{i\_1}$. Neglecting the terms containing $a_2$'s (see Chapter 6 for justification) and only keeping the $a_{1\_1}^3$ term other terms are of higher orders, the coefficients of the linear and cubic terms in the expansion of $v_{o\_2}(v_{i\_1})$ are as follows (minus the algebra):

$$a_{1\_sys} = a_{1\_1} a_{1\_2}$$
$$a_{3\_sys} = a_{1\_2} a_{3\_1} + a_{3\_2} a_{1\_1}^3$$

Then, using (8.7), we have

$$\frac{1}{\text{IP}_{1dB\_sys}} = -\frac{1}{0.145} \frac{a_{3\_sys}}{a_{1\_sys}} = \frac{1}{\text{IP}_{1dB\_1}} + \frac{G_1}{\text{IP}_{1dB\_2}} \qquad (8.8)$$

where $G_1 = a_{1\_1}^2$ is the first-stage power gain. The result expressed in a reciprocal term is advantageous: The reciprocal of IP$_{1dB\_sys}$ is simply the sum of the reciprocals of $IP_{1dB\_1}$ and $IP_{1dB\_2}$ with the second term multiplied by the first-stage gain. The functional form of (8.8) is analogous to the formula for the equivalent capacitance of two capacitors in series connection. Equation (8.8) can be generalized for an N-stage cascaded amplifier as

$$\frac{1}{IP_{1dB\_sys}} = \frac{1}{IP_{1dB\_1}} + \frac{G_1}{IP_{1dB\_2}} + \cdots + \frac{G_1 \cdots G_{N-1}}{IP_{1dB\_N}} \tag{8.9}$$

In practice, $OP_{1dB}$ is often more useful. From (8.9), we have

$$\frac{1}{OP_{1dB\_sys}} = \frac{1}{G_2 \cdots G_N \cdot OP_{1dB\_1}} + \frac{1}{G_3 \cdots G_N \cdot OP_{1dB\_2}} + \cdots + \frac{1}{OP_{1dB\_N}} \tag{8.10}$$

where the 1-dB reduction in total gain at the 1-dB compression point is neglected.

While (8.9) and (8.10) allow for calculations of the cascaded $P_{1dB}$ if the gain and $P_{1dB}$ of each stage are known, they are, in practice, more useful in designing a cascaded system. This is because power amplifiers with higher $P_{1dB}$ are generally more expensive, and the $P_{1dB}$ specification of each stage should be properly selected based on the required system specification. For instance, for a specified cascaded $OP_{1dB\_sys}$, (8.10) implies that the $P_{1dB}$'s of early stages should be sufficient (but not unnecessarily high), so that the cascaded $OP_{1dB\_sys}$ is dominated by the last stage ($OP_{1dB\_N}$).

#### 8.3.2.2 IP₃

$IIP_3$ is only different from $IP_{1dB}$ by a constant factor (Chapter 6); therefore, it is obvious that the cascaded $IIP_3$ has the same functional form as (8.9) and is given by

$$\frac{1}{IIP_{3\_sys}} = \frac{1}{IIP_{3\_1}} + \frac{G_1}{IIP_{3\_2}} + \cdots + \frac{G_1 \cdots G_{N-1}}{IIP_{3\_N}} \tag{8.11}$$

As discussed in Chapter 6, $IP_3$ is a measure of intermodulation. Sections 8.4 and 8.5 show how intermodulation affects the performance of transmitters and receivers differently.

We note that these equations on cascaded linearity specifications should generally be considered semiquantitative in that they provide a useful insight into the relationship under consideration, but the actual performance data of a system should be determined from its measurements.

## 8.4 Specifications for RF Transmitters

This section considers the performance requirements for transmitters. Our focus is on the RF portion of the system, particularly the system specifications that are noise- and nonlinearity-related. As a result, we do not cover some other important but relatively self-evident specifications such as gain, output power, and efficiency.

### 8.4.1 Peak-to-Average Power Ratio

As we have discussed, the nonlinearity of an RF system is characterized in terms of one or two single-tone signals. $P_{1dB}$ and IP3 are two examples. These specifications are always provided in average powers, although it is the peak power that physically causes the nonlinear effects (see Chapter 6). This is because the average power is much easier to accurately measure, and the peak power is simply related to the average power by a factor of 2. (Note that the term peak power can have different meanings in different conditions; it is used here in the context of a sinusoidal wave where $V_{peak} = \sqrt{2}V_{rms}$.) The situation changes remarkably when we are concerned with a passband RF signal for which peak-to-average power ratio (PAPR), also known as the peak-to-average ratio (PAR), is no longer a fixed value. Instead, PAPR varies widely with the wireless protocol. In some high-data throughputs wireless systems PAPR can be more than 10 dB. A high PAPR has a serious implication for the system design and characterization, as will be discussed next.

Figure 8.9(a) shows waveforms for two functions of $V_1(t) = \sin(2\pi f_1 t)$ and $V_2(t) = 1/\sqrt{2} \sin(2\pi f_1 t) + 1/\sqrt{2} \sin(2\pi f_2 t)$. In the plot $f_2 = 2f_1$. The reader can verify that $V_1$ and $V_2$ have the same average power by integrating the square of the functions over one period, $1/f_1$. A feature of interest in the plots is that the maximum swing of $V_2$ is 1.24 as opposed to 1 for $V_1$. The physical origin for this increased peak value is simple: When two sinusoidal waves of different frequencies are superimposed, at a certain moment, due to the favorable phase condition, the superimposed waveform has a larger peak value than a single tone signal even if they have the same average power. In terms of PAPR, signal $V_2$ has a higher PAPR than $V_1$ by 1.87 dB ($1.24^2$). If we allow the phase of the second term ($f_2$) to be a random variable, then in a certain moment the peaks of the two sinusoidal waves are exactly overlapping, resulting in a $V_2$ peak value of $\sqrt{2}$. Therefore, in this case the maximum possible PAPR is 3 dB. We can see without working out the detail that this phenomenon becomes more prominent as more sine functions are added to $V_2$. This line of thinking can be extended to a passband signal, which may be approximated by a group of sinusoidal waves with a small frequency spacing, as sketched in Figure 8.9(b). In fact, a special RF transmission technique known as orthogonal frequency division multiplexing ((OFDM), [2, 8]), which is widely used in Wi-Fi and mobile communications systems, actually consists of a large number of subcarriers. For example, per the 802.11a standard [9], the Wi-Fi OFDM signal contains 64 subcarriers (52 of which are used or populated) with a spacing of 312.5 kHz. Although these subcarriers are generated using a digital signal processing (DSP) method called the fast Fourier transform (FFT), for the purposes of our discussion, they can be treated as a group of individually modulated sinusoidal waves. A wider passband generally contains more such subcarriers. This is the physical origin for the large PAPR values for those high data rate wireless protocols. (Recall the reciprocal relationship between the data rate and the bandwidth.) Due to the random nature of the subcarriers, the PAPR value for a specific protocol is usually provided in a statistical manner [8, 10]. Typically, the PAPR of a Wi-Fi 20MHz signal is considered to be in the range of 5–10 dB.

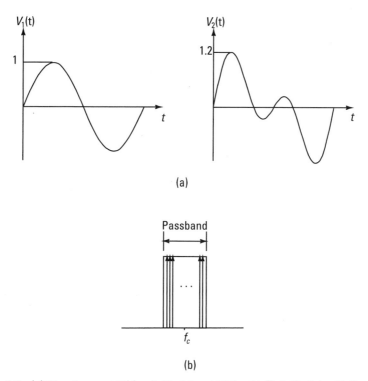

**Figure 8.9** (a) Waveforms of $V_1(t) = \sin(2\pi f_1 t)$ and $V_2(t) = 1/\sqrt{2} \sin(2\pi f_1 t) + 1/\sqrt{2} \sin(2\pi f_2 t)$ and (b) passband approximated by a group of discrete frequencies.

**248** RF Circuits and Applications for Practicing Engineers

The PAPR of a signal is determined at the baseband by the protocol and cannot be reduced by any RF means. The significance of a high PAPR in RF designs is that a nonlinearity specification measured at an average power is no longer sufficient in system characterization. This leads to the increasing popularity of EVM as a means to characterize wireless transmitters because EVM is specified using actual modulated signals. Section 8.4.3 details EVM.

### 8.4.2 Phase Noise

Before our discussion on EVM, we should reexamine the constellation diagram in Figure 8.4. As explained earlier, each point on the diagram represents the amplitude and phase of an RF carrier. Like any electrical circuit parameter, both the amplitude and phase in an actual circuit implementation can be considered variables with a certain amount of fluctuation and inaccuracy due to noise, and possibly distortion. This variability of a $QAM$ signal can be graphically represented by a cluster of sampling points on the diagram, as shown in Figure 8.10(a). The variation in the angular direction is what is called phase noise. Phase noise affects RF system performance in a wide variety of ways. Robins provided an excellent treatment on the subject [11]. We narrow our discussion to its impact on the EVM specification.

Section 8.2 shows that a $QAM$ carrier can be either characterized by $A_i$ and $\theta_i$ as in (8.3) or $A_{iI}$ and $A_{iQ}$ as in (8.4). The equivalence of (8.3) and (8.4) reveals the origin of the phase noise in this scheme: It is from the amplitude

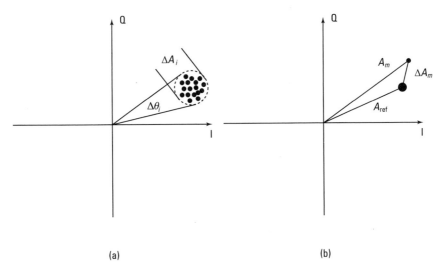

**Figure 8.10** (a) Sampling points on the constellation diagram, and (b) error vector

noise of the in-phase and quadrature signals. Chapter 10 provides more detail on how phase noise is generated. Readers can also find more in-depth discussion in [11]. The point we make here is that the phase noise of an electrical signal has the same origin as the amplitude noise of an electrical signal. From these equations we also can see that both the carrier and baseband can contribute to phase noise. Also, phase noise is a critical specification for both transmitters and receivers. A detailed account can be found in [1].

### 8.4.3 EVM

The plot in Figure 8.10(a), in fact, resembles an actual result of a device in an EVM measurement where a known signal is sent to the device input, and the output signal from the device is fed back to the tester. The detected signal is displayed on the constellation diagram as a sampling point. $\Delta A_i$ and $\Delta \theta_i$ in Figure 8.10(a) represent the respective variation range in the amplitude and angle directions. EVM is a standard parameter used as a figure of merit for the fidelity of a QAM signal when transmitted by a device. It is defined as follows.

Let $A_{ref}$ be the vector representing an ideal point (without the device) on the constellation diagram and $A_m$ be the vector for an actual sampling point, as sketched in Figure 8.10(b). The error vector is

$$\Delta A_m = A_m - A_{ref} \tag{8.12}$$

Then the average magnitude of this error vector normalized to the amplitude of the ideal vector is defined as EVM; that is,

$$\text{EVM} = \frac{\sqrt{\langle |\Delta A_m|^2 \rangle}}{|A_{ref}|} \tag{8.13}$$

Here $\langle \cdots \rangle$ denotes the ensemble average. There are some variations in the definition of EVM in the literature, but the essence of EVM remains the same: It is a normalized root mean square (RMS) of the error magnitude defined in (8.12).

In practice, EVM measurement is more than a number. The measured pattern, such as the one sketched in Figure 8.10(a), provides rich information on system performance. In addition to the tightness of the distribution, EVM measurements can reveal a number of system errors. One example is that if the system is driven into saturation (the gain is compressed), the distribution will shift toward the center. It is also clear now why, for a given average power, a high-PAPR signal tends to have a poor EVM, since the system gain is more likely to be compressed with a high-PAPR input signal. More information on how to use EVM in the system analysis and trouble-shooting can be found in [10] as well as in application notes by EVM equipment manufacturers. Numerically, EVM is provided in either the percentage or decibels, and they are related by

$$\text{EVM}(\text{dB}) = 20\log(\text{EVM}(\%))$$

As the $M$ value in a $M$-$QAM$ scheme increases, the required EVM value for signal detection becomes more demanding. Generally, 1% (−40 dB) is considered an excellent specification, although some of the latest wireless systems require even better performance.

### 8.4.4 Spectral Regrowth

Recall the intermodulation of a two-tone measurement introduced in Chapter 6: When two RF signals, $f_1$ and $f_2$ ($f_2 > f_1$), are applied to a nonlinear system, a third-order mixing generates two intermodulation signals at $f_1 - \Delta$ and $f_2 + \Delta$; here $\Delta = f_1 - f_1$. This concept can be extended, by the same reasoning for the passband PAPR, to a passband signal. That is, the third-order mixing of interband frequencies will generate two sidebands next to the passband, as depicted in Figure 8.11. Note that $f_1$ and $f_2$ must be within the passband. The passband bandwidth is usually much smaller than the carrier frequency; that is, $\Delta \ll f_1, f_2$, which is the condition assumed in Chapter 6.

The slanted shape of the sidebands is due to the fact that as the sideband frequency, denoted as $f_s$ in Figure 8.11(a), moves closer to the edges of the passband, a wider frequency range within the passband can contribute to $f_s$. Quantitatively, let $\Delta$ be the distance from $f_s$ to the lower passband edge (we only need to consider one side) and $\Delta'$ be the frequency range within which a frequency can have a pair to generate an intermodulation at $f_s$ [$f_L$ and $f_U$ in Figure 8.11(a) are such a pair]. We can show (without the accompanying algebra) that

$$2\Delta' = B - \Delta$$

It is clear from the equation that when $\Delta = 0$ ($f_s$ at the band edge) the entire passband contributes to $f_s$, while for $\Delta = B$ ($f_s$ at the far end of the sideband) the contribution to $f_s$ is only from two frequencies at the low- and high-passband edges.

The sidebands are often referred to as spectral regrowth or adjacent channel power (ACP). Like all nonlinear effects, they grow with input power.

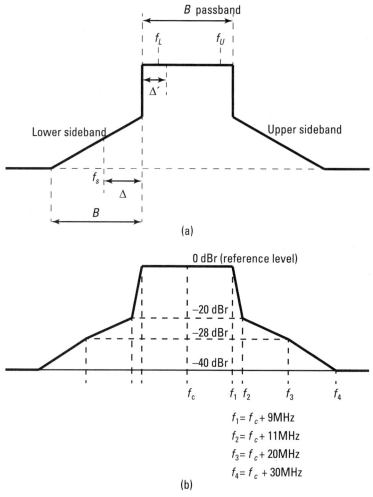

**Figure 8.11** (a) Spectral regrowth, and (b) spectral mask requirement for a 2.4-GHz 20-MHz Wi-Fi signal.

In practice, there are two major consequences of this phenomenon: First, it interferes with the neighboring channel signals. To limit this interference to below a certain level, communication protocols for wireless applications usually specify a channel spectral mask. Figure 8.11(b) illustrates such a requirement for a 20-MHz 2.4-GHz Wi-Fi channel [9]. Second, when a channel is close to the boundary of the allocated band, for instance, a 20-MHz Wi-Fi channel centered at 2.412 MHz (see Figure 8.7), the spectral regrowth spills into a restricted frequency band (2.31–2.39 GHz in the United States) designated by the regulatory. This regulatory specification can be more restrictive than the spectral mask requirement.

For a well-designed system, the spectral regrowth should be dominated by the last-stage RF power amplifier. In practical designs, the power amplifier must be chosen based on the output power specified for the compliance with the spectral mask. For the regulatory compliance, a high rejection filter, such as a SAW filter, for the specific band is usually used to ease the difficulty in meeting the requirement. In some cases, the output power levels for the channels at the band edges may still have to be lowered to pass the regulatory compliance tests.

We have considered two nonlinearity related performance specifications for RF power amplifiers, namely, EVM and spectral regrowth. Both have critical impacts on system performance for the following reason. For any wireless systems, two critical figures of merit, coverage distance and data throughput, are dependent, other things being equal, on the transmit power and the modulation scheme, respectively. Most modern wireless protocols allow adjustments on the transmit power and the throughput speed (e.g., a Wi-Fi system). The EVM at low-power levels of the RF amplifier sets the limit for the highest throughput (assuming that the protocol supports the speed). As the output power of the RF amplifier increases, at a certain point, the EVM starts to degrade. As a result, the data throughput must be reduced. This is why higher throughputs can only be maintained over relatively shorter distances. This process continues until it reaches the limit imposed by the spectral regrowth. In general, the EVM specification sets the highest throughput while the spectral regrowth determines the coverage distance.

## 8.5 Specifications for RF Receivers

The specifications for wireless receivers are primarily determined by the receiver IC [10]. The goal of the front-end RF circuit design and system integration is to optimize the system-level receiver performance based on the requirements for a specific application environment. This section focuses on several issues related to this design goal.

Figure 8.12 shows a block diagram for a wireless receiver. In practice, actual receiver architectures vary widely and constantly evolve. Nevertheless, this block diagram captures some basic concepts that are pertinent to our discussion.

Figure 8.12 suggests that there are three functional blocks in a wireless receiver: (1) amplification of the received RF signal; (2) frequency

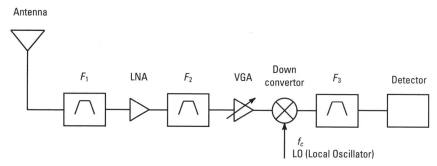

**Figure 8.12** Block diagram for a wireless receiver.

down-conversion from the RF carrier to a lower frequency usually known as intermediate frequency (IF), which can be, but is not necessarily, the baseband; and (3) signal detection and data recovery.

We start with the last block. A detector itself is not an RF circuit, but it imposes two critical requirements on the RF amplifier block. First, the signal level at the detector needs to be in a relatively narrow window for the detector to function properly. In an application environment, however, the received RF power at the antenna port has an extremely wide range. To accommodate this wide gap in dynamic ranges between the two ends of the RF amplifier block, the RF gain not only has to be sufficiently high but also adjustable. Hence, a variable gain amplifier (VGA) is required. It is usually implemented in the transceiver IC. In some cases, a bypass mode is also included in the front-end LNA to allow a further gain adjustment. The second requirement of the detector is adequate SNR. The minimum SNR as a specification for a detector, which is denoted as $(S/N)_{min}$ in our discussion, is defined in terms of the thermal noise. $(S/N)_{min}$ is generally fixed by the communication protocol and the detector's characteristics. The design challenges for the RF portion of a wireless receiver are to minimize any added broadband noise (noise figure) and to limit any interfering signals to an acceptable level. Sections 8.5.1 and 8.5.2 consider these separately.

### 8.5.1 Receiver Sensitivity and Noise Figure

Sensitivity of a wireless receiver is defined as the minimum signal power at the antenna port that can be detected by the receiver. Note that, in casual conversation, the phrase "higher sensitivity" actually implies a sensitivity that, when used as a specification, has a lower numerical value. This section discusses the effect of the noise figure of an RF block on the sensitivity.

Recall from Chapter 4 that the system bandwidth does not affect its noise figure, because it is a ratio of two S/N (SNR) terms. When considering receiver sensitivity, it is the S/N itself at the detector input, rather than a ratio, that determines the minimum detectable signal. The bandwidth of the receiver has no effect on the received signal as long as it is larger than the signal bandwidth, whereas the noise power (such as thermal noise) is generally proportional to the receiver bandwidth. Hence, in this case, S/N is no longer meaningful without specifying the bandwidth. To avoid this ambiguity, a different term, carrier-to-noise ratio (CNR), which is defined as the SNR within the signal band, is sometimes used in the literature [10]. In practical circuit implementations, a bandpass filter with a bandwidth that is close or equal to the signal bandwidth ($F_3$ in the block diagram in Figure 8.12) is usually placed before the detector. It is clear that the bandwidth of this filter, $B_s$, should be the parameter for noise calculation, and the noise at the detector is then given by: $kTB_sFG$, where $F$ and $G$ are the system noise figure and gain respectively. Then, the minimum detectable signal at the antenna port is

$$S_m = \left(\frac{S}{N}\right)_{min} \cdot kTB_sF \qquad (8.14)$$

Using the numerical result in Section 4.3, $S_m$ can be expressed in decibel terms as

$$S_m(\text{dB}) = \left(\frac{S}{N}\right)_{min}(\text{dB}) - 174\ \text{dBm} + 10\log B_s(\text{Hz}) + F(\text{dB}) \qquad (8.15)$$

Equation (8.15) indicates that the order of magnitude of $S_m$ is mainly determined by the thermal noise floor −174 dBm at room temperature (290K) and $(S/N)_{min}$ and $B_s$; both are related to the communication protocol and the device technology. Numerically, if we take $(S/N)_{min} = 1$ (0 dB) (note that a signal is coherent while a noise is random, which allows various algorithms to recover the signal even if the signal level is comparable or lower than noise), $B_s$ = 20 MHz; then we have a baseline for $S_{min}$ at −101 dBm. Considering that the maximum transmit powers of a Wi-Fi system are typically in the range of 10–20 dBm, this sensitivity implies that a transmitted signal can still be recovered even after a propagation loss of more than 100 dB.

Equation (8.15) also shows that the noise figure numerically increases (making it worse) $S_{min}$ proportionally (decibel for decibel). In a typical environment where a wireless system is deployed, especially in mobile applications, the received signal at an antenna port fluctuates widely (easily on the order of 20 dB or higher) due to the radio propagation characteristics (with multipath fading [12] being one of the major causes). Therefore, a change by a dB or so in the noise figure will hardly yield any discernible impact on practical user experience. For this reason, while the improvement in noise figure is still beneficial, front-end LNAs are often employed mainly for reducing various external interferences. This leads to our discussion in Section 8.5.2.

### 8.5.2 Interference in Receivers

In practical applications, various interfering signals, also known as spurious signals (or simply spurs), may mix into a receiving channel through some unintended means. The presence of these spurious signals, even at a remarkably low level, can drastically degrade the sensitivity. For this reason the sensitivity in (8.15) is also referred to as spurious-free sensitivity. There are almost countless scenarios in which interference can happen. A number of cases are considered in [10]; most of them are related to the architecture and circuit performance of a receiver IC. There are roughly two categories of interference. One consists of interference from the transceiver itself, and the other stems from the external environment. Sections 8.5.2.1 and 8.5.2.2 explain these interference problems, considering one case for each category.

#### 8.5.2.1 Interference from a Clock Signal

A wireless transceiver requires a highly accurate frequency reference [10], usually known as a clock signal, for both analog (RF) and digital circuits. It is typically in the 10-MHz range. One of the key requirements for a clock signal is a sharp rise/fall edge [13]. (It is usually generated by a high-$Q$ resonator, such as a crystal oscillator [14].) As a result, a clock signal can have extremely wide harmonic components. In practice, a well-designed transceiver IC has sufficient on-chip isolation to protect the receiving signal from corruption by the clock signal. However, a harmonic of the clock signal may leak to a receiving channel through a path on the PCB, if the board layout is not properly designed. The leakage can occur in a variety of ways; improper grounding is one of them, as briefly discussed in Section 8.6.

Note that the figure of −101 dBm cited earlier for the sensitivity floor is the total power over the bandwidth while the clock and its harmonics are at discrete frequencies. The capability to detect a modulated signal at the presence of a single-tone interferer depends on the detection technique and the acceptable level of error. Generally, an interfering signal at a level that is much lower level than the spurious-free sensitivity may still cause detection problem. Consider the example used above for the Wi-Fi OFDM 20MHz signal.

Since there are 52 subcarriers, on average, each subcarrier has a power of 17 dB below the total power. When an interferer falls into one of the subcarriers, the signal detection is invariably affected. If we assume that the system tolerates an interferer that is 10% (or 10 dB lower) of the subcarrier, then it is simple algebra to show that the presence of an interferer at a level of 10 dB below the sensitivity floor (i.e., −111 dBm) raises the minimum detectable signal by 17 dB.

In practice, it is not always straightforward to determine whether clock signal interference is the root cause for sensitivity degradation. If the receiver has a baseband test port, the presence of a clock signal is evidenced by a spurious signal in the baseband spectrum whose frequency (after converting back to the RF receiving channel) happens to be an exact harmonic of the fundamental clock frequency. If this option is not available, the fact that the sensitivity degradation only happens at certain receiving channels is an indication that the interference from the clock may be the culprit. This is because the clock leakage is usually frequency-selective, if the circuit board is reasonably well-designed.

In addition to the interference from the on-board clock signal, transmission power may affect the receiving sensitivity. Some cases are discussed in [10].

#### 8.5.2.2 Interference from Intermodulation of External Signals

The interference from a clock signal is unintended and generally cannot be accurately predicted. In contrast, the external interferences are known (in the statistical sense), and the challenge is to design a front-end circuit that can effectively block the interferers without compromising the noise figure.

Wireless systems are ubiquitous in modern life. Wireless devices are often used in environments where multiple wireless signals coexist. Consider the case of a receiver for a satellite communication system. This type of system usually requires a sensitivity that is near the thermal noise floor because of the low power levels of the incoming signals. If such a device is used in proximity to other mobile devices, the transmitting powers of the mobile devices, which are usually more than 100 dB higher than the receiver sensitivity, can cause severe interference to the receiver. There are several mechanisms for this type of interference [10]. We use one example here to illustrate the potential challenges in dealing with this kind of interference.

Figure 8.13 shows a condition where two wireless devices with respective transmission frequencies of $f_1$ and $f_2$ are closely located to a wireless receiver with a receiving frequency of $f_r$. Recall from Chapter 6 that the third-order intermodulation of $f_1$ and $f_2$ occurs at $2f_1 - f_2$ and $2f_2 - f_1$. Then after the first-stage LNA, we expect five frequency components indicated in Figure 8.13.

Since most popular wireless bands are crowded in the low-gigahertz range, there is a good chance that one of the intermodulation frequencies falls in the receiving channel; that is, $2f_1 - f_2 = f_r$ or $2f_2 - f_1 = f_r$. This phenomenon is known as out-of-band intermodulation interference. It is so called because $f_1$ and $f_2$ are obviously outside the receiving band, but their intermodulation is within the desired band. The presence of this interference may severely degrade receiver sensitivity. Quantitatively, if we assume that the maximum transmission powers are in the range of 30 dBm and the propagation loss is 50 dB, then in a worse-case condition (with the strongest interference), there are two interfering signals at a power level of −20 dBm that are coexisting with the intended signal at the receiver antenna. At this input power level, even if the resulted intermodulation is as low as 100 dB, the interference may still be significantly higher than the spurious-free sensitivity discussed above. This is why the rejection of out-of-band intermodulation is a critical system specification for applications in such a wireless environment.

It may appear that a simple solution to this interference problem is to place a sharp bandpass filter before the first LNA ($F_1$ in Figure 8.13) to substantially attenuate the interferers. However, a high-rejection filter always means a high insertion loss, which is directly translated into a degradation in noise figure, as discussed in Section 8.3. While the worst case interference may happen rarely, the degradation in noise figure affects the system performance more regularly. A practical solution usually consists of a filter specifically designed according to the system requirements and an LNA with excellent linearity performance. In terms of specifications, a typical nonlinear parameter such as $P_{1dB}$ or $IP_3$ is clearly insufficient in this case. There is no standard

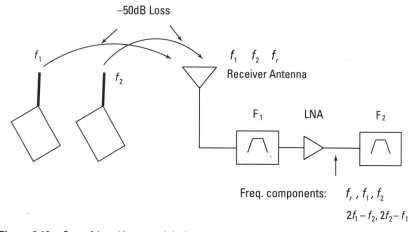

**Figure 8.13** Out-of-band intermodulation.

specification method for the out-of-band intermodulation. In practice, it is usually provided in decibels below a reference level under a set of frequency and power conditions.

Any leakage signals of $f_1$ and $f_2$ after the first stage can be sufficiently suppressed by a high-rejection filter ($F_2$ in Figure 8.13) without any concern about its impact on the noise figure. Therefore, contrary to the general understanding that the system intermodulation is mainly determined by the nonlinearity of the last stage [see (8.11)], the nonlinearity of the first stage is most critical in this particular case.

## 8.6 Grounding in RF Circuits

EMC control of a wireless product is usually the responsibility, at least partially, of RF engineers. In addition to signal-integrity related issues such as the clock interference, an EMC task also includes compliance with various emission limits regulated by government agencies. Most EMC-related problems are unintended and are hard to quantify, which often makes EMC control a difficult task for practical engineers. There is rich literature on EMC, including [15–17]. A complete treatment of the subject is beyond the purview of this book. This section considers the important topic of grounding, which, when poor, accounts for a significant portion of practical EMC problems.

The mechanical structure of a ground in an electrical circuit generally does not play a role in circuit performance. In contrast, from the EM-wave viewpoint, an RF signal on a PCB is a propagating EM field that is confined in a space defined by two conductors. For a microstrip transmission line, one of the conductors is ground (with the other the signal trace), as illustrated in Figure 8.14(a). Any structural change on either conductor will cause a disturbance to the RF signal. What is unique about ground, in the context of EM emissions, is that the ground is usually a large piece of metal extended to other parts in the system. Whenever an imperfection occurs to the ground, it forces the EM field to stray from the intended RF path to the other parts of the circuit, causing potential undesired emission. Consider, for example, the structure in Figure 8.14(a); according to EM field theory, there exists an RF current pattern in the ground (not shown) that is associated with the magnetic field. It is then convenient to use this current to visualize the EM field propagation.

Figure 8.14(b) shows a case where an open slot is required on the ground plane to allow two vias to go through without shorting them. If an RF trace happens to be above the opening, the ground for the RF signal is no longer

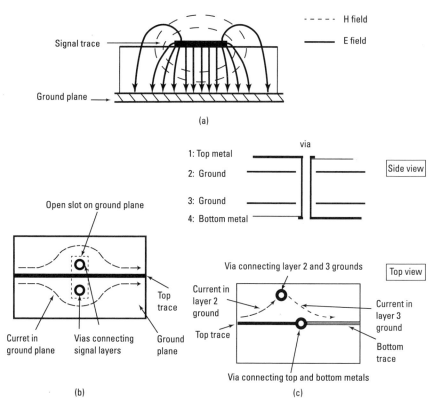

**Figure 8.14** (a) EM field pattern of microstrip line, (b) open slot in ground, and (c) changing plane of RF signal

continuous. This is a classical example in EMC analyses. We repeat it here for readers who are new to the field.

In another example, consider a PCB design of a four-layer board that has RF traces on both sides of the board. At a certain location, the signal trace needs to change from one side to the other, as illustrated in Figure 8.14(c). The key point here is that the signal trace and the ground must be paired. That is, when the top trace transitions from layer 1 to layer 4, the transition for ground from layer 2 to layer 3 has to follow the signal path closely. In Figure 8.14(c) the via for the ground transition is placed too far away from the via for signal trace transition. This imperfection may potentially cause an EMC problem.

These two examples show how a poor grounding scheme may cause an RF signal to stray away from the intended path, causing potential emission problems. The same grounding problem can cause unintended coupling, too, where an external signal can be more easily mixed into an RF path. We can understand this by realizing that the two processes are reciprocal if there are no unilateral components such as amplifiers involved. In general, good grounding practice is important for both emission control and prevention of unintended signal coupling.

There are other issues concerning the grounding designs in an RF system, particularly in a mixed-signal circuit. For a more detailed discussion, interested readers may consult [15].

# References

[1] Razavi, B., *RF Microelectronics*, Prentice Hall, 1998.

[2] Lathi, B. P., and Z. Ding, *Modern Digital and Analog Communication Systems*, Fourth Edition, Oxford University Press, 2009.

[3] Haykin, S., *Communication Systems*, Third Edition, John Wiley and Sons, 1994 (and later editions).

[4] Ziemer, R. E., and W. H. Tranter, *Principles of Communications*, Seventh Edition, John Wiley and Sons, 2015.

[5] Vendelin, G. D., A. M. Pavio, and U. L. Rohde, *Microwave Circuit Design Using Linear and Nonlinear Techniques*, John Wiley and Sons, 1990.

[6] Pozar, D. M., *Microwave Engineering*, Second Edition, John Wiley and Sons, 1998.

[7] Alexander, C. K., and M. N. O. Sadiku, *Fundamentals of Electric Circuits*, McGraw-Hill, 2000.

[8] van Nee, R., and R. Prasad, *OFDM for Wireless Multimedia Communications*, Norwood, MA: Artech House, 2000.

[9] IEEE 802.11g/a standards for WLAN.

[10] Q. Gu, *RF System Design of Transceivers for Wireless Communications*, Springer-Verlag, 2005.

[11] Robins, W. P., *Phase Noise in Signal Sources*, IEE, 1984.

[12] Rappaport, T. S., *Wireless Communications*, Second Edition, Prentice Hall, 2002.

[13] Dally, W. J., and J. W. Poulton, *Digital Systems Engineering*, Cambridge University Press, 1998.

[14] Matthys, R., *Crystal Oscillator Circuits*, Krieger Publishing Company, 1992.

[15] Holzman, E., *Essentials of RF and Microwave Grounding*, Norwood, MA: Artech House, 2006.

[16] Weston, D. A., *Electromagnetic Compatibility*, Marcel Dekker, 2001.

[17] Williams, T., *EMC for Product Designers*, Third Edition) Newnes, 2001

# 9
# Passive Components in RF Circuits

This chapter covers passive components used in RF applications, a category that encompasses a wide variety. The discussion in this chapter is not intended as a comprehensive review on the subject. Rather, our focus is on those components that are most commonly used in PCB-based circuits as well as in RF measurement setups. We further narrow the scope of our discussion to the component characteristics that have direct impacts on RF (and DC in some cases) performance. The chapter does not cover component technologies and some other component attributes that can be critically important in practical designs, such as tolerance, thermal, and mechanical properties, in any detail. Readers may obtain further information on these subjects in [1–4].

## 9.1 Capacitors, Inductors, and Resistors

Up to this point, we have used these passive components in our circuit analysis under the assumption that they function in an ideal fashion. This section discusses the actual circuit behaviors of a physical device that unavoidably include all nonideal (or parasitic) effects. Recall the concept of the parasitic resistance associated with the $Q$ factor of capacitors and inductors, briefly discussed in Section 3.4.1. This chapter expands our discussion to other parasitic effects and explains why these effects should be part of design considerations under

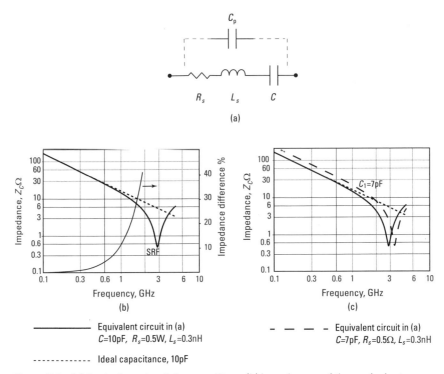

**Figure 9.1** (a) Equivalent circuit for capacitors; (b) impedances of the equivalent circuit in (a) and ideal capacitance; and (c) compensation for the parasitic effect.

certain conditions. This topic is particularly important for RF engineers since most parasitic effects only become problematic at higher frequencies. Among the three components, capacitors and inductors are much more of a concern in this regard because of being reactive (thus lossless) and frequency-dependent.

In the component industry, there is a category called RF or high-frequency components whose parasitic effects are minimized. Then there is a subcategory intended for ultra high-frequency (e.g., > 10 GHz) or ultra low-loss applications. Components in this subcategory employ special technologies that further reduce parasitic effects. However, the cost of those special components may be prohibitively high for many commercial applications. A quantitative understanding of the parasitic problem will help users determine whether these special components are required for a specific design task.

As is customary, this chapter uses an equivalent circuit to describe the parasitic effects of a physical component.

## 9.1.1 Capacitors

In RF circuit applications, the two major concerns of a physical capacitor are the parasitic inductance associated with the component leads and the power dissipation due to the dielectric loss. In practice, these two parasitic effects are represented by an inductance $L_s$ and a resistance $R_s$ in a simple RLC circuit shown in Figure 9.1(a). (Ignore the parallel capacitance $C_p$ for now.) Some manufacturers also offer much more sophisticated SPICE (Simulation Program with Integrated Circuit Emphasis) models that can have as many as 20 circuit elements. While a good SPICE model can improve accuracy in simulation, the RLC equivalent circuit allows for a simple circuit representation of the measurable characteristics of physical capacitors. Specifically, the impedance of a capacitor can be measured through S-parameters. The presence of the parasitic inductance is evidenced by the resonance characteristic of the impedance plot. The RF component industry refers to resonance frequency, which is usually provided by the manufacturers, as self, or series, resonance frequency (SRF). From the *SRF* specification, $L_s$ can be calculated by $L_s = 1/[(2\pi \cdot SRF)^2 C]$. Similarly, the parasitic resistance $R_s$, which was introduced in Section 3.4.1 as *ESR*, is simply the real part of the capacitor impedance. Manufacturers usually provide $R_s$ versus frequency or $Q$ versus frequency plots. $Q$ and $R_s$ are related by $Q = 1/(\omega C R_s)$. [See (3.4.5).]

The parasitic inductance is determined by the lead structure. As such, $L_s$ is independent of frequency and roughly remains unchanged with capacitance for a specific series of capacitors. Therefore, for a given series, *SRF* is approximately proportional to $1/\sqrt{C}$. In comparison, $R_s$ is typically an increasing function of frequency, but not always monotonically. Furthermore, in general, $R_s$ varies considerably among a number of factors such as capacitance values, capacitor types, and manufacturers.

In certain types of capacitors, the impedance plot exhibits a parallel resonance (a dip in the $S_{21}$ measurement) at a sufficiently high frequency. This effect can be represented by a parallel capacitor $C_p$ in Figure 9.1(a). $C_p$ will be ignored in our analysis, because the parallel resonance frequency is usually so high (several times higher than *SRF*) that its impact is inconsequential within the usable frequency range. Nevertheless, if a capacitor is intended for applications at high gigahertz and beyond, its frequency characteristics should be fully examined. Sections 9.1.1.1–9.1.1.3 consider three specific cases.

### 9.1.1.1 Capacitors in RF Matching Networks

As outlined in Chapters 2 and 3, the function of a capacitance in a matching network lies in its reactance $-j/\omega C$ in series connection or susceptance

$j\omega C$ in parallel connection. For a practical capacitor intended for RF applications, at an operation frequency considerably below its *SRF* the parasitic resistance $R_s$ is always much smaller than the reactance, and hence $R_s$ can be ignored in this frequency range. Then the main parasitic effect is from the inductance $L_s$, which effectively reduces the reactance from $-j/\omega C$ to $-j(1/(\omega C) - \omega L_s)$. It is clear that this effective reactance continuously decreases as the frequency increases until the *SRF* point. To visualize how the impedance of the equivalent circuit in Figure 9.1(a) changes with frequency, Figure 9.1(b) shows an impedance plot for $C = 10$ pF in a log-log scale. In this example, two parasitic parameters are taken as $L_s = 0.3$ nH and $R_s = 0.5\Omega$, which are typical for a general purpose 0402 (1.0 mm × 0.5 mm) ceramic capacitor in this frequency range. For comparison, Figure 9.1 also plots the impedance for the ideal 10-pF capacitance.

The impedance curve (in this case, it is almost identical to the reactance curve with the exception of the frequency region close to the resonance frequency) of the equivalent circuit exhibits a resonance frequency *SRF* (around 3 GHz) whereas the impedance of the ideal capacitance is a straight line with a negative slope, both of which are expected. What is important to us is the frequency point where the two curves start to move apart. In Figure 9.1(b) this point is about 600 MHz. After this point the physical capacitor no longer behaves in a circuit exactly in accordance with the ideal behavior. To quantify this effect, the percentage difference between the two curves is plotted in the same figure. If 10% (an arbitrary choice) is taken as a cutoff criterion, we conclude that this specific 10-pF capacitor ($L_s$ is given) can be used, without significant correction for the parasitic effects, up to a frequency approximately one third (1 GHz) of the *SRF* value. In fact, this conclusion is valid for any capacitance $C$ as shown in (9.1). Let $\omega_{0.1}$ be the angular frequency when the parasitic term $\omega L_s$ is 10% of the ideal value $1/(\omega C)$; that is,

$$\frac{\omega_{0.1} L_s}{\frac{1}{\omega_{0.1} C}} = 10\%$$

Then, we obtain

$$\frac{\omega_{0.1}}{2\pi \cdot SRF} = \sqrt{0.1} \approx \frac{1}{3} \quad (9.1)$$

In practice, the restriction on the usable frequency range of a capacitor is not as severe as (9.1) indicates, because the parasitic effects can be compensated, to a certain extent, by reducing the nominal capacitance. Figure 9.1(c) illustrates this concept graphically, showing that a 7-pF capacitor provides the same impedance as an ideal 10-pF capacitor at 2 GHz. Mathematically, let $C_1$ be the nominal capacitance of a capacitor that provides the same reactance as that of an ideal capacitance $C_i$ at frequency $\omega_i$; then we have

$$-\frac{1}{\omega_i C_i} = -\frac{1}{\omega_i C_1} + \omega_i L_s \qquad (9.2)$$

The solution of (9.2) is $C_1 = C_i/(1 + \omega_i^2 L_s C_i)$. Clearly, $C_1 < C_i$. This process is in fact often carried out in practice through bench tuning by engineers without detailed knowledge of $L_s$. From Figure 9.1, we also notice that if the frequency is pushed higher (e.g., the 10-pF capacitance used at 3 GHz), the slope of the corresponding $C_1$ curve at the intercept becomes steeper. The physical implication is that the circuit will be increasingly sensitive to the variation of component parameters as the frequency increases. Consequently, despite the fact that the solution to (9.2) always exists, this technique has a limitation in practice.

Finally, note that the *SRF* of a capacitor used for a matching network will not be arbitrarily low with respect to the operation frequency, because it is the reactance (or susceptance) rather than the capacitance that counts in a matching circuit. For example, the reactance of 100 pF is almost always too low at 5 GHz to be used for a matching network at that frequency. This can be quantitatively analyzed as follows.

Let $X_0 (= 1/(2\pi f C))$ be the required reactance for a matching network, the ratio of the *SRF* value of the required capacitor to the operation frequency $f$ is given by

$$\frac{SRF}{f} = \frac{1}{2\pi f \sqrt{CL_s}} = \sqrt{\frac{X_0}{2\pi f L_s}} \qquad (9.3)$$

According to (9.1), when the ratio in (9.3) is >3, the parasitic effect can virtually be ignored. As the frequency increases the ratio drops at a rate of $1/\sqrt{f}$. The factor $X_0$ in (9.3) implies that matching at low impedance levels (note that the resistance and reactance are scaled in a matching network as discussed in Section 3.4 on the Q factor) is more difficult to avoid the parasitic effects. Thus, the design of a high-power amplifier (which, as discussed in Chapter 6, requires matching at low impedance) at a high frequency imposes significant challenges in terms of dealing with the *SRF* effect. Equation (9.3) offers two clues in this regard: First, design transmission lines and shunt capacitors properly to avoid any matching network that uses a unnecessarily large capacitance. (For example, choose two small capacitances instead of one large capacitance.) The Smith chart technique introduced in Chapter 3 turns out to be remarkably useful for this purpose. Second, lower $L_s$ by using a special type of ultra low SRF capacitors.

### 9.1.1.2 DC Blocking Capacitors

As mentioned in Section 3.5.3, it is convenient in many cases to have a DC blocking capacitance that is large enough to be transparent (have no effect on matching networks) at the operation frequency. The same concept can be realized in circuits by choosing a capacitor whose *SRF* is close to the operation frequency. Theoretically, the *SRF* approach is most ideal because it is the condition where the impedance of the capacitor is at a minimum. In practice, however, this approach only yields a tangible benefit in a limited condition, as explained next.

DC blocking capacitors are most often placed at the circuit input and output where the impedance is 50Ω. Under this condition, we may treat the 50-Ω source plus the DC blocking capacitor as an effective source as shown in Figure 9.2(a) and study how much the reflection coefficient of this effective source deviates from the perfect condition $\Gamma_s = 0$. As noted in Chapter 1, a return loss of 20 dB is practically a very good specification. Let's use $\Gamma_{\text{eff}} = 0.1$ ($S_{11} = -20$ dB) as a condition for the DC blocking capacitor that is still considered transparent. Figure 9.2(b) shows $S_{11}$ (decibels) for five capacitances over almost five decades in a frequency range from 10 MHz to 10 GHz. The frequency for the lowest point in $S_{11}$ (decibels) of each curve is the *SRF* of the corresponding capacitor. It is immediately clear from the plots that matching the capacitor's *SRF* with the operation frequency is not necessary in most parts of the spectrum considered in Figure 9.2, since there is a wide capacitance range that yields a virtually unnoticeable effect ($S_{11}$ better than −20 dB). It is also clear that a capacitor can work well even after it becomes inductive, as long as the impedance $\omega L_s$ (also plotted in Figure 9.2) is still sufficiently low. The plot of $\omega L_s$ shows that it crosses the −20-dB line around 6 GHz. It is when the operation frequency is close to or beyond this point that the *SRF* method may be employed, as illustrated by the 2.2-pF curve. The same plot also shows that this method only works for a relatively narrow band. For wideband applications, if the operation frequency is beyond 6 GHz, a different series of capacitors with smaller parasitic inductance must be found.

While this analysis is useful in illustrating the notion that the *SRF* method may offer some benefits in certain conditions, the assumption of constant $R_s$ used in Figure 9.2(a) is too simplistic for these plots to be used as guidance in practical design. In reality, at high frequencies, $R_s$ can be significantly

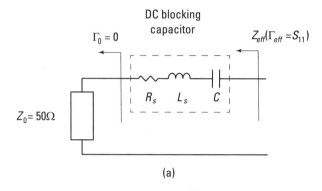

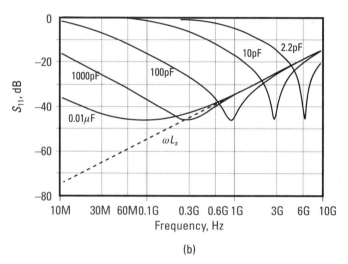

**Figure 9.2** (a) Effective source when the DC blocking capacitor is included, and (b) $S_{11}$ of the effective sources.

higher than 0.5Ω and is generally dependent on capacitance. For this reason, the feature in Figure 9.2(b) that all capacitance curves converge to the $\omega L_s$ curve may not correspond to real-world situations. In practice, for a narrowband application, a frequency-appropriate DC blocking capacitor is normally recommended. For example, in a 5-GHz Wi-Fi circuit, a DC-blocking capacitor in the 10pF range is expected to work properly while a 1-$\mu$F capacitor may or may not function well in that frequency band.

Ultimately, for broadband applications where insertion loss or return loss is critical, the most reliable method in selecting the DC-blocking capacitor is to experiment with different capacitance values.

At higher frequencies (≳10 GHz), some special types of ultra-broadband capacitors, which have relatively flat frequency characteristics, are generally recommended.

We have considered the imperfections (parasitic effects) of a capacitor in a perfect termination condition. Sometimes, in practice, it is found in a wideband circuit, typically in RF switch circuits, that the return loss—and insertion loss—in the middle of the frequency band are sensitive to the DC-blocking capacitance, which seems to contradict our conclusion. Often, this is caused by accidental matching, meaning that the imperfection of the device ($\Gamma \neq 0$) happens to be matched by a specific capacitor. This observation should not be confused with the *SRF* effect.

### 9.1.1.3 Power Dissipation

There are three possible areas of concern in terms of the power dissipation of a capacitor: (1) the amount of power dissipation; (2) its effect on circuit performance; and (3) the maximum power rating.

Regarding the first question, for a given $R_s$, the power dissipation of a capacitor can be calculated from either the RF current or voltage. But neither quantity is directly measurable. In principle, once a complete equivalent circuit network is established, the power dissipation of any resistive element in the network can be estimated using an EDA simulation. However, this type of simulation is not always possible. (In the case of transistor matching networks, the availability of the transistor model is an obvious issue.) In fact, it is not necessary, in most practical cases, because of the relatively small impact on the circuit. For example, consider the case of the DC-blocking capacitor in a 50-Ω condition. This is a series connection, and the power dissipation in the capacitor is actually the insertion loss of the circuit. The ratio of the power dissipated to the power delivered to the load is simply the resistance ratio of 0.5Ω/50Ω, which implies that 99% of the power is delivered to the load in this case. Thus we have an insertion loss of 0.04 dB. If this capacitor is on the input side of an LNA, this power dissipation degrades the noise figure by 0.04 dB (see Chapter 8), an amount of degradation that is hardly noticeable in practice. When the operation frequency is in the high gigahertz range or beyond, $R_s$ can be considerably higher than 0.5Ω. In that case, a special low-*ESR* capacitor may yield a more tangible improvement in the noise figure.

A similar analysis can be applied to a shunt capacitor if we convert the impedance of the equivalent circuit to a shunt admittance. Then the power dissipation of the shunt capacitor can be estimated by the ratio of the conductance of the resulted admittance to that of the load. We leave the details to the reader. Note that the result depends on both the frequency and the capacitance.

In general, the power dissipation of a capacitor cannot be accurately calculated without a detailed knowledge of the network where the capacitor is placed. Nonetheless, at low to medium powers (> several watts), it is usually ignored in practice.

In high-power applications, the power dissipation in capacitors becomes a concern. In addition to the difficulty in estimating the power dissipation, the maximum power rating is not consistently provided by manufacturers. For ultra high-power circuits (>10W), when uncertain, the best practice is to follow the manufacturer's bill of materials (BOM), because, generally, the components on the BOM have proven reliable for that particular circuit condition.

### 9.1.2 Inductors

We are mainly concerned with SMT inductors. While the manufacturing technologies have evolved significantly since the introduction of SMT technology several decades ago, the basic component structures for inductors remain largely unchanged. Inductors commonly used in RF circuits can be roughly grouped into three categories: multilayer, wire-wound (with ceramic core), and air-core (also known as spring inductor). More detailed descriptions for each type of component can be found in [3–5].

For many RF applications, multilayer inductors are cost-effective and have adequate specifications. Wire-wound inductors generally have better specifications in terms of parasitic parameters, which will be discussed in more detail later in this section. An air-core inductor does not have any core material, which is the reason for both its advantages (minimal parasitic effects) and disadvantages (larger size and limited inductance values). Another special type of inductor, the ferrite bead, mentioned in Chapter 5 in the context of bias circuit designs, is detailed in Section 9.1.2.4.

As in the case of capacitors, our discussion of the electrical characteristics of inductors is also focused on the parasitic parameters and their effects on RF circuit applications. The parasitic elements of an inductor can be represented by the simple equivalent circuit shown in Figure 9.3(a). In Figure 9.3(a), $L$ is the nominal inductance, and $R_L$ accounts for the power loss inside the component; $C_L$ represents the parasitic capacitance between the terminals of the inductor. It is an easy exercise to show that the equivalent circuit in Figure 9.3(a) has the same resonance frequency as that of the classical RLC parallel circuit; that is, $f_0 = 1/(2\pi\sqrt{LC_L})$. As with capacitors, this resonance frequency is also referred to as *SRF* and is usually provided on the datasheet.

In general, power dissipation in inductors is higher than that in capacitors. Contributions to $R_L$ are from two key elements of an inductor: (1)

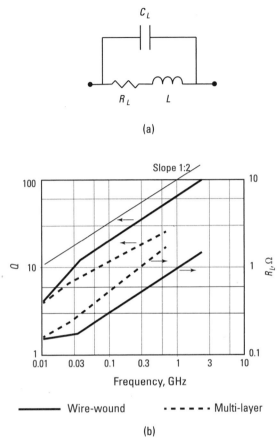

**Figure 9.3** (a) Equivalent circuit for inductors, and (b) $Q$ and $R_L$ plots for wire wound and multilayer inductors of 10 nH.

conducting wire and (2) insulating core material. At the frequency range of typical RF circuits, power loss inside a conducting wire is due to the skin effect [6, 7]. This effect becomes significant when the skin depth, which is inversely proportional to $\sqrt{\sigma f}$ (where $\sigma$ and $f$ are conductivity and frequency, respectively), is comparable with the wire diameter. For copper this occurs around 20–30 MHz. After that point, $R_L$ continuously rises with frequency, following a $\sqrt{f}$ dependence. At lower frequencies, the power loss of the core material is usually dominant. Figure 9.3(b) depicts $R_L$ as a function of frequency for a 10-nH wire-wound inductor along with that for a 10-nH multilayer one. The curves in Figure 9.3(b) are for illustration purposes only. In practice it is usually the $Q$ factor, which is also shown in Figure 9.3(b), rather than $R_L$

that is provided on the datasheet. The conversion from $Q$ to $R_L$ is simply by [see (3.17)]:

$$Q = \frac{\omega L}{R_L(\omega)}$$

In the skin-effect region, $R_L(\omega) \propto \sqrt{\omega}$; thus $Q$ is also proportional to $\sqrt{\omega}$, as illustrated in Figure 9.3(b). Furthermore, because both $L$ and $R_L$ are proportional to the wire length, the $Q$ curves for a series of inductors are approximately independent of the $L$ value.

Sections 9.1.2.1–9.1.2.4 consider the inductor characteristics in several RF circuit applications.

### 9.1.2.1 Inductors in RF Matching Networks

The parasitic effects of an inductor in matching circuits can be analyzed in a similar manner as in the treatment of capacitors. In the case of inductors, the impedance increases with frequency and reaches a maximum at *SRF*. Two major conclusions (reached for capacitors in Section 9.1.1) for 10% deviation and the tuning process for compensation of the parasitic effect are perfectly applicable to inductors. We do not repeat them here.

### 9.1.2.2 Power Dissipation

Recall that power dissipation in inductors is usually more significant than that in capacitors in RF applications. Moreover, unlike the case of DC-blocking capacitors, it is difficult to provide a general estimate for the power dissipation in an inductor even if $R_L$ is known due to the unknown nature of the impedance of the network where the inductor is placed. For a low-noise circuit operated at a few gigahertz or higher, a 0.1-dB or so improvement in noise figure can often be achieved by replacing a general-purpose multilayer inductor with a wire-wound one in the input-matching network.

For high-power applications, in addition to added insertion loss, RF power dissipation inside an inductor can also be a reliability concern. Similar to the situation for capacitors, it is difficult to quantify a design criterion in this regard, due to the lack of both a specification by the manufacturer and knowledge of the actual power dissipated in the inductor. For an operation frequency of a few hundred megahertz or higher, transmission lines are normally more suitable as a matching component for their better thermal characteristics and convenience in tuning (see Section 3.5.3). At lower frequencies, air-core inductors offer the best results with regard to this performance requirement. Again, design examples by the manufacturer are always a good starting point.

### 9.1.2.3 Inductors in Bias Circuits and Bias-Tees

As outlined in Section 5.2.3, there are three key design considerations in selecting an inductor for a bias circuit: (1) DC ratings, (2) RF isolation, and (3) resistive load at low RF frequencies. Among them, the DC requirements in current ratings and DC resistance (usually denoted as *DCR*) are always specified on the datasheet. For RF isolation, an inductor should have a large enough nominal inductance and an *SRF* that is still above the operation frequency. Practically, it is not difficult to find an inductor whose *SRF* is as high as 10 GHz. Beyond 10 GHz, a quarter-wavelength transmission line is more often employed in practical designs. Therefore, it is generally not difficult to implement a bias circuit for narrowband applications, whether at low or high frequencies as long as the current rating is moderate (on the order of amperes). In most practical cases, the bias inductance does not need to be a perfect RF choke; its reactance only needs to be larger than the network impedance where the inductor is connected by a factor ranging from 3 to 10. Similar to the situation of DC-blocking capacitors, the difficulty is in broadband bias circuits. Special types of ultra broadband inductors are available at a premium cost.

A component commonly used as a bias device for broadband RF measurements is the bias-tee, also known as the bias-T owing to its circuit shape as shown in Figure 9.4. A bias-tee allows both the DC supply voltage (at the $V_{cc}$ port) and the RF signal (at the *RF* port) to be present at the *RF* and *DC* ports without interference with each other. In terms of specifications, in addition to its DC current capability, frequency range and insertion loss are two key specifications for a commercial bias-tee. It should be noted, however, that the frequency range of a bias-tee is usually specified at a DC current lower than

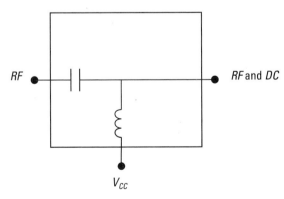

**Figure 9.4** Bias-tee.

the DC current rating. For an application that extends to very low frequencies such as below 1 MHz, a bias-tee that appears adequate based on the frequency specification may suffer a significant degradation in insertion loss at the low frequency end once the DC current reaches a level that is still well within its DC current rating. This is because the inductance, which needs to be sufficiently high to support the specified isolation at low frequencies, decreases as the DC current increases. For this reason if the frequency range of a bias-tee is not specified at the intended current levels, the designer should verify the performance independently.

The third requirement for the inductor in a bias circuit is to provide an adequate resistive load for the stability at a frequency range from a few tens to a few hundreds of megahertz. This is often realized through a special component, a ferrite bead, discussed in Section 9.1.2.4.

### 9.1.2.4 Ferrite Beads

Section 5.2.3 discusses the basic concept of ferrite beads in the context of stability. Ferrite refers to a special type of magnetic materials that exhibit ferrimagnetism [6, 8]. Ferrite materials find uses in a variety of RF devices [1, 6], and the ferrite bead is one of them. A ferrite bead is essentially an inductor with a ferrite core. When used in a bias circuit, the distinguishing property of the ferrite core, in comparison with the nonmagnetic core used in regular inductors, is its increasing power loss with frequency, making the device a natural high-frequency rejection filter. In fact, in terms of circuit functions, ferrite beads are most widely utilized as filters in the applications of electromagnetic interference (EMI) control and signal integrity. The former means preventing RF signals of the circuit from radiating through the DC bias lines, and the latter implies blocking noise from the DC source to the amplifier [2, 4]. For this reason, a ferrite bead is also commonly known as an EMI filter or an RF choke. Here, our focus is on using ferrite beads for stability improvement applications.

We start with the equivalent circuit for ferrite beads. In this case, the power loss at high frequencies is a feature rather than a parasitic effect. It turns out this feature can be conveniently represented by a parallel RLC circuit as shown in Figure 9.5(a). The DC resistance $R_{DC}$ in Figure 9.5(a) is equivalent to $R_L$ in the equivalent circuit for inductors in Figure 9.3(a). As a DC bias device, the $R_{DC}$ value of a ferrite bead is critically important in component selection, but it has no direct impact on the high-frequency characteristics of the device. Consequently, we do not discuss it here.

The parallel circuit that is comprised of three constant components in Figure 9.5(a) can be viewed as merely a convenient way of representing a

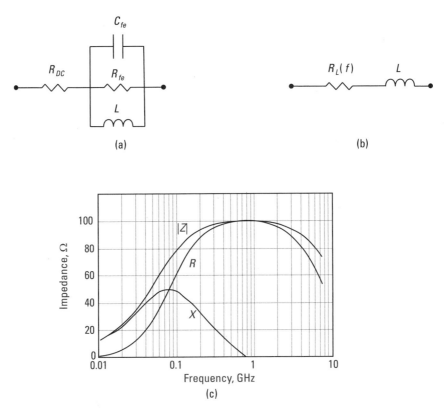

**Figure 9.5** Equivalent circuits and characteristics of ferrite beads: (a) RLC parallel circuit representation, (b) frequency-dependent resistance representation, and (c) impedance and its real and imaginary parts of the circuit in (a), $R_{DC} = 0$, $R_{fe} = 100\Omega$, $C_{fe} = 0.2$ pF, and $L = 200$ nH [see Figure 5.9(d)].

frequency-dependent series resistance $R_L(f)$ in Figure 9.5(b). The series circuit of $R_L(f)$ and $L$ actually bears some similarity with the equivalent circuit for inductors shown in Figure 9.3(a). The omission of $C_L$ is justifiable, because ferrite beads are typically used in a frequency range where the effect of $C_L$ can be neglected. In Figure 9.5(c) the amplitude of impedance $|Z|$ along with its real and imaginary parts $R$ and $X$ of the equivalent circuit in Figure 9.5(a) are plotted against frequency. As a matter of fact, in addition to $R_{DC}$ and the DC current rating, the other key datasheet specification for a ferrite bead is impedance at 100 MHz, usually along with a plot of impedance $|Z|$ versus. frequency. Data of the equivalent circuit parameters is available only in some cases. The curves shown in Figure 9.5(c), which are calculated based on the equivalent circuit used in our stability analysis in Figure 5.9(d), are realistic representations of the frequency dependences of the impedance $|Z|$ and its real and imaginary parts $R$ and $X$.

Details of the $|Z|$, $R$, and $X$ curves shown in Figure 9.5(c) vary greatly among different ferrite beads. Generally, at low frequencies the impedance is dominated by the inductance. The resistance component in the impedance increases with frequency until reaching a maximum at the resonance frequency. For the purpose of stability improvement, the characteristics around 100 MHz are most critical. An impedance in the range of 100Ω with a significant resistance component is usually ideal. There is no definitive guidance in selecting a ferrite bead for stability purposes. In some cases, experimentation might be required.

We finally note that the impedance curves (including $R$ and $X$ curves) are strongly affected by the DC current. Most of the ferrite bead specifications we discuss here are provided under the conditions of zero current. In real applications, the circuit performance, therefore, must be verified at the required operating current.

### 9.1.3 Resistors

Parasitic effects, especially parasitic inductance, also exist in a resistor. However, in RF circuit applications, they are generally not as great a concern as they are with capacitors and inductors. This is chiefly because resistors are rarely used for matching networks where precise component characteristics are critical. Also, at a moderate frequency (a few gigahertz), the impedance of a parasitic element is usually smaller than the nominal resistance. One exception is a 0-Ω resistor being used as a jumper for bridging a gap between two PCB transmission lines. Since the nominal resistance is zero in this case, the component behavior in the circuit is dominated by the parasitic effects. At sufficiently high frequencies, the 0-Ω resistor can cause significant discontinuity in the transmission line, particularly if the widths of the resistor and the transmission line are considerably mismatched. While this mismatch can be minimized by choosing a resistor size comparable to that of the transmission line, the discontinuity in the vertical direction is unavoidable due to the component structure. If the transmission line integrity is critical, a better, but slightly less convenient, method is to use a copper (or brass) shim with a width cut exactly as that of the transmission line.

When a resistor is used for termination or power dissipation in a high-power application, its power rating has to be adequate for the application.

## 9.2 Devices for RF Power Wave Manipulations

Throughout this book, we emphasize that an RF signal is essentially an electromagnetic propagating wave. As such, the traveling direction, just as its magnitude and phase, can be a critical attribute for an RF signal in certain applications. Additionally, the book emphasizes that the media for RF signal transmissions, mostly transmission lines, have to be uniform to avoid any unwanted reflection. These characteristics of RF circuits require special devices for signal manipulations such as redirecting, combining, and dividing. This section introduces several such devices.

All of the devices discussed in Sections 9.2.1–9.2.5 have at least two ports. Their circuit behaviors, in principle, are fully characterized by their corresponding N-port S-parameters [6, 9]. Our description of these devices, however, is entirely focused on the datasheet specifications and their implications in typical applications. Furthermore, we only consider those specifications that are unique to each specific device and generally do not detail other important but self-explanatory specifications such as frequency range, return loss and insertion loss (except in special cases), amplitude and phase imbalance (when power splitting is involved), and power and temperature ratings.

### 9.2.1 Directional Couplers

Section 1.3.2 notes that two opposite traveling waves in the same transmission line can be measured individually. The key component for that measurement is the directional coupler. The basic function of a directional coupler is to sample the incident power with a known factor (coupling) without disturbing the "through" path, such as in the cases of VNA and power measurements discussed in Chapter 10. The reason for "directional" will be clear at the end of this discussion. Figure 9.6 shows a general configuration for a four-port

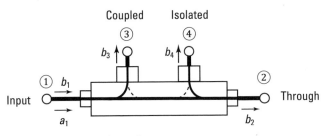

**Figure 9.6** Four-port directional coupler.

**Table 9.1**
Specifications of the Four-Port Directional Coupler

| Input\Output Ports | 1 | 2 | 3 | 4 |
|---|---|---|---|---|
| 1 | Return loss | Through | Coupling | Isolation |
| 2 | Through | Return loss | Isolation | Coupling |

coupler, where port 1 is assumed to be the input port. Practical devices are usually designed to be symmetric with port 1 and port 2. We only need to consider the case of port 1 being the input. Assuming all ports are terminated with 50Ω, then only output going power waves, designated as $b_1 - b_4$ in Figure 9.6, need to be considered. Table 9.1 summarizes the terms used to specify the output power $b_i$ ($i = 1 - 4$) with respect to the input power $a_1$ or $a_2$.

Using the argument of conservation of energy from (1.37), we have the following relationship for the power waves:

$$|a_1|^2 = |b_1|^2 + |b_2|^2 + |b_3|^2 + |b_4|^2 + D \qquad (9.4)$$

All $b_i$ terms in (9.4) can be directly measured (S-parameters). $D$ is the power dissipation inside the coupler. In practice, however, not all datasheet specifications are defined in these direct measurable quantities. Specifically, four most common specifications for a directional coupler are:

- Coupling factor (decibels):

$$-10\log\frac{|b_3|^2}{|a_1|^2} = -S_{31}(\text{dB})$$

- Return loss (decibels):

$$-10\log\frac{|b_1|^2}{|a_1|^2} = -S_{11}(\text{dB})$$

(The return loss specification is also often in the VSWR format on datasheets.)
- Directivity (decibels):

$$-10\log\frac{|b_4|^2}{|b_3|^2} = -10\log\frac{|S_{41}|^2}{|S_{31}|^2}$$

- Insertion loss (decibels):

$$-10\log\frac{D}{|a_1|^2}$$

The first two specifications are directly measurable while the last two are not. Directivity can be calculated from the isolation, which is defined as:

- Isolation (decibels):

$$-10\log\frac{|b_4|^2}{|a_1|^2} = -S_{41}(\text{dB})$$

Note that a (−) sign is used in all the above definitions so that the parameter is a positive number in decibels.

It is straightforward to show that

$$\text{Isolation (dB)} = \text{Directivity (dB)} + \text{Coupling factor (dB)}$$

The significance of directivity in applications can be better understood if we consider the output power wave at port 3, $b_3$, when the incident power is applied at port 2. Assuming that the device has perfect symmetry with respect to port 1 and port 2, we have $|S_{41}| = |S_{32}|$. Then directivity can also be written as:

$$\text{Directivity} = -10\log\frac{|S_{32}|^2}{|S_{31}|^2} \tag{9.5}$$

In typical applications (see the example for power measurement described in Section 10.5), the power coupling from port 1 to port 3 is the desired function and is usually referred to as forward coupling. In contrast, as explained in Chapter 10, it is undesirable to have any power coupling at port 3 if the incident power wave is from port 2, which is the isolation specification and is also called backward coupling. The directivity expressed in (9.5) is then a ratio of the backward coupling to the forward coupling. An ideal directional coupler should maintain a consistent forward coupling and have no backward coupling. Therefore, directivity is essentially a figure of merit for the coupler's capability to discriminate the backward coupling. The directional coupler is so called precisely because of its discrimination for different directions of power waves. In practice, a coupler always has a finite isolation. If an application demands an exceptional isolation, some other means may be employed to achieve the required system performance.

The last parameter is through loss, which is usually not a datasheet specification, but can be measured by:

$$-10\log\frac{|b_2|^2}{|a_1|^2} = -S_{21}(\text{dB})$$

With all four S-parameters $S_{i1}$ ($i = 1 - 4$) known, the insertion loss $D$ can be calculated using (9.4).

Here we assume that all ports are perfectly terminated and hence that no reflections need to be considered. In real measurement setups, there is always some imperfection in a termination. A 30-dB return loss is generally an excellent specification. On the other hand, both the coupling factor and directivity can be very small (20 dB or lower). As a result, the coupling powers to be measured may be too small compared with the unwanted reflection powers. In that case, some special measurement techniques would be required. One example is described in [6].

In practice, three-port directional couplers, which have the second coupling port internally terminated, are more common.

## 9.2.2 Circulators

As the name implies, the incident wave of a circulator travels in a circular manner as illustrated in Figure 9.7. Two key specifications of a circulator are insertion loss and isolation. In our labeling scheme, when all ports are terminated, $S_{21}$, $S_{32}$, and $S_{13}$ are the insertion loss, and $S_{12}$, $S_{23}$, and $S_{31}$ are the isolation. In a typical application, ports 1 and 2 are connected to a source and a load, respectively, and port 3 is the detection (or measurement) port. Then the circulator reduces the power reflection from the load to the source by the amount of isolation, provided the third port is perfectly terminated. For example, for an isolation of 20 dB, even when port 2 is an open or a short circuit, the return loss at port 1, in theory, would be at least 20 dB. In addition, a circulator allows an accurate measurement of the reflected power wave at

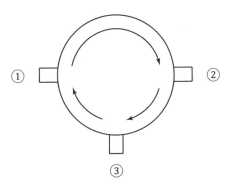

**Figure 9.7** Circulator.

port 3. (A coupler has a similar capability, but, as explained in Section 9.2.1, the coupling port only takes a fraction of the reflected power).

#### 9.2.2.1 Isolator

An isolator is simply a circulator with a built-in termination at the third port. It serves the isolation function only.

### 9.2.3 Power Dividers and Combiners

In some circuit applications, an RF signal needs to be applied to two or more loads. This circuit function obviously cannot be realized by simply splitting a transmission line into two branches without causing a major disturbance to the circuit. A special device known as a power divider is required to implement this function. Similarly, to combine two RF signals, a special device known as a power combiner must be used if the circuit needs to be properly matched. Section 9.2.3.1 discusses power dividers, and then Section 9.2.3.2 explains how the same devices may be used for power combining.

#### 9.2.3.1 Dividers

In the RF component market, another device name, power splitter, is also often seen for the power-dividing function. In most cases, dividers and splitters only differ by name, although some manufacturers do make a distinction between them in specifications. In component selection for power dividing, rather than solely basing the choice on the catalog names, the user should pay attention to the performance specifications, which we discuss here.

We limit our discussion to two-way equal-split dividers. Information on the more general case of $N$-way and unequal-split dividers, along with

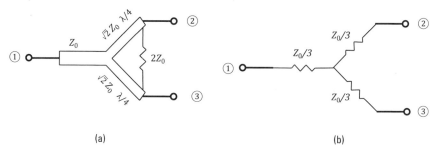

**Figure 9.8** (a) Wilkinson power divider, and (b) resistive power divider.

the dividers' structure and operation principles, can be found in [6, 9]. There are two major types of dividers commonly used in practice: the Wilkinson divider and the resistive divider. Their basic circuit structures are sketched in Figure 9.8.

Details of the parameters in Figure 9.8 can be found in [6, 9]. Practical devices usually have a more complex structure for enhanced performance specification (often at the expense of some other parameters). Both types of dividers have a theoretically perfect return loss at all three ports when terminated with 50-Ω loads. The Wilkinson divider utilizes transmission lines to achieve the required impedance conditions. This circuit configuration yields a lossless power division (note there is no power dissipation in the internal resistor if the two output ports are matched), but it has a limited frequency range. The resistive divider shown in Figure 9.8(b) is comprised of three resistive branches and does not have a limit at the low frequency end. The downside of a resistive divider is that it has half the input power dissipated internally. Table 9.2 summarizes some key specifications of both types of dividers. (Refer to Figure 9.8 for port designations.)

**Table. 9.2**
Specifications of Power Dividers

| Device\Specs | Frequency range | $S_{11}, S_{22}, S_{33}$ | $S_{21}, S_{31}$ | $S_{23}$ |
|---|---|---|---|---|
| Wilkinson | Limited bandwidth | Return loss | 3-dB+Ins loss* | Isolation |
| Resistive | Down to DC | Return loss | 6-dB+Ins loss* | 6 dB |

* In practice, there are two different definitions used by manufacturers to specify insertion loss for a divider: In the first definition, used in Table 9.2, insertion loss is the unintended internal power loss of the device. In this definition, insertion loss is usually a fraction of a decibel. In the second definition, insertion loss is the direct measured power ratio $S_{21}$ (and $S_{31}$), which includes 3-dB and 6-dB losses for the Wilkinson and resistive dividers, respectively, due to power division. In fact, when the type of a divider is unknown, this measurement is the quickest way to determine it.

With regard to isolation, the Wilkinson divider has an infinite isolation in theory, as opposed to 6 dB of the resistive divider. (A higher isolation can be achieved, at the sacrifice of insertion loss, with a resistive divider in a different configuration.)

### 9.2.3.2 Combiners

Unlike the directional devices discussed in Section 9.2.2, practical devices for power dividing and combining are reciprocal; that is, the input and output can be exchanged. Then it appears that the same device for power dividing can be used as a power combiner as illustrated in Figure 9.9(a, b). In fact, in many cases manufacturers do not distinguish dividers and combiners by simply labeling the product as divider/combiner. There is, however, a subtle difference between the two cases.

We note that the reciprocal property ensures that $S_{ij} = S_{ji}$, but it does not imply that the power combination is simply a reverse of power division. Take the Wilkinson divider as an example. When used as a combiner, for signal $f_1$ [see Figure 9.9 (b)], the power loss from port 2 to port 1 is also 3 dB, which is consistent with the conclusion that $S_{12} = S_{21}$. However, the other half of the input power dissipates in the internal resistor. The same is true for signal $f_2$. Hence, we conclude that, in general, for power combining, half of the input power is lost inside the device whereas for power dividing no power loss occurs. This difference can be explained if we consider the signals at port 2 and 3

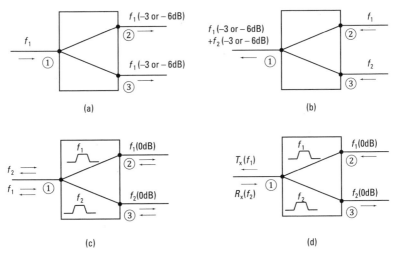

**Figure 9.9** (a) Divider, (b) combiner, (c) diplexers, and (d) duplexer. The decibel numbers in the parentheses are nominal through loss.

for the two cases: (1) For power dividing, the two signals are identical; and (2) for power combining, the two signals are generally incoherent (no fixed phase relationship). In fact, if we truly reverse the power dividing process by applying two identical signals at ports 2 and 3, there will be no internal power dissipation because the voltage across the internal resistance is always zero. The other extreme condition is when two coherent signals with 180-degree phase difference are applied at ports 2 and 3. In that case, the input power completely dissipates internally. These conclusions are intuitive, and we leave the rigorous proof to [6].

In summary, power dividers and combiners are exchangeable in terms of functionality. When the input powers are at a low or medium level (usually ~watts), the same device can be used for both functions. At higher powers, the power rating of the device has to be considered, particularly for combiners. For this reason, many manufacturers of high-power components have two separate categories for power dividers and power combiners.

A subcategory of power dividers/combiners is the so-called diplexers/duplexers. Again, there is no standard naming convention in the industry for these terms. The essence of the devices in this category is frequency selectivity, which is realized through an internal filter array. The difference between a diplexer and a duplexer is mainly in application. A device used in the configuration shown in Figure 9.9(c) is commonly called a diplexer, which separates or combines signals from two frequency bands [labeled $f_1$ and $f_2$ in Figure 9.9(c)]. Typical examples are a 2-GHz–5GHz dual-band Wi-Fi system and a mobile phone system that operates in both the cellular band (800 MHz) and the PCS band (1,800–1,900 MHz). In these applications, within each band the signals are bidirectional. In comparison, a device used in a frequency division multiplexing (FDM) system is typically called a duplexer, with the transmitting (Tx) and receiving (Rx) signals as shown in Figure 9.9(d). In this case, the separation of the two frequency bands is usually small. For example, for the cellular band, it is only 20 MHz (between 849 MHz and 869 MHz). While the difference in frequency band spacing may require different filter design techniques, from Figure 9.9(c, d), we can see there is no difference between a diplexer and a duplexer in terms of functionality. In some cases, however, a duplexer may employ a circulator to realize the separation of the Tx and Rx signals. In that case, the duplexer is different from a diplexer.

### 9.2.4 90-Degree Hybrid Couplers

A 90-degree hybrid coupler (also known as a hybrid or a quadrature hybrid) is a four-port device that can be represented symbolically by the diagram shown in Figure 9.10. In our scheme, port 1 is designated as the input, port 4 as an isolation port, and ports 2 and 3 are output ports with a 90-degree phase difference. In typical applications, a hybrid is used with the following two termination conditions:

- Port 4 is terminated with a 50-Ω load.
- Ports 2 and 3 are terminated with a pair of identical loads (generally not equal to 50Ω).

Under these two conditions, the hybrid has the following properties:

- Property 1: The outgoing wave at port 3 has a 90-degree phase shift with respect to that at port 2;
- Property 2: Port 1 has a perfect return loss;
- Property 3: The reflected powers from ports 2 and 3 dissipate in the load at port 4. When total reflection occurs at ports 2 and 3 (open or short circuit), power dissipation at port 4 equals the incident power at port 1;
- Property 4: $S_{22}$ and $S_{33}$ are perfect; that is, the impedances looking into ports 2 and 3 are 50Ω.

Since there is no internal resistance to dissipate power, properties 2 and 3 are equivalent (i.e., one must lead to the other). They are stated separately here, since each property has its own importance in applications (e.g., the implication of property 3 in selection of the resistor at port 4 in balanced amplifier; see Section 5.4.1). It may appear from Figure 9.10 that a hybrid, when port 4 is terminated, has the same functions as a power divider/combiner. It is property 2 (or 3) that sets a hybrid apart from a regular divider/combiner: Property 2 ensures a perfect input return loss as long as the loads at the two outputs are

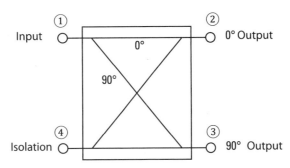

**Figure 9.10** 90-degree hybrid coupler.

identical whereas the perfect input return loss is only true for a divider when the two outputs are terminated with 50Ω.

### 9.2.5 Low–Passive-Intermodulation Components

Chapter 8 explains the importance of intermodulation, along with other nonlinearity specifications. While, in principle, every component in a signal path contributes to the system's linearity specifications, traditionally, intermodulation is considered only relevant to active components because the nonlinearity of passive components is generally substantially lower than that of active ones. This is no longer true, however, for today's wireless systems, as intermodulation requirements become increasingly demanding. To meet the market demand, the component industry has introduced a new category: low-passive-intermodulation (PIM) components. Typically, components in this category are specified by a two-tone measurement (see Chapters 6 and 10) with 20W (43 dBm) for each tone. Under this condition, a low-PIM device should have an intermodulation of at least −100 dBm (−143 dBc) and may have one as low as −120 dBm (−163 dBc).

In practice, as explained in Chapter 8, the intermodulation specification for a system is usually written for a specific operation condition. Also mentioned in Chapter 8 is that intermodulation specifications cannot be reliably converted from one condition to another. For these reasons, it is not always immediately clear whether a low-PIM component is suitable for a specific application. If in doubt, the user should test the selected component under the required condition first before using it in the application.

# References

[1] Wallace, R., and K. Andreasson, *Introduction to RF and Microwave Passive Components*, Norwood, MA: Artech House, 2015.

[2] Smolskiy, S. M., L. A. Belov, and V. N. Kochemasov, *Handbook of RF, Microwave, and Millimeter-Wave Components*, Norwood, MA: Artech House, 2012.

[3] Riddle, A., "Passive Lumped Components" in *RF and Microwave Passive and Active Technologies*, M. Golio and J. Golio (eds.), CRC Press, 2008.

[4] Bahl, I., *Lumped Elements for RF and Microwave Circuits*, Norwood, MA: Artech House, 2003.

[5] Meeldijk, V., *Electronic Components*, John Wiley and Sons, 1995.

[6] Pozar, D. M., *Microwave Engineering*, Second Edition, John Wiley and Sons, 1998.

[7] Ramo, S., J. R. Whinnery, and T. Van Duzer, *Fields and Waves in Communication Electronics*, Third Edition, John Wiley and Sons, 1993.

[8] Spaldin, N. A., *Magnetic Materials: Fundamentals and Applications*, Second Edition, Cambridge University Press, 2010.

[9] Collin, R. E., *Foundations for Microwave Engineering*, Second Edition, McGraw-Hill, 1992.

# 10

# RF Measurements

RF measurements are an integral part of RF designs. This chapter is devoted to several measurement techniques, commonly used for the characterization of RF circuits and systems. We focus mainly on the practical aspects of these techniques in applications, with special attention paid to the operation of the pertinent measurement instrument.

## 10.1 Measurement Techniques of S-Parameters

### 10.1.1 The VNA and Calibration Techniques

The VNA, an instrument for the S-parameter measurement, measures both the magnitude and phase, whereas a scalar instrument measures only magnitude.

A VNA measurement usually requires several steps of calibration. This section outlines the operation principle of the VNA, from which we can see why these calibration steps are necessary.

Figure 10.1 illustrates conceptually a VNA for a two-port S parameter measurement, showing only the setup for the forward measurements ($S_{21}$ and $S_{11}$). The measurements of $S_{12}$ and $S_{22}$ can be simply configured by a switch (inside the instrument and not shown) between ports 1 and 2. The detector used in a VNA is capable of detecting the vector ratio of two power waves, labeled as $a'$ and $b'$ in Figure 10.1. (Note that a power wave is characterized by its amplitude and phase; see Section 1.3.1.) A double-pole double-throw (DPDT) switch allows two measurement configurations for $S_{21}$ and $S_{11}$, respectively. The configuration shown in Figure 10.1 is for $S_{21}$, which is the vector ratio of $b_2$ to $a_1$ at the DUT. In this scheme, if we assume that the system is consistent (that is, it does not change from measurement to measurement, other than random noise), we can see that $a_1$ is related to $a'$ with a fixed but unknown factor, and so is $b_2$ to $b'$. The question then is how the desired result $b_2/a_1$ can be derived from $b'/a'$, which is what the detector actually measures. The answer lies in calibration.

The calibration process for this case is simply a measurement without the DUT [i.e., $(b_2/a_1)_{cal} = 1$ (or $S_{21} = 1$]. This step of calibration is usually

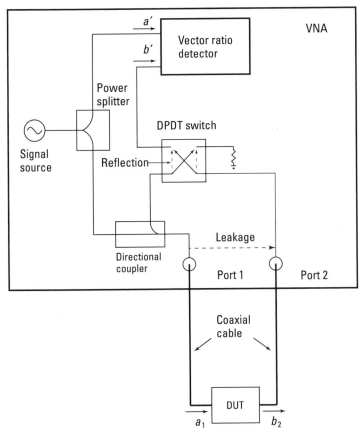

**Figure 10.1** Illustration of the operation principle of vector network analyzer.

labeled as "transmission" or "thru"(short for through). The measured result by the detector, which is denoted as $(b'/a')_{cal}$, is the calibration data, stored in the VNA. After this calibration, we make a second measurement with the DUT inserted. Now we have $(b_2/a_1)_{DUT} = S_{21}$, and the result by the detector is $(b'/a')_{DUT}$. Then the two measurement results are related by

$$\frac{\left(\frac{b'}{a'}\right)_{DUT}}{\left(\frac{b'}{a'}\right)_{cal}} = \frac{\left(\frac{b_2}{a_1}\right)_{DUT}}{\left(\frac{b_2}{a_1}\right)_{cal}} = S_{21}$$

Thus, the unknown factors are removed by a calibration process.

In practice, the conditions "without DUT" and "with DUT" are implemented by a set of two coax thru standards (also called adaptors) with different genders (male and female) but identical electrical length. One case is illustrated in Figure 10.2(a).

In the above calibration scheme, we, in fact, implicitly assume that there are no system errors in the setup. If there were, the calibration would no longer be accurate. Consider a possible leakage from port 1 to port 2 as shown in Figure 10.1. In such a case, $b'$ has two components, from the leakage path and the DUT ($b_2$) respectively. A possible method to eliminate this leakage effect is to add another step in calibration (called isolation by some VNA manufacturers). In this step, the measurement is made without any connection, hence no contribution from $b_2$. Then the measured result by the detector is the contribution from the leakage alone. Without getting into detailed mathematics, we still can conceptualize a process that utilizes the two calibration data to extract the desired $S_{21}$ from the DUT measurement.

In reality, the leakage in the $S_{21}$ measurement usually can be neglected, due to the high quality of the components used in modern VNAs. Nonetheless, this simple case provides some insight into how a calibration process, which usually consists of several measurements (each under a known condition such as "transmission" and "isolation" in the above cases), is used to eliminate the unknown factors and the undesired couplings in the S-parameter measurements of a DUT.

For the $S_{11}$ measurement, the DPDT switch is in the reflection position (see Figure 10.1), and the rest of the analysis is the same as that for the $S_{21}$ measurement. In the reflection measurement, however, some undesired couplings between the incident wave and reflected wave are almost unavoidable. A noticeable example is the leakage of the incident wave to the coupling

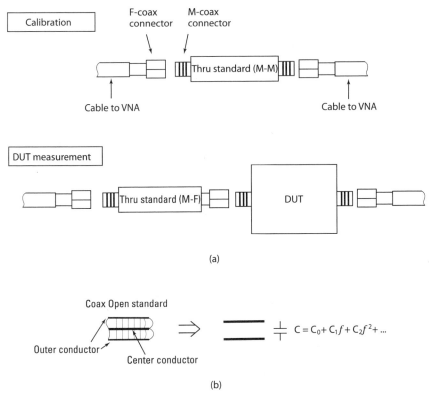

**Figure 10.2** (a) A set of two thru standards and (b) a circuit model for an open standard.

port of the coupler due to the finite isolation specification (see Section 9.2.1). As a result, more error corrections are generally necessary in the reflection measurement. The standard mathematical tool used by VNA manufacturers is the so-called error model represented by the signal flow graph technique. Detailed accounts of the error model and the signal flow graph technique can be found in [1, 2] and [3, 4], respectively. We simply describe the results here.

There are three error terms in this model for the reflection measurement. To solve them in a set of equations, three measurements are performed using three terminations known as the short (S), open (O), and load (L) standards. The standard practice by VNA manufacturers is to provide a calibration kit that consists of coaxial terminations of S, O, and L standards along with the thru (T) standards already discussed. A calibration using this set of standards is referred to as a SOLT calibration (also known as a full two-port calibration).

None of the physical standards are perfect in terms of circuit characteristics. Take the open standard as an example. There is always an EM field coupling between the two conductors at the opening, and this coupling increases with frequency. To take this effect into account, an open can be modeled as being terminated by a capacitor, which is expressed as a polynomial series in frequency as indicated in Figure 10.2(b). Actual implementations on the standard modeling vary among different VNAs. Some vendors provide a specific data file for the standard set; others allow a user's input for the model parameters (such as $C_0$ and $C_1$, etc.), or ignore this effect completely. Generally, for a typical wireless product at several gigahertz, this level of accuracy may not be necessary; that is, the user can use these standards without any modeling parameters.

The DC bias for the DUT is typically provided by one or two bias-Ts (see Section 9.1.2) in the measurement. The same bias-Ts must also be used in all calibration steps so that their effects are included in the calibration data.

### 10.1.2 TRL Calibration

Another calibration technique supported by almost all commercial VNAs is the thru, reflect, and line (TRL) method. The reflect standard can be either open or short, and the line standard is essentially a transmission line with a specific electrical length. The TRL method is generally considered more accurate than the SOLT method as the TRL standards do not require a sophisticated device modeling. A complete mathematical treatment on the TRL calibration technique can be found in [5].

In principle, TRL standards can be either coaxial components or implemented on a PCB (coax to PCB connectors are still required). In practice, however, other than some sophisticated fixturing applications, the PCB-based TRL method is almost exclusively used because of its convenience in the measurements for the surface-mount device. Although the TRL design normally falls outside the activities of a product development engineer, it is occasionally desirable to have accurate S-parameter data for a specific device. Since prototyping a PCB circuit is a routine technique for today's RF engineers, a set of custom TRL standards can be made with only moderate effort at most RF laboratories. The following is a step-by-step description of the design process:

1. A test board designed for the selected DUT. Figure 10.3(a) shows two examples. The board length only needs to be convenient. The transmission line impedance should be the same as the system reference impedance. In theory [5], a different impedance is allowed for

some standards. To simplify the process we use the same reference impedance for all the transmission lines utilized in our design. Figure 10.3(a) shows only the signal traces. Other necessary PCB elements such as grounds for DUT and pads for coax to PCB connectors are omitted, since they are not relevant to the calibration process.
2. A THRU board, whose length must be the sum of the two signal traces on the test board. (See the discussion on the reference plane in Step 3).
3. A board for open reflect standards (ports 1 and 2) as shown in Figure 10.3(b). It is critically important to keep the trace lengths of the test board and the reflect standard identical (at both ports) as indicated in Figure 10.3(b). The separation of the two reflects should be large enough to prevent any coupling between them. [The board layout in Figure 10.3(b) is for illustration only.]

This brings us to an important concept in the S-parameter measurement, namely, the reference plane. Chapter 1 determines that the effect of a section of transmission line, in terms of S-parameters, is a phase shift. In an TRL calibration, the VNA uses the end points of the two reflects, referred to as reference planes in the industry, as the reference for the phase calculation at ports 1 and 2. Essentially, the reference planes are the boundaries for specific S-parameter data. Consider two layouts of the test board, sketched in Figure 10.4(a, b). (Only the port 1 side is considered.) In both cases, the trace length is the same as the reflect standard (length "a" in Figure 10.3), as required. The difference is in the reference plane. It is clear that the measured S-parameter of the DUT in the case of Figure 10.4(b) includes the effect of an extra lead, labeled as $\delta$, in comparison with that in Figure 10.4(a). From this example, we can see the reference planes are determined by the test board layout. There is no correct way of choosing a reference plane, although it is commonly chosen to be the point where the device lead is first in contact with the PCB trace as the case in Figure 10.4(a) or simply the edges of the device. Additionally, a thru standard is treated by the VNA as having zero length between the two reference planes, which is the reason for the requirement in step 2.

For the user of S-parameter data, it is important to pay attention to the reference planes and to design the application circuit accordingly. Otherwise some unexpected results may occur. In addition to the case just discussed, let's examine another example to show this point.

# RF Measurements

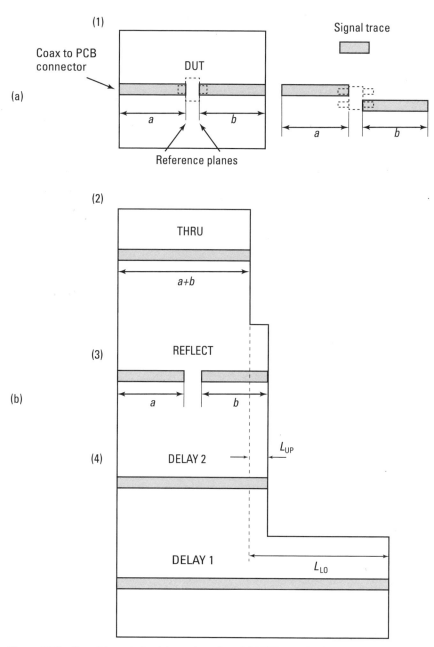

**Figure 10.3** Board layouts for (a) test board, and (b) TRL standards.

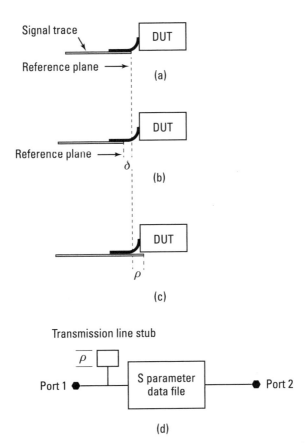

**Figure 10.4** (a, b) Two reference planes, (c) board layout with extra trace length, and (d) equivalent circuit for case (c).

In this case, the S-parameters are measured with the port 1 reference plane shown in Figure 10.4(a), but in the actual circuit board the trace extends underneath the DUT, as illustrated in Figure 10.4(c). This extra trace is equivalent to a short stub in parallel with the actual S-parameters as illustrated in Figure 10.4(d). At sufficiently high frequencies (>10 GHz), this extra stub may cause some observable effects in terms of comparison between the simulation (without the stub) and the measurement results (with the stub).

The concept of the reference plane applies to coax components too. The reference planes in that case must be in the connectors. But their exact locations are generally not a concern because there is no arbitrariness allowed in coaxial connection. (It has to be tight.)

4. The design of the line standard(s) is slightly more complicated. A line standard must be longer than the thru standard. The propagation delay due to the extra length is the parameter for the line standard and is required in the VNA calibration. For this reason, the line standard is often referred to as the delay line. Chapter 1 shows that the delay of a transmission line $\Delta T_d$ is uniquely determined by its physical length $L$ and the effective dielectric constant $\varepsilon_r'$ as

$$\Delta T_d = \frac{L\sqrt{\varepsilon_r'}}{c} \quad (10.1)$$

This delay line introduces an extra phase shift, $\Delta\theta$, relative to the thru standard at the phase detector. $\Delta\theta$ is related to $\Delta T_d$ by

$$\Delta\theta = 2\pi f \Delta T_d \quad (10.2)$$

It is usually stated in the VNA manual that the optimal range of $\Delta\theta$ for accurate measurement results is from 20° to 160°. Using this criterion, the corresponding usable frequency range for a given delay standard is bound by a lower limit $f_{LO}$ and an upper limit $f_{UP}$:

$$f_{LO} = \frac{20°}{360°} \frac{1}{\Delta T_d} \quad (10.3)$$

and

$$f_{UP} = \frac{160°}{360°} \frac{1}{\Delta T_d} \quad (10.4)$$

Obviously, $f_{UP}$ and $f_{LO}$ always follow the 8:1 ratio. If the desired frequency range exceeds this ratio, most VNAs allow a second (or more) line standard.

The critical parts of a TRL design are to determine the delay parameter and the physical length based on the required frequency range. We can start with either $f_{LO}$ or $f_{UP}$, depending on the application, and let the other limit set by the 8:1 ratio. In the following example, we first select $f_{LO}$. Then the delay $\Delta T_d$ can be calculated by (10.3), and the corresponding physical length $L$ is calculated by (10.1). Table 10.1 shows the numerical results for the two design parameters: $\varepsilon_r' = 1.8$ and $f_{LO} = 0.3$ GHz.

In this design, a second delay is included to extend the upper limit of the frequency range. In Figure 10.3(b), all TRL standards are on the same PCB. This arrangement is convenient for our illustration as well as for certain

**Table 10.1**
Parameters for Delay Lines

|  | $f_{LO}$ (20°), gigahertz | $f_{UP}$ (160°), gigahertz | Delay, ps | Physical length, millimeters |
| --- | --- | --- | --- | --- |
| Delay 1 | 0.3 | 2.4 | 185.1 | $L_1 = 41.4$ |
| Delay 2 | 2.4 | 19.2 | 23.1 | $L_2 = 5.17$ |

$\varepsilon'_r = 1.8$, $f_{LO} = 0.3$ GHz. $Z_0 = 50\,\Omega$.

practical applications. The key factors in the board design are the operation frequency and the frequency characteristics of the coax-to-PCB connector. In a frequency range of 10 GHz or lower, a typical RF coax-to-PCB connector (also known as a PCB end launch connector) can work well, which allows flexibility in the board design. For a significantly higher frequency, a special precision connector must be used to ensure a high-quality coax-to-PCB transition. In that case the PCB design must accommodate the form factor required by the connector.

On the lower frequency limit, we note that for a moderate $f_{LO} = 300$ MHz, the total PCB length of the delay 1 standard would be more than 5 cm. The physical length can become unrealistically long if a significantly lower limit is required unless a different substrate material with a much higher $\varepsilon'_r$ is used. VNAs usually allow measurements below $f_{LO}$. The results are less accurate but may still provide useful information for the circuit/component evaluation.

### 10.1.3 Port Extension Technique

We now describe a method that allows the S-parameter measurements of the DUT using the same test board shown in Figure 10.3(a), but without the TRL calibration. Consider a case where we will make a measurement of the test board shown in Figure 10.3(a) using a coax SOLT calibration. The reference planes, in this case, are at the input and output coax to PCB connectors (not shown) respectively. If we can move the reference planes from the connectors to the device [indicated in Figure 10.3(a)], we will be able to use the measured data on the test board to derive the desired S-parameter of the device. This way, we no longer need the TRL calibration. This is the concept behind the port extension.

The key parameter for changing the reference plane is the delay of the signal trace between the two reference planes. It can be determined by an $S_{11}$

measurement of the test board with the device removed. (Note that the reference plane is at the connector at this moment.) This is essentially a measurement of a transmission line with an open termination at the far end. Let $S_{11\_c}$ and $S_{11\_open}$ be the reflection coefficients at the connector and open respectively, then according to (1.18), we have

$$S_{11\_c} = S_{11\_open} e^{-j2\beta\ell} = S_{11\_open} e^{-j4\pi f \Delta T_d} \tag{10.5}$$

Since $S_{11\_open} = 1$ (see Section 1.3), the delay $\Delta T_d$ can be derived as

$$\Delta T_d = \frac{\theta_c}{4\pi f} \tag{10.6}$$

where $\theta_c$ is the phase of $S_{11\_c}$. (Note that the negative sign is ignored.) A VNA screen shot is shown in Figure 10.5(a) where the phase of a measured $S_{11\_c}$ is plotted over a frequency range from 50 MHz to 10 GHz. From the slope, we obtain: $\Delta T_d = 62.8$ ps.

Most VNAs are equipped with the port extension capability, which allows the user to add a delay to the measured data. We can use this capability to determine the delay without a calculation as follows: Gradually increase the delay in the port extension mode while observing the phase versus frequency plot. When this plot becomes flat over the entire frequency range as shown in Figure 10.5(b), the corresponding delay is the desired parameter for the transmission line under test. In Figure 10.5(b), it is 64.8 ps, which is in good agreement with the above calculated result. The same process can be repeated for port 2. Once the delays are determined, the VNA can extract the desired S-parameters (with the reference planes at the device) from the measured data at the connectors.

This technique works reasonably well at moderate frequencies (roughly several gigahertz, depending on the quality of the open transmission line). However, there are two major limitations on this technique: (1) it does not remove the effect of the trace loss, which can be significant at higher frequencies, particularly if the DUT is a low-gain or no-gain device such as an RF switch; and (2) at higher frequencies, the determination of the delay becomes somewhat arbitrary, because the phase versus frequency plot is not exactly a straight line, as evidenced by the expanded view in Figure 10.5(c). The uncertainty is due to various imperfections in the PCB structure. In practice, this technique is usually employed for a quick estimate of the device performance. For a more serious device characterization, the TRL technique is still recommended.

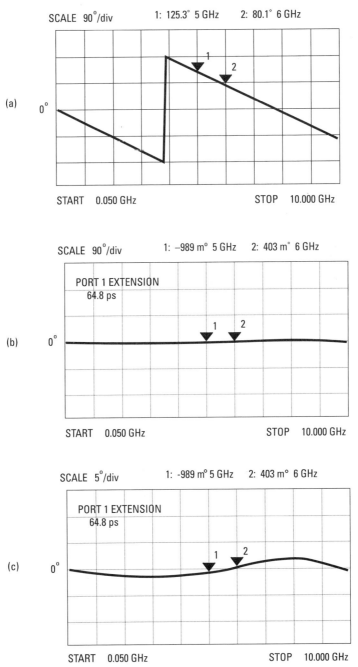

**Figure 10.5** (a) Phase plot of $S_{11\_c}$; (b) after the delay correction; and (c) expanded view of (b).

### 10.1.3.1 Pigtail Technique

The port extension technique can also be used in an actual circuit for trouble shooting purposes. However, it is limited to cases where the desired test point is directly connected to the coax-to-PCB connector through a transmission line. In many practical situations, the desired measuring point is in the middle of a circuit. An example is illustrated in Figure 10.6(a), where we need to measure $S_{11}$ at the input of the second stage. In principle, a VNA with a probing capability [2] is an ideal solution for this case. However, such equipment is mostly used for the on-wafer measurement because of its high cost and difficult calibration process. For a typical PCB application, the pigtail technique turns out to be handy for a measurement task in the few gigahertz range (may still be usable up to 5–6 GHz). A pigtail is a short piece of thin semirigid coax cable with one end attached to a coax connector, as illustrated in Figure 10.6(b). Its function is essentially the same as the open transmission line on the test board we discussed earlier in this section. While a pigtail is even less perfect as an open standard because of the extrusion of the center conductor, it proves to be a rather effective tool for probing a circuit condition, provided there is enough space and the availability of a ground pad near the probing location. We use the case in Figure 10.6(a) to show the process, outlined as follows:

1. Cut the trace, if necessary. (If there is a series component near the probing point, removing the component can also be an option.)
2. Perform the $S_{11}$ calibration using coax SOL standards. If a DC block or a bias-T is required to block the DC voltage at the test point, they should be included in this calibration.
3. Connect the pigtail to the VNA cable connector and use the port extension technique to move the reference plane to the tip of the pigtail.
4. Solder the pigtail, both the tip and ground, to the probing point as shown in Figure 10.6(c). At this point, the VNA result is usually a good approximation of the $S_{11}$ at the probing point.

We complete this section with some practical tips and comments on the VNA measurement, listed as follows.

- Set the power level properly. Most VNAs have a default power level of 0 dBm. This power level exceeds the saturation levels for many low-noise and small-signal amplifiers, resulting in false data in the gain measurement. On the other hand, if the power is set lower than necessary, the measurement accuracy may not be optimized.

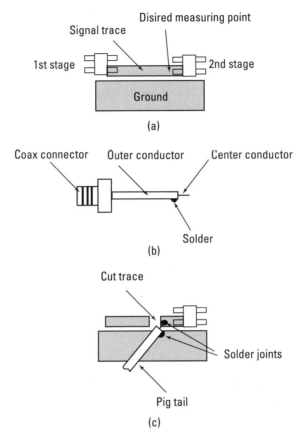

**Fig. 10.6** VNA measurement using a pigtail: (a) measuring point, (b) pigtail, and (c) soldering the pigtail to the measuring point after port extension.

- Ensure a good connection. Erroneous measurement data often results from faulty cable/component connections, either too loose (more often) or too tight. A torque wrench is a useful tool for tightening coax connectors.
- Check the calibration. Poor calibration data affects all subsequent measurements. It is a good practice to check the calibration using a known component other than the standards used in the calibration. Any deficiency in these measurements, such as an abnormally large fluctuation or any irregularities, is a strong indication that the calibration was not done properly.
- Consider whether a bias-T or a DC block is required in the measurement. If yes, they should be included in the calibration.

Many advanced VNAs offer a wide variety of functions and parameters to be selected in measurements and calibrations. Study the user's manual closely before selecting or changing these functions and parameters. It is always prudent to reset the instrument to the default setting before a new session of calibration and measurement.

Time domain reflection (TDR) is a useful technique in certain situations for the RF circuit/system characterization. Many VNAs are equipped with TDR capability. As this subject is not covered in this book, interested readers are referred to [1, 2].

## 10.2 Spectrum Measurements

Spectrum measurement of an RF signal is a measurement in the frequency domain. The instrument for the spectrum measurement is called a spectrum analyzer. The counterpart of the spectrum analyzer is the oscilloscope, which is used in time domain waveform measurement. In the early days of RF technology, the spectrum measurement was the only means to study an RF signal in any detail because the speed of oscilloscopes was simply not fast enough to capture an RF waveform. Situations are drastically different in today's market. The bandwidth of a top-line oscilloscope has exceeded 100 GHz and will certainly further advance. Technology aside, measurement in the frequency domain is still the primary tool in RF circuit and system characterizations. Recall, for example, that the wireless protocols and regulatory requirements discussed in Chapter 8 are all specified in terms of the frequency spectrum.

In principle, if an RF waveform can be digitized at a sufficiently high rate, according the Nyquist theorem on sampling rate, (see, for example, [6, 7]), its spectrum can be revealed through a discrete Fourier transformation. In fact, many contemporary oscilloscopes are equipped with a fast Fourier transformation (FFT) capability, and almost all modern spectrum analyzers employ an analog-to-digital converter (ADC) at some point in the signal path. A review of recent developments in spectrum measurement techniques can be found in [2, Chapter 4].This section focuses on some basic concepts that are important in understanding the measurement results regardless of the system architecture and the circuit implementation (digital or analog) of the instrument.

Figure 10.7 shows the basic architecture of an all-analog spectrum analyzer. This architecture is still rather informative for the purposes of our discussion despite the fact that a number of the components in the diagram have been replaced by digital devices in modern analyzers.

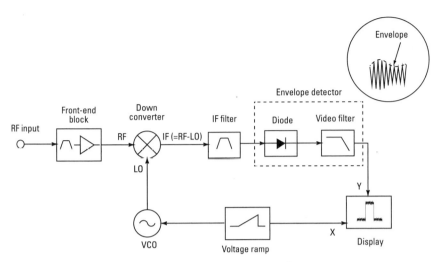

**Figure 10.7** Block diagram of an all-analog spectrum analyzer.

The front-end block, which consists of various amplifiers and filters, is mainly for signal conditioning and dynamic range. This block is critical to the overall performance of an analyzer but has no effects in terms of the system's operation principle.

Then, effectively, the first functional component is the frequency down converter. The LO is from a voltage-controlled oscillator (VCO), which in turn is controlled by a voltage ramp generator. The same generator sets the X coordinate of the instrument display. Thus, the frequency sweep on the display is synchronized with the LO.

On the Y coordinate, the intermediate frequency (IF) signal is filtered by the IF filter, which has a fixed center frequency but an adjustable bandwidth. The output of the IF filter is detected by an envelope detector (also known as power detector) consisting of a diode detector and a video filter. The envelope (see the insert in Figure 10.7) is scaled with the input RF power and is used as the input for the Y coordinate of the display.

A modern spectrum analyzer has a remarkably wide range of measurement capabilities. Each capability is configured through a series of instrument settings. This section mentions only four settings that constitute a very basic spectrum measurement: (1) the input power limit, which is usually called the reference level or magnitude on the instrument; (2) the frequency range, either in the format of center and span or start and stop; (3) the resolution bandwidth (RBW); and (4) the video bandwidth (VBW). The first two items are self-explanatory and will not be further discussed. We will focus on items (3) and (4), particularly on how they affect the measurement results.

## 10.2.1 Resolution Band Width (RBW)

RBW is essentially the bandwidth of the IF filter. As the name implies, this setting determines the analyzer's resolution, which is the minimum separation of two discrete frequencies that can be resolved by the instrument. To see why this is the case, consider a situation where a single-tone [or continuous-wave (CW)] signal is to be measured. In this case, as the LO sweeps, the IF follows with a constant amplitude, as IF = RF − LO. The output of a filter is simply a product of the input and the filter response; therefore, the spectrum at the IF filter output (which is what is displayed on the analyzer) is a duplicate of the IF filter response. This process is depicted in Figure 10.8(a). The simplistic filter response in Figure 10.8(a) is for illustration only, and the determination of the RBW in an analyzer is more complex than what appears [9]. Nonetheless, it becomes clear from this analysis that when the separation of two

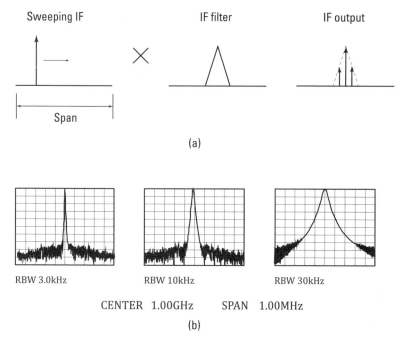

**Figure 10.8** (a) A single-tone signal at the IF filter, and (b) actual measurements with three RBWs.

single-tone signals are comparable to the IF bandwidth, the analyzer can no longer accurately resolve the two signals. This is why the IF bandwidth is the resolution bandwidth of an analyzer.

Figure 10.8(b) shows three cases of a single-tone signal measured with three different RBWs. We can see that while the spectrum width increases with RBW, the measured power value remains constant. Also, the width of the filter response scales with the RBW.

Next, we consider the measurement of a signal with a spread spectrum. Such a signal is specified by its power spectral density (PSD). We can still use the same technique (i.e., let the IF signal sweep through the IF filter) to visualize the shape of the spectrum at the IF filter output. There are two major effects of the RBW on the measurement results.

First, the RBW affects the fidelity of the measurements: We can view the IF sweep as a sampling process with a window of the RBW on the PSD, as illustrated in Figure 10.9(a). For the simplicity of discussion, we assume a rectangular filter response. Then the grey area within the RBW window is proportional to the power level to be displayed on the analyzer. It is clear that the case of RBW2 more accurately represents the original spectrum at a spot frequency [$f_m$ in Figure 10.9(a)] than RBW1. In general, the RBW has to be narrow enough in comparison with the signal bandwidth. Figure 10.9(b) shows examples of measured spectra on an input signal with two RBW settings.

Second, the RBW affects the power reading of the measurement: The question here is how to convert the power reading on the display to the actual signal power spectral density at the input. Upon reexamination of Figure 10.9(a), we can see that the power measured on the analyzer $P_{SA}$ is simply

$$P_{SA}(f) = PSD(f) \cdot RBW$$

In terms of decibels, the power spectral density at a specific frequency can be calculated by

$$PSD(f)(\text{dBm/unit frequency}) = P_{SA}(f)(\text{dBm}) - 10\log(RBW) \quad (10.7)$$

Once $PSD(f)$ is known, the total power is an integration over the signal band. For a signal with a near rectangular spectrum, the signal bandwidth can be easily estimated. Let it be $B$; then the total power is

$$P_{\text{total}}(\text{dBm}) = PSD(\text{dBm/unit frequency}) + 10\log(B) \quad (10.8)$$

Equations (10.7) and (10.8) provide the basics on how to interpret the measured power on an analyzer. Most analyzers are able to calculate the signal

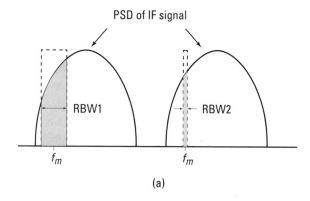

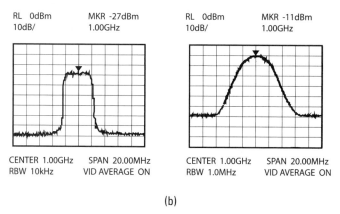

**Figure 10.9** (a) Use of the RBW window to sample the PSD of IF signal, and (b) measurement results with different RBWs.

bandwidth (usually called the occupied bandwidth) and the spectral density as well as the total power. Manual calculations of these parameters are generally not necessary.

The spectrum measured in Figure 10.9(b) is a 3G cellular signal. The input signal power in the CW mode (without modulation) is adjusted to be 0 dBm and its bandwidth is approximately 4 MHz. For RBW = 10 kHz, the measured $P_{SA}$ at the middle of the band is −27 dBm. Using (10.7) and (10.8), we get: $PSD = -27$ dBm $- 40$ dB $= -67$ dBm/Hz and $P_{total} = -67$ dBm/Hz $+ 66$ dB $= -1$ dBm, which is in a reasonable agreement with the CW measurement. At a higher RBW (1 MHz), both the power measurement and the spectrums are distorted.

If the *PSD* in Figure 10.9(a) is a noise spectrum, it is clear that the noise level measured by the analyzer is proportional to the RBW. This is confirmed in Figure 10.9(b) where the noise floor rises by 20 dB when the RBW is increased from 10 kHz to 1 MHz.

As a final note on RBW, we point out that the sweep time is inversely proportional to the RBW, which is expected given the general relationship between the filter responses in the time and frequency domains. In practice, the sweep time may need to be part of the considerations when selecting the RBW for the measurement.

### 10.2.2 Video Band Width

The video filter is a lowpass filter. Its bandwidth controls the noise level on the display. Another method to reduce the noise level is video average, which is an average of multiple sweeps. Figure 10.10 illustrates how the RBW and VBW affect the noise level on a measurement.

As in the case of RBW, the sweep time is inversely proportional to the VBW.

In most modern spectrum analyzers, an ADC takes place either at the envelope detector or at the IF output of the down-converter. In these analyzers, despite the absence of physical filters, the settings of RBW and VBW still exist and have the same functionality as we outlined above.

## 10.3   Phase Noise and Frequency Stability

This section revisits the concept of phase noise, which is briefly discussed in Chapter 8 in the context of its impact on EVM. Here, we treat the subject in a

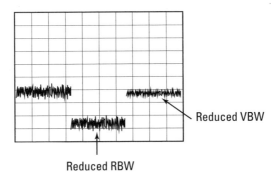

**Figure 10.10**   Effects of RBW and VBW on noise levels in measurements.

slightly more rigorous manner and then describe how phase noise is measured. In addition, the section briefly covers the related topic of frequency stability.

### 10.3.1 Types of Modulation

A single-tone signal, referred to as a carrier in this discussion, can be generally expressed as follows

$$V(t) = A(t)\sin\phi(t) \tag{10.9}$$

where $V(t)$ is a linear quantity, and $A(t)$ and $\phi(t)$ are its amplitude and phase, respectively. The angular frequency of $V(t)$ is related to $\phi(t)$ by

$$\omega(t) = \frac{d\phi}{dt} \tag{10.10}$$

Note that $\phi$ is always in radians in (10.10). For an ideal signal, the amplitude and the changing rate of the phase, are constant; that is,

$$A(t) = \text{constant} = A_0\ ;$$

and

$$\frac{d\phi}{dt} = \text{constant} = \omega_0$$

Hence, for an ideal signal, (10.9) becomes

$$V(t) = A_0 \sin\omega_0 t \tag{10.11}$$

which is the usual sinusoidal expression for an RF carrier. Our concern here is how to mathematically describe, and eventually how to measure, a noise that is added to the carrier given in (10.11). While noise, by nature, is random and generally has a broad spectrum, it is often intuitive and convenient to use a small sinusoidal perturbation with a known amplitude to study the system's response. The result from such an analysis can usually be generalized to noise, which is characterized by its spectral density. We follow this approach in our discussion here.

Recall from Chapter 8 the concept of carrier modulation as a means of generating an information-bearing RF signal. By the same token, a perturbation can be considered a modulation with a small amplitude. There are three types of modulation: amplitude modulation (AM), phase modulation (PM), and frequency modulation (FM). Sections 10.3.1.1 and 10.3.1.2 consider them separately.

### 10.3.1.1 AM

The perturbation term is in on the amplitude and is expressed as $\Delta A \sin\omega_m t$, where $\Delta A$ and $\omega_m$ are much smaller than $A$, and $\omega_0$. $\omega_m$ is the modulating frequency, analogous to the baseband frequency discussed in Chapter 8. Then (10.9) becomes

$$V(t) = \left(A_0 + \Delta A \sin\omega_m t\right) \sin\omega_0 t \qquad (10.12)$$

Using the trigonometry identity, we obtain

$$V(t) = A_0 \sin\omega_0 t - \frac{\Delta A}{2}\cos(\omega_0 + \omega_m)t + \frac{\Delta A}{2}\cos(\omega_0 - \omega_m)t \qquad (10.13)$$

Equation (10.13) indicates that an AM signal has three frequency components: the original carrier at $\omega_0$ and two sidebands at $\omega_0 + \omega_m$ (upper band) and $\omega_0 - \omega_m$ (lower band), respectively.

### 10.3.1.2 PM

For PM, the perturbation is inserted on the phase as

$$\phi(t) = \omega_0 t + \Delta\phi \sin\omega_m t \qquad (10.14)$$

And $V(t)$ in (10.9) is given by:

$$V(t) = A_0 \sin\left(\omega_0 t + \Delta\phi \sin\omega_m t\right) \qquad (10.15)$$

Again, using the trigonometry identity yields

$$V(t) = A_0 \left[\sin\omega_0 t \cos\left(\Delta\phi \sin\omega_m t\right) + \cos\omega_0 t \sin\left(\Delta\phi \sin\omega_m t\right)\right] \qquad (10.16)$$

The two factors in (10.16), $\cos(\Delta\phi\sin\omega_m t)$ and $\sin(\Delta\phi\sin\omega_m t)$, can be expressed in a series of Bessel functions, which are the so-called special functions of mathematical physics [10, 11]. Practical values of $\Delta\phi$ are always very small; that is, $\Delta\phi \ll 1$. Hence, $\Delta\phi\sin\omega_m t \ll 1$. Therefore, usual approximations $\cos x \approx 1$ and $\sin x \approx x$ for $x \ll 1$ can be applied to these factors. We justify this approximation in a moment. Then we have

$$V(t) = A_0 \left[\sin\omega_0 t + \Delta\phi \sin\omega_m t \cos\omega_0 t\right] \qquad (10.17)$$

Thus,

$$V(t) = A_0 \sin\omega_0 t + \frac{A_0 \Delta\phi}{2}\sin(\omega_0 + \omega_m)t - \frac{A_0 \Delta\phi}{2}\sin(\omega_0 - \omega_m)t \quad (10.18)$$

Equation (10.18) indicates that the PM has the same frequency composition as the AM. The two sidebands in (10.18) are in quadrature to those in (10.13) and have +/− signs reversed. From (10.18), the ratio of a sideband to the carrier is $\Delta\phi/2$. For a rather poor specification of −40 dBc, $\Delta\phi = 2 \times 10^{-2}$. Thus, our approximation $\Delta\phi \ll 1$ is almost always valid in practice.

#### 10.3.1.2 FM

In this case, the perturbation is inserted on the frequency as

$$\omega(t) = \omega_0 + \Delta\omega \sin\omega_m t \quad (10.19)$$

The corresponding $\phi(t)$ is

$$\phi(t) = \int_0^t \omega(t)dt = \omega_0 t - \frac{\Delta\omega}{\omega_m}\left[\cos\omega_m t - 1\right] \quad (10.20)$$

Comparing (10.20) with (10.14), we can see that other than an inconsequential constant phase shift and an exchange of cosine and sine functions, the FM is the same as the PM, provided $\Delta\phi = \Delta\omega/\omega_m$. Therefore, PM and FM are indistinguishable from a measurement viewpoint. In comparison, when an AM is mixed with a PM, the effect can be detected. See [11] for details.

### 10.3.2 Phase Noise

Figure 10.11(a) summarizes the modulation process (either AM or PM). Now if we replace the deterministic perturbations in (10.12) and (10.15) with a noise that is characterized by its spectral density, we can conceptually see a situation where the carrier has two continuous sidebands as depicted in Figure 10.11(b). In Figure 10.11(b), the noise spectrum starts from DC and has a $1/f$ frequency dependence up to the corner frequency, as briefly discussed in Chapter 4. The frequency dependence of the sideband is expected to be, in some way, related to the noise spectrum. The plot in Figure 10.11(b) is for illustration only. This noise modulation process is the origin for phase noise. The frequency measured from the carrier frequency, which is usually referred to as offset frequency, is an equivalent parameter of $\omega_m$.

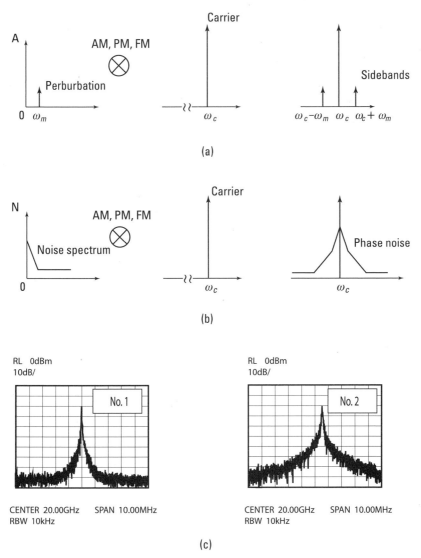

**Figure 10.11** (a) Modulation by a small perturbation signal, (b) modulation by a noise spectrum, and (c) phase noise of two 20-GHz oscillators.

We now briefly consider the mechanism of noise modulation. It is often stated in the literature (see, for example, [1, 12]) that the AM noise of a good oscillator can be neglected. This can perhaps be justified, if we realize that the amplitude of an oscillator is determined by the saturation state of the active device (unity loop gain). In other words, we can think of the amplitude being held steady by a negative feedback mechanism. In contrast, the phase or frequency of the oscillator is free to vary over a relatively large range (unless the resonator's Q factor is extremely high). As a result, the phase noise is generally dominant in terms of noise contribution to the sidebands. In the case of a VCO, the noise modulation is obvious since any noise on control voltage directly translates into frequency noise. For this reason, the noise power in the sidebands is used as a figure of merit for the phase noise of an oscillator.

The exact relationship between the noise in an oscillator circuit and the resulting phase noise is generally complex. For a more elaborate discussion on the topic, see [11, 13, 14] and the references therein. From an application perspective, we are mainly concerned with how the phase noise is measured and specified.

In practice, the phase stability of a signal source is most commonly characterized by the so-called single-sideband (SSB) phase noise measured with a spectrum analyzer. An SSB phase noise is the spectral density at a specified offset frequency normalized by the carrier power. Thus, the unit for phase noise is dBc/Hz. Figure 10.11(c) shows the measurement results of two 20-GHz oscillators that are made of different transistors. We can see that oscillator 2 still has significant noise power at rather far offset frequencies. This explains why in addition to EVM, phase noise is also critical to adjacent channel power and spectrum mask (see Chapter 8) specifications.

Clearly, in the phase noise measurement using a spectrum analyzer, the phase noise of the internal oscillator of the analyzer (LO in Figure 10.7) must be considerably below the noise levels of the signal to be measured. It should also be noted that in such a measurement the displayed noise spectral density may be lower than the actual value by about 2.5 dB [9]. For device comparison purposes, this correction factor is not important as long as all measurements are done using the same instrument. However, if the phase noise specification is required in an absolute term, a spectrum analyzer equipped with the phase noise capability is a better choice because all the correction factors are included in the measurement results.

When the phase noise of the signal is comparable to that of the analyzer, or the power density of the sideband is beyond the dynamic range of the instrument, a different measurement method can be employed. In this method, the carrier is first eliminated by a mixing process; then the sideband noise spectrum can be directly measured at the offset frequency. Details on this technique can be found in [1, 15].

### 10.3.3 Frequency Stability

We have noted that phase noise is provided at a specific offset frequency. In principle, there is no lower limit on the offset frequency. Practically, however, it is difficult to use a spectrum analyzer to measure phase noise at an offset frequency much lower than 1 Hz. Given the relationship between the phase and frequency in (10.10), a slow varying phase fluctuation (which is the physical meaning of a phase noise at low offset frequency) can be characterized by the long-term frequency stability, which can generally be measured without difficulty by a frequency counter [16].

The long-term frequency stability is a critical parameter for a signal source when used as a time standard or an independent clock reference. It is generally not a concern in wireless systems because the clock of a wireless device is always synchronized with a remote reference through a special phase lock sequence.

A well-established approach in characterization of the long-term stability of a signal source is the Allan variance (also known as two-sample variance). Readers can find more in-depth discussion on the Allan variance in [1, 14].

## 10.4 Noise Figure Measurement

There are several techniques for the noise figure measurement. We focus on the Y-factor method, the technique still most commonly used in practice. A different technique known as cold-source method will be briefly described in Section 10.4.3. Spectrum analyzers can also be used as a convenient but much less accurate method for noise figure measurement [2,17].

### 10.4.1 Y-Factor Method

The building blocks for a noise figure measurement using the Y-factor method are a special noise source and a noise receiver, as shown in Figure 10.12(a). Traditionally, this measurement architecture is implemented in a dedicated instrument called a noise figure meter. However, increasingly, spectrum analyzers are equipped with Y-factor–based noise figure capability as one of the measurement options.

Figure 10.12(b) shows details of the function blocks of the noise source and the receiver system. The noise source has two states, corresponding to two output noise powers: (1) the cold state, in which the noise power is simply the thermal noise of the resistor $R_s$ at the operation temperature, denoted as $T_c$; and (2) the hot state, in which the switch is connected to the biased diode.

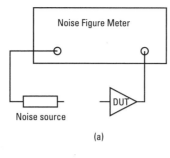

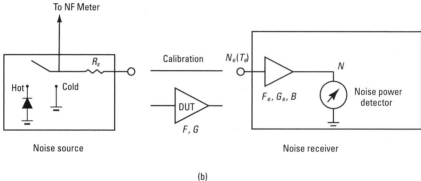

**Figure 10.12** (a) Instrument of the Y-factor based noise figure measurement, and (b) block diagrams of the noise source and the receiver.

The noise power at the source output is the sum of the noise generated by the diode and the thermal noise of the resistor. Let $T_h$ and $T_{\text{diode}}$ be the equivalent temperatures for the output noise at the hot state and the diode noise respectively (see Section 4.5.2); then we have

$$T_h = T_{\text{diode}} + T_c \qquad (10.21)$$

A noise source is specified by its excessive noise ratio (ENR) defined as

$$\text{ENR} = \frac{T_h - T_c}{T_0} \qquad (10.22)$$

where $T_0 = 290$K is the standard temperature for noise figure (Section 4.5.2). Substitution of (10.21) into (10.22) leads to

$$\text{ENR} = \frac{T_{\text{diode}}}{T_0} \qquad (10.23)$$

The diode-based noise source is not the only type of noise source, but it is the most common in practice. The ENR definition in (10.22) is applicable to any noise source, while (10.23) is a special case for the diode-based noise source. We use only (10.23) in our discussion. Commercial noise sources are provided with an ENR table (as the newer ones are usually in an electronic file), calibrated against frequency at the standard temperature $T_0$. Equation (10.23) is significant in that the ENR data is independent of $T_c$; hence the standard ENR data is valid at any operation temperature. This is not true if $T_h$ can be set independently. In that case, ENR varies with $T_c$, as implied in (10.22). This is the reason some different formulas are found for temperature correction in the literature.

The noise receiver unit has a gain-filter block followed by a noise power detector. The bandwidth of the gain block, $B$, should be smaller than that of the DUT. Then, according to $N = kTB$ (see Section 4.2), the bandwidth used in the noise power calculation is determined by $B$.

In the calibration mode, where the noise source is directly connected to the noise receiver, the measured noise powers, denoted as $N_c$ and $N_h$ for the cold and hot states, are

$$N_c = kBG_eT_c + G_eN_e \qquad (10.24)$$

and

$$N_h = kBG_eT_h + G_eN_e \qquad (10.25)$$

where $N_e$ is the equivalent input noise power due to the receiver [see (4.15)], and $G_e$ is the receiver gain. Equations (10.24) and (10.25) can be considered two special points in a general relationship between the equivalent temperature of the noise source $T_s$ and the measured noise power $N$:

$$N = kBG_eT_s + G_eN_e \qquad (10.26)$$

Equation (10.26) can be graphically represented as shown in Figure 10.13(a).

The Y-factor is defined as

$$Y = \frac{N_h}{N_c}$$

Then, by noting that $N_e = kBT_e$, where $T_e$ is the noise temperature of the receiver (Section 4.5), and using (10.25) and (10.26), we derive

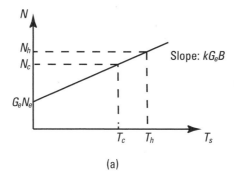

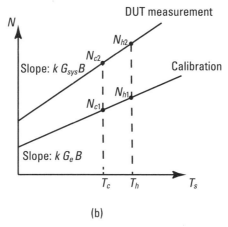

**Figure 10.13** (a) Graphic representation of (10.24) to (10.26), and (b) calculation of the receiver gain using two slopes.

$$Y - 1 = \frac{T_h - T_c}{T_e + T_c} \quad (10.27)$$

Equation (10.27) shows the key advantage of the Y-factor method: It does not include the receiver parameters, $G_e$ and $B$; both are difficult to accurately determine. For the case $T_c = T_0$, it is straightforward to see that the noise figure of the receiver $F_e$ can be calculated from Y and ENR:

$$F_e = \frac{\text{ENR}}{Y - 1} \quad (10.28)$$

Here $F_e = 1 + T_e/T_0$ (4.45) and (10.22) are used.

Now we insert the DUT and repeat the cold-hot measurement on the cascaded system (DUT +receiver). The measurement yields the system noise figure $F_{sys}$. Using (8.6), the DUT's noise figure, $F_d$ is calculated by

$$F_d = F_{sys} - \frac{F_e - 1}{G_d} \tag{10.29}$$

$G_d$ is the DUT gain. It can be determined from the two measurement data, as illustrated in Figure 10.13(b).

The two slopes determined from the measurements "calibration" and "DUT measurement" are:

$$kG_e B = \frac{N_{h1} - N_{c1}}{T_h - T_c}$$

and

$$kG_{sys} B = \frac{N_{h2} - N_{c2}}{T_h - T_c}$$

Since $G_{sys} = G_d G_e$, we obtain

$$G_d = \frac{N_{h2} - N_{c2}}{N_{h1} - N_{c1}} \tag{10.30}$$

Thus, both the gain and noise figure of the DUT are determined after two consecutive cold-hot measurements.

### 10.4.2 Uncertainty and Temperature Correction

There are a number of sources in the measurement setup that may cause uncertainty in the measurement results using (10.29). The most frequently mentioned is the mismatch between the DUT and the noise source. Section 5.1 discusses the possible variation of the mismatch factor due to the unknown phase relationship between the two components. A comprehensive review of this topic in the context of noise figure measurement can be found in [18]. Also, various tools, usually in the spreadsheet format, for noise figure uncertainty calculation are available online or from noise figure equipment manufacturers. A feel for uncertainty in the noise figure measurement is especially important for today's RF engineers, because the noise figure of many low-noise devices are so good that they are almost comparable with the measurement uncertainty.

In addition to uncertainty, there are a few factors that affect the measurement accuracy. Operation temperature is one of them. We briefly consider it as follows.

When $T_c \neq T_0$, strictly speaking, (10.28) is no longer accurate. We will work out the correction factor for this case. From that, we can see that for a change of a few degrees, the correction can be neglected for most practical purposes.

Using (10.22) in (10.28), we have

$$Y - 1 = \frac{T_h - T_c}{T_e + T_c} = \frac{T_0 ENR}{T_e + T_c}$$

Solving for $T_e$,

$$T_e = \frac{T_0 ENR}{Y - 1} - T_c \qquad (10.31)$$

On the other hand, $T_e$ is related to $F_e$ by

$$T_e = T_0 (F_e - 1) \qquad (10.32)$$

From (10.31) and (10.32), we obtain

$$F_e = \frac{ENR}{Y - 1} + \left(\frac{T_0 - T_c}{T_0}\right) \qquad (10.33)$$

Comparing with (10.28), the second term in (10.33) is the correction term.

Again, we emphasize that the ENR in (10.33) is the standard value, independent of $T_c$. In addition, (10.33) is only valid for diode-based (also called solid-state) noise sources.

For a $T_c$ that is 10°C different from $T_0$, the correction is 0.03, which corresponds to about a 0.1-dB correction for a noise figure of 0.5 dB ($F_e$ = 1.12). Most instruments allow the user to input the operation temperature (or detect it automatically) and display the results after the temperature correction.

### 10.4.3 Cold-Source Method

Section 10.4.2 discusses the uncertainty associated with the DUT gain measurement using the Y-factor method. A different noise figure–measurement technique, called the cold-source method, has become more readily available from equipment manufacturers in recent years. The key feature of this technique is that in the DUT measurement, only the cold noise source is used for the noise power measurement. The DUT gain is determined from a full S-parameter measurement. As a result, any mismatch factors can be removed. The cold-source method is usually offered as an option of a VNA, as it requires an S-parameter measurement. An instrument equipped with the cold-source technique is generally specified with higher accuracy than those using the Y-method but also costs more. More detailed information on the cold-source technique can be found in [2] and in the application notes of the equipment manufacturers (for example, [19]).

## 10.5 Power and Nonlinearity Measurements

### 10.5.1 Power Measurements

In typical RF product development, RF power measurements are relatively straightforward. Instruments commonly used are the spectrum analyzer and power meter. The latter is generally more accurate. Complications in power measurement are usually associated with the time characteristic of the signal to be measured. Pulsed signals and certain types of modulated signals are two common examples. For certain applications, accuracy can also be of critical importance. For a more elaborate discussion, readers can refer to [1, 2]. This section describes a measurement setup for power amplifier characterization. It is particularly suitable for automatic test equipment (ATE) applications.

Figure 10.14 illustrates the basic idea of this power measurement system. Actual implementation may vary somewhat. In this setup, the input and output powers of a DUT are recorded at the same time as the signal generator sweeps (usually at a discrete step) through a power range. This scheme completely eliminates any measurement uncertainty associated with the repeatability problem of the signal generator and the driver amplifier. The measurement accuracy is essentially limited by the power meters.

The directional coupler, discussed in Chapter 9, is the main component in this measurement scheme. The input power is recorded by the power meter 1 at the coupling port 1. To convert a measurement of the power meter 1 to the input power, a calibration is first performed by directly connecting the power meter 2 to the isolator 2 (the point where the input power is). The calibration data can be either a fixed offset value or a calibration table.

After the calibration, the DUT, along with the attenuator 1 (most power meters only allow a maximum of 20 dBm), is inserted. The attenuation of a high-quality attenuator can generally be considered a constant over the power. Then the recorded data at the power meters 1 and 2 are, after corrections, the

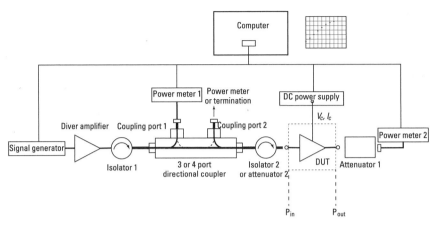

**Figure 10.14** Power measurement setup for power amplifier characterization.

input and output powers, respectively. In the same power sweep, the supply current and voltage may be recorded as well for the calculation of the power added efficiency (see Chapter 7).

This setup also allows for the measurement of the large-signal input return loss. In that case, the isolator 2 should be replaced by the attenuator 2, and the coupling port 2 is connected to a power meter. This is also a case that illustrates why the directivity of the coupler is important in such an application. Consider a simplified situation without the attenuator 2. Assume that the directivity is 10 dB, and the input return loss of the DUT is 5 dB. Then using (9.5), the power from the reflected power at the coupling port 1 is 15 dB (10+5) below the expected power level. This corresponds to a potential 0.13-dB error, depending on the phase relationship of the two power waves. In certain applications, this level of error is unacceptable.

In PCB-based applications, sometimes a power detector is required for the monitor or control purposes. A common device for this function is a diode-based IC, which integrates the required filtering and bias circuits. In certain special cases, mostly for reasons of frequency range or time-response characteristics, a discrete power detection circuit may be needed. Readers can find detailed design equations for this circuit in [4, 20].

### 10.5.2 Nonlinearity Measurements

The measurement setup for a nonlinearity parameter usually can be conceived based on its definition. In practice, however, some extreme conditions may require special arrangements in the measurement setups.

### 10.5.2.1  $P_{1dB}$

$P_{1dB}$ can be determined from a plot of $P_{out}$ versus $P_{in}$. (See the insert in Figure 10.14.) In addition, an $S_{21}$ measurement in the power sweep mode of a VNA can be used to measure the $P_{1dB}$. This method may be convenient in some cases but is generally less accurate than the measurement by a power meter.

For very high-power measurements, self-heating, which usually reduces the DUT's gain, may affect the measurement accuracy. Should that happen, a pulse measurement can be employed to shorten the measurement time.

### 10.5.2.2  Intermodulation

Intermodulation is usually specified by a two-tone measurement that involves two input signals and a measurement at a different frequency resulted from mixing of the input signals. Figure 10.15 shows a typical setup. The attenuator before the DUT is for reducing any mismatch effects.

In the IP3 measurement, the IP3 point is determined by an extrapolation from the IM3 versus $P_{in}$ plot, as explained in Chapter 6. In practice, this extrapolation is often done manually, which inevitably leads to a repeatability problem. As a result, the IP3 data is generally for reference only.

The input power range for the IP3 measurement should be selected properly. It needs to be sufficiently low so that no higher-order effects beyond third-order effects significantly impact the results, but it should not be so low as to cause difficulty in the measurement due to an extraordinarily large dynamic range. The trend of modern wireless systems in the intermodulation specification is to use a set of predefined power and frequency conditions so that there is no data processing involved.

For high-power measurements and high-dynamic measurements (where the intermodulation level is much lower than the carrier), the measurement

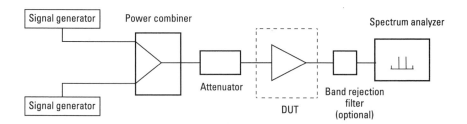

**Figure 10.15**  Measurement setup for a two-tone intermodulation.

setup must be verified first, without the DUT, to ensure that the baseline is below the target specifications (see Section 9.2.5).

Finally, in a case where the intermodulation frequency is separated far enough from the input signals, a rejection filter, along with an attenuator (to ensure an adequate termination at the DUT output) may be used to ease the requirement on the dynamic range of the analyzer, as illustrated in Figure 10.15. This technique can be used for harmonic measurements as well. A careful calibration must be performed in such a setup.

# References

[1] Bryant, G. H., *Principles of Microwave Measurements*, IEE, 1993.

[2] *Modern RF and Microwave Measurement Techniques*, V. Teppati, A. Ferrer, and M. Sayed (eds.), Cambridge University Press, 2013.

[3] Gonzalez, G., *Microwave Transistor Amplifiers Analysis and Design*, Second Edition, Prentice Hall, 1997.

[4] Pozar, D. M., *Microwave Engineering*, Second Edition, John Wiley and Sons, 1998.

[5] Engen, G. F., and C. A. Hoer, "Thru-Reflect-Line: An Improved Technique for Calibrating the Dual Six-Port Automatic Network Analyzer," *IEEE Trans. Microwave Theory and Techniques*, Vol. 27, No. 12, 1979.

[6] Lathi, B. P., and Z. Ding, *Modern Digital and Analog Communication Systems*, Fourth Edition, Oxford University Press, 2009.

[7] Haykin, S., *Communication Systems*, Third Edition, John Wiley and Sons, 1994 (and later editions).

[8] Da Silva, M., "Real-Time Spectrum Analysis and Time-Correlated Measurements Applied to Nonlinear System Characterization," in *Modern RF and Microwave Measurement Techniques*, V. Teppati, A. Ferrero, and M. Sayed (eds.), Cambridge University Press, 2013.

[9] Spectrum Analysis Basics, HP Application Note 150, 1989.

[10] Mathews, J., and R. L. Walker, *Mathematical Methods of Physics*, Addison-Wesley, 1970.

[11] Robins, W. P., *Phase Noise in Signal Sources*, IEE, 1984.

[12] Rhea, R. W., *Oscillator Design and Computer Simulation*, Second Edition, Noble Publishing Corporation, 1995.

[13] Lee, T. H., and A. Hajimiri, "Oscillator Phase Noise: A Tutorial," *IEEE J. Solid-State Circuits*, Vol. 35, No. 3, 2000.

[14] Rubiola, E., *Phase Noise and Frequency Stability in Oscillators*, Cambridge University Press, 2009.

[15] Understanding and Measuring Phase Noise, Hewlett-Packard Application Note 207, 1976.

[16] Fundamentals of the Electronic Counters, Hewlett-Packard Application Note 200, 1997.

[17] Fundamentals of RF and Microwave Noise Figure Measurements, Agilent Application Note 57-1, 2000.

[18] Noise Figure Measurement Accuracy: The Y-Factor Method, Keysight Technologies Application Note, 2020.

[19] High-Accuracy Noise Figure Measurements Using the PNA-X Series Network Analyzer, Keysight Technologies Application Note, 2014.

[20] Irvine, "Detectors and Mixers," in *Microwave Solid State Circuit Design*, I. Bahl and P. Bhartia (eds.), John Wiley and Sons, 1988.

# 11

# RF Switches

RF switches are increasingly utilized in today's wireless products to support multiprotocol and multiband operations. While many wireless transceiver ICs have internal RF switches, discrete RF switches still offer the most flexibility in terms of system configuration and performance selection. As a result, they are widely used in various RF systems. This chapter discusses the applications of these discrete RF switches in circuit and system designs.

For the completeness of coverage, we first mention that in general there are two major groups of RF switches: (1) electromechanical switches including micro-electromechanical system (MEMS) devices and (2) solid-state (or semiconductor) switches. The switching process in the first group of devices is a mechanical movement, as the name indicates, whereas the devices in the second group utilize two distinctly different electrical states to achieve the desired switching function. The devices we consider in this chapter all belong to the second group. Among them, there are three main categories for discrete RF switches: Si and compound semiconductors (mainly GaAs and GaN) FET-based devices (almost always as a MMIC component), and PIN diode switches.

Sections 11.1–11.3 outline the basics of these solid state RF switches to illustrate the underlying operation principles for some of the RF switch characteristics commonly considered in applications. The depth of these discussions is generally not sufficient for actual MMIC designs. Much more detailed information on RF switch IC designs and a review of various semiconductor technologies for RF switches can be found in [1–3] and information on compound semiconductor device physics in [1, 4, 5].

## 11.1 Realization of Switching Functions Using Two-Impedance Devices

We use a single-pole double-throw (SPDT, or SP2T) switch for illustration purposes in our discussion. Figure 11.1(a) shows the switch symbol. The commonality of the devices considered in this section is that they have two distinctively different impedance states, high and low, that are controllable by a set of external logics. Figure 11.1(b) illustrates how these two states are utilized to realize the switching function. In Figure 11.1(b), all three ports of the switch are explicitly terminated at $Z_0 = 50\Omega$, which is the condition an RF switch is usually specified. The path through the low impedance (labeled $Z_{on}$) is the "ON" path, and the other path through the high impedance $Z_{off}$ is the "OFF" path. $Z_{on}(Z_{off})$ must be at least an order of magnitude smaller (larger) than $50\Omega$. The loss of the ON path is the insertion loss (IL) defined in Section 5.1, and the loss through the OFF is the isolation (ISO). Both can be calculated by the same formula (with the derivation left to the reader):

$$\text{IL(ISO)} = -20\log\left|\frac{2Z_0}{2Z_0 + Z_{on}(Z_{off})}\right| \quad (11.1)$$

Figure 11.2 shows the plots of the function in (11.1) for $Z_0 = 50\Omega$, from which we can estimate IL = 0.4 dB for $Z_{on} = 5\Omega$, and ISO = 15 dB for $Z_{off} = 500\Omega$. In Figure 11.2, we assume that both $Z_{on}$ and $Z_{off}$ are a pure resistance, which is generally true for $Z_{on}$. $Z_{off}$ is usually a capacitive impedance, as seen in Sections 11.2 and 11.3. Since we are only interested in the condition $|Z_{off}| \gg Z_0$, the plot in Figure 11.2 is still a reasonable approximation for the isolation. The circuit configuration in Figure 11.1(b) also implies that $Z_{off}$ affects not only the isolation but may also increase the insertion loss when it is too small. This is because a portion of the source power would be diverged to the OFF path. For instance, for $|Z_{off}| = 500\Omega$, the insertion loss of the ON path is still 0.1 dB even if $Z_{on} = 0$.

Compared with a mechanical switch, a major disadvantage of the solid state switches is the limited isolation due to the finite value of $Z_{off}$, especially at higher frequencies. To improve the isolation of a switch, a shunt $Z_{on}$ can be added to the OFF path as shown in Figure 11.1(c). For $Z_{on} \ll Z_0$, the voltage

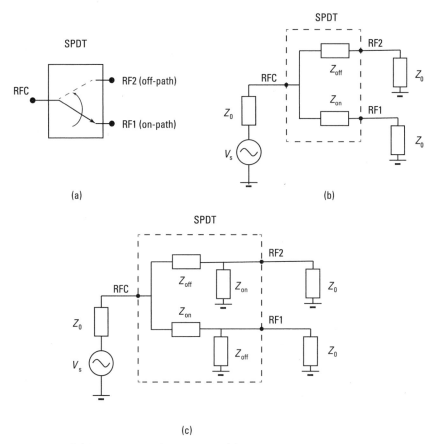

**Figure 11.1** (a) Symbol for the SPDT switch, (b) circuit realization using two-impedance devices, and (c) SPDT switch consisting of four devices.

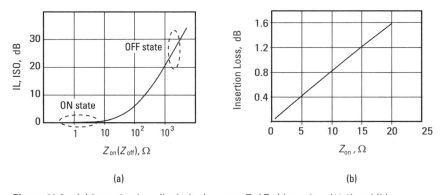

**Figure 11.2** (a) Insertion loss (isolation) versus $Z_{on}(Z_{off})$ based on (11.1) and (b) expanded view of $Z_{on}$ region

on the load $Z_0$ is determined by $Z_{on}$; hence the isolation is improved by a factor of $(Z_0/Z_{on})^2$. For the ON path, a shunt device in OFF state does not significantly affect the insertion loss as long as $|Z_{off}| \gg Z_0$.

The circuit in Figure 11.1(c) is the basic architecture for the RF switches discussed in this chapter. Sections 11.2 and 11.3 consider how the switch circuit is implemented using two types of switching devices, namely, PIN diode and FETs.

## 11.2 PIN Diode RF Switches

The name of PIN diode is derived from its device structure (a P region, an intrinsic region, and an N region) as shown in Figure 11.3(a). The semiconductor used for a PIN diode can be either GaAs(GaN) or Si. The operation of a PIN diode is quite straightforward. When forward-biased, the intrinsic region becomes conductive, and the equivalent resistance of the diode decreases as the forward current increases. At reverse bias or no bias, the diode functions like a capacitor with the intrinsic region as the insulating layer. By controlling the bias voltage we can achieve two impedance states required for the switching function, as illustrated in Figure 11.3(b). The component values in Figure 11.3(b) can vary by an order of magnitude in either direction in actual devices.

Figure 11.3(c) shows the basic switch circuit [Figure 11.1(b)] with a PIN diode implementation. This circuit clearly indicates that $Z_{off}$ becomes smaller as the frequency increases. In the literature, a figure of merit for the frequency characteristic of a switching device, known as the cut-off frequency $f_c$, is defined as the frequency when $Z_{off} = Z_{on}$; that is,

$$f_c = \frac{1}{2\pi R_F C_R} \qquad (11.2)$$

For $R_F = 2\Omega$ and $C_R = 0.1$ pF, $f_c$ is about 800 GHz. In some cases, the factor $R_{on}C_{off}$ is directly used for the same purpose. Figure 11.4 shows the S-parameter simulation results for the two-switch configurations with $R_F = 2\Omega$ and $C_R = 0.1$ pF. The labeling in Figure 11.4(b) follows the convention often seen in the manufacturers' datasheets, even though, strictly speaking, the insertion loss and isolation should be positive numbers in decibels. Figure 11.4(b) confirms the amount of improvement in the isolation by the series-shunt configuration, which is about 28 dB in this case. In actual devices, the isolation deteriorates at considerably faster rate with frequency due to other effects that are not included in the highly simplified equivalent circuit model.

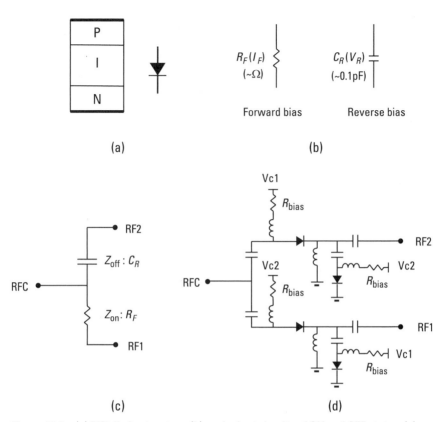

**Figure 11.3** (a) PIN diode structure, (b) equivalent circuits of ON and OFF states, (c) a switch using two PIN diodes, and (d) a complete switch circuit with series-shunt configuration.

Figure 11.3(d) shows a complete switch circuit using four PIN diodes. We observe that multiple inductors and DC-blocking capacitors are used to provide adequate isolation between the RF and DC paths. The need for inductors limits the use of a PIN diode at the low-frequency end in a practical application. The resistance $R_{bias}$ is selected according to the required bias current for the ON state. In general, PIN diode–based RF switches are made from discrete components, although in some cases the entire switch circuit may be built inside a module, which appears as an SMT IC. This circuit architecture is an advantage of the PIN diode switch's flexibility. There is a wide range of PIN diodes available on the market. On top of that, the user can improve, to a certain extent, some specifications such as the switching speed

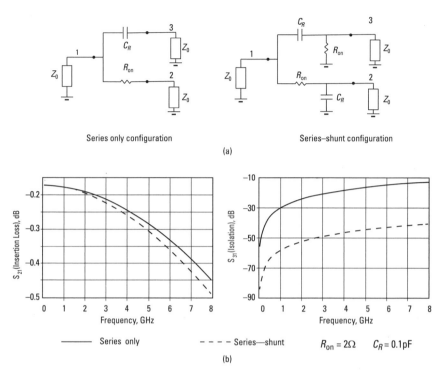

**Figure 11.4** S-parameter simulation of insertion loss and isolation: (a) circuit setups and (b) simulation results.

and the insertion loss by adjusting the bias conditions. On the other hand, this circuit architecture is also the reason for the PIN diode switch not being used as widely as its FET counterparts; it is difficult to design, requires more PCB space, and has a higher BOM cost.

For a discussion on the other device characteristics of the PIN diode switch, such as nonlinearity and switching speed, see [1, 2]. For practical design of PIN diode RF switches, application notes by the specific PIN diode manufacturers are usually good references.

## 11.3 FET-Based RF Switches

FET-based RF switches do not require any inductor for RF/DC isolation, and hence they are usually implemented in an integrated circuit.

Recall that there are two major types of FET-based RF switches, Si and GaAs. The silicon-on-insulator (SOI) technology currently offers the best performance for discrete Si RF switches, while the pHEMT is the dominant technology for GaAs and GaN devices. The operation principle of the RF performance is essentially the same for the two types of devices. The main difference is the DC bias configuration. Section 11.4 provides a brief comparison between the Si and GaAs(GaN) RF switches in practical component selection. Sections 11.3.1 and 11.3.2 focus on RF characteristics and some DC bias–related issues, respectively.

### 11.3.1 RF Performance of FET Switches

Figure 11.5(a) depicts how an FET is used in a switch circuit. Unlike the PIN diode, the FET is a three-terminal device. As shown in Figure 11.5(a), the drain and source are used for the signal paths, and the gate voltage sets two impedance states for the switching function. This three-terminal scheme separates

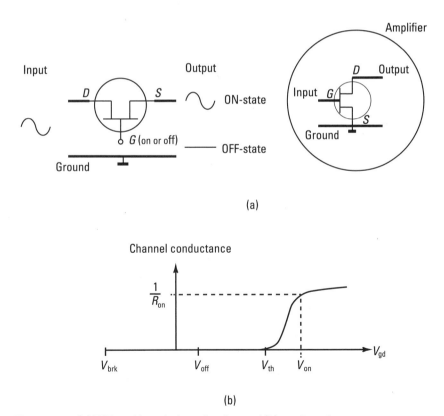

**Figure 11.5** (a) FET used in switch application, and (b) two impedance states controlled by $V_{gd}$.

the control path from the signal path, eliminating the need for an RF choke. For comparison, Figure 11.5(a) (in the insert) also shows the configuration of an FET used in RF amplifiers.

The operation of the impedance switching can be understood if we recall that the drain-source channel becomes conductive once the gate-source voltage $V_{gs}$ is above the threshold voltage $V_{th}$ (see Figure 6.2). However the transfer function, $I_{ds}$ versus $V_{gs}$, is not suitable for our discussion here, because $I_{ds}$ in the transfer function is specified with the drain-source voltage $V_{ds}$ being in the active region (Figure 6.2). For the switch applications, the ohmic region is where the FET operates. Conceptually we can use a parameter channel conductance to replace $I_{ds}$ to represent the two impedance states as shown in Figure 11.5(b). As mentioned in Chapter 6, the source of an FET for amplifier applications is always the terminal connected to ground, which makes device and lead structures of the source drastically different from those of the drain. In contrast, the drain and source of a switch FET is generally symmetric with respect to the gate. Figure 11.5(b) uses $V_{gd}$, instead of $V_{gs}$, as the control voltage, simply because the drain is usually taken as the signal input side in RF switch applications. In linear (or small signal) operation the drain-source voltage ($= I_{ds}R_{on}$) is small enough for $V_{gd}$ and $V_{gs}$ to be considered virtually equal in terms of channel condition. While the detail of the curve of Figure 11.5(b) is not important to our discussion, we can make two observations: (1) $R_{on}$ in Figure 11.5(b) is the same as that in Figure 6.2, and (2) $V_{gd}$ must be below $V_{th}$ in order to keep the transistor in the OFF state. The second point is critical to understanding the power-handling capability discussed next. Note we did not specify whether $V_{th}$ is positive or negative in Figure 11.5(b), a topic that will be considered in Section 11.3.2.

The equivalent circuit shown in Figure 11.6(a) can be used for both impedance states. The channel resistance for the ON state $R_{on}$ is much smaller than the impedances of $C_{gs}$ and $C_{gd}$, whereas $R_{off}$ for the OFF state is much larger than the capacitive impedances. Hence we can simply use $R_{on}$ for the ON state and $C_{gs}/2$ for the OFF state (assuming $C_{gs} = C_{gd}$). $R_{bias}$ in Figure 11.6(a) is used to set the DC bias point (note that the ground is AC ground) and should be large enough to be considered floating (or of no effect) to the RF signals in the designed operation frequency range. Then the analysis of the insertion loss and isolation for the PIN diode can be applied to the FET-based switches as well, with $R_F$ replaced with $R_{on}$ and $C_R$ with $C_{gs}/2$. More complex equivalent circuits with other circuit elements, along with explanations of their physical origins in the device structure, can be found in [1, 2].

Next we consider the power-handling capability of an FET switch. Among the four possible device conditions in Figure 11.1(c), the shunt FET in

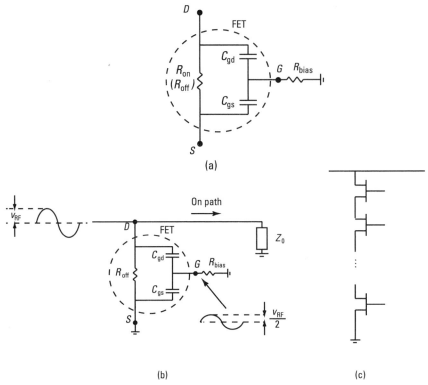

**Figure 11.6** (a) Equivalent circuit of FET in RF switch applications, (b) power-handling capability, and (c) stack of *n* FETs in shunt configuration.

the OFF state normally has the most impact on the power-handling capability of a switch. We analyze this case in the following. The RF power has effects on the FET operation condition in the other three device configurations as well. For a complete consideration, the reader is referred to [6].

We use Figure 11.6(b) to facilitate our discussion. Let $v_{RF}$ be the amplitude of RF voltage at the drain (with reference to ground), as shown in Figure 11.6(b). Then $v_{RF}$ is also the RF voltage across the drain and source. Since the gate is the midpoint between the two terminals, the RF voltage at the gate is $v_{RF}/2$. Recall that the DC bias point of $V_{gd}$ is set at $V_{off}$ [see Figure 11.5(b)], the total voltage across the gate-drain junction is $V_{off} - v_{RF}/2$. Then it is clear that when $v_{RF}$ reaches the point:

$$v_{RF\_max} = 2(V_{th} - V_{off}) \qquad (11.3)$$

the transistor is no longer in the OFF state. Section 11.3.2 explains that $V_{\text{off}}$ is always slightly lower than the supply voltage (except for the case when an internal voltage converter is employed). Then the corresponding RF power at $v_{\text{RF\_max}}$ can be conveniently expressed as

$$P_{\text{max}} = \frac{v_{\text{RF\_max}}^2}{2R_0} = \frac{2(V_c - V_{\text{th}})^2}{R_0} \tag{11.4}$$

where $V_c$ is the supply voltage. A quick estimate shows that for a 1-W switch with a 50-$\Omega$ load, $V_c$ needs to be more than 5V. A common technique to boost the power capability is to stack multiple devices, as depicted in Figure 11.6(c). For a stack of $n$ FETs, each device only needs to support $1/n$ of $v_{\text{RF}}$. Thus the maximum power in this case is given by

$$P_{\text{max}} = \frac{2\left[n(V_c - V_{\text{th}})\right]^2}{R_0} \tag{11.5}$$

Equation (11.4) clearly indicates the benefit of increasing the supply voltage in terms of power-handling capability. However, the breakdown voltage of an FET in the OFF state ($V_{\text{brk}}$ in Figure 11.5) sets the maximum supply voltage.

So far in our analysis, we have assumed that $R_{\text{bias}}$ has no effect on the RF signal. As the frequency decreases, the impedance of $C_{\text{gd}}$ (and $C_{\text{gs}}$) becomes larger. At a certain point, $R_{\text{bias}}$ starts to shunt the RF path, resulting in a reduced RF amplitude ($<V_{\text{RF}}/2$) at the gate. At this point, the maximum RF power starts to decrease.

The power-handling capability is usually manifested as an increased insertion loss once the input power reaches a certain point. Similar to the gain compression in RF power amplifiers, the input power where the insertion loss of a switch is increased by 1dB is defined as the $P_{\text{1dB}}$ compression point of the switch. Unlike RF power amplifiers, the $P_{\text{1dB}}$ point of an RF switch is usually very close to or even exceeds the maximum power rating. As a result, the 0.1-dB compression point $P_{\text{0.1dB}}$ is also used as a power specification.

The discussion of RF power amplifiers in Chapter 6 shows that some nonlinearity effects, such as third-order intermodulation and EVM, affect the circuit performance, at an input power level significantly lower than the onset of the gain compression. A similar conclusion applies to the FET switches. Analogous to the nonlinear relationship between the input and output variables of an amplifier [see (6.8)], the nonlinear effects of an RF switch at input powers considerably lower than the compression point are generally associated with the $I_{ds}$ versus $V_{ds}$ characteristics of the ON state (it may not be the only contributing factor). This can be seen by examining the simple circuit in Figure 11.7(a): If $R_{on}$ is a constant, the circuit is perfectly linear; otherwise it is not. Since $R_{on} = V_{ds}/I_{ds}$, the linearity of the switch circuit is determined by the relationship between $I_{ds}$ and $V_{ds}$.

There are a variety of equivalent circuits for FETs designed for RF switch applications. We use the following formula [1] to illustrate the concept:

$$I_{ds} = \beta\left(V_{gs} - V_{th}\right)^2 \left(1 - \lambda V_{ds}\right)\tanh\left(\alpha V_{ds}\right) \qquad (11.6)$$

where $\alpha$, $\beta$, and $\lambda$ are device parameters. According to [1], $\lambda \sim 0.01$ and $\alpha \sim 2$. For the ON state, $V_{ds}$ is less than 1V in practical cases; thus $1 - \lambda V_{ds} \approx 1$. Then the $V_{ds}$ dependence of $I_{ds}$ is all in the factor of $\tanh(\alpha V_{ds})$. We note that the linear term of the Taylor expansion of $\tanh(x)$ is $x$, which implies that at a small $V_{ds}$, $I_{ds} \propto V_{ds}$, leading to a constant $R_{on}$. Recall that the gain

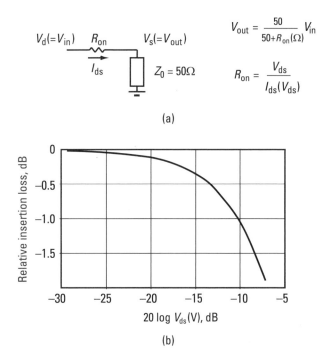

**Figure 11.7** (a) Equivalent circuit for ON state in linearity analysis and (b) relative insertion loss versus $V_{ds}$, $\alpha = 2$ [see (11.7)].

compression of an amplifier is specified by the amount of deviation from the linear gain, the insertion loss compression can be examined by the relative insertion loss defined as:

$$\text{relative insertion loss} = 20\log\left[\tanh(\alpha V_{ds})\right] - 20\log(\alpha V_{ds}) \quad (11.7)$$

The plot of relative insertion loss versus $20\log(V_{ds})$ is shown in Figure 11.7 for $\alpha = 2$. While the compression curve of an actual device includes other factors (including the power handling capability discussed in this section), the main purpose of (11.6) in our discussion is to demonstrate the origin of the nonlinearity of an ON-state FET in the switch application. We note that the nonlinear relationship between the input and output variables in this case (the tanh function) only has odd terms in the Taylor expansion. This implies that any nonlinear effects related to the third-order intermodulation (see Section 6.3) must be evaluated carefully.

### 11.3.2 DC Bias Networks for FET Switches

Section 5.1 briefly discusses the difference in the gate-source bias requirement between the enhancement and depletion modes FETs for amplifier applications. For the switch application, the same conclusion holds. That is, the threshold voltage of the Si MOSFET is positive, and the device is in the OFF at state zero bias, whereas the threshold voltage of the GaAS (GaN) pHEMT is negative, and the OFF state requires a negative gate-drain voltage. The scheme in Figure 11.8(a) is used to illustrate how the DC bias network is set up for a series-shunt FET switch.

The DC bias scheme of a Si MOSFET is quite straightforward because the device is in the OFF state at $V_{gd} = 0$, and the gate-drain(source) junction is essentially insulating. Consider the schematic, if V1 = H (the supply voltage) and V2 = L (ground), the transistors $Q_1$ and $Q_3$ are in the ON state, while $Q_2$ and $Q_4$ are in the OFF state. The RF ports, which are connected to the drain or the source, are always at ground potential. This is why for the Si MOSFET switch there is no need for DC-blocking capacitors as long as there is no external voltage present on the RF ports. Another convenience of the Si switch, from the user's perspective, is the single control logic. (Most GaAs RF switches require two complementary control voltages.) This is because an inverter circuit (shown in Figure 11.8) can be easily integrated into a Si switch IC. Similarly, almost all Si RF switches have integrated ESD protection circuits. As a result, the Si RF switch generally has much higher ESD rating than the GaAs RF switch.

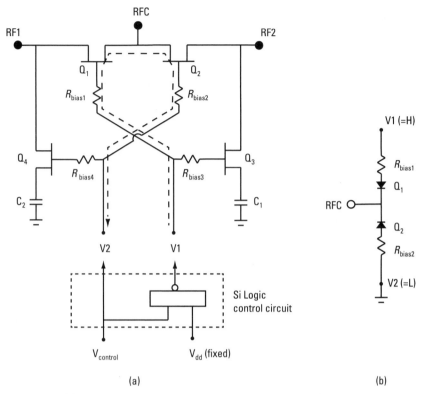

**Figure 11.8** (a) DC bias network for series-shunt FET switch, and (b) DC equivalent bias circuit.

For GaAs RF switches, the DC bias network is considerably more complex because of the negative threshold voltage. To see how the DC voltages are established in this case, we can model the gate-drain and gate-source junctions as a diode with the gate serving as an anode [1]. For the case of V1 = H and V2 = L, the DC bias conditions of each transistor can be analyzed by following the current paths from V1 to V2. Figure 11.8(b) shows an equivalent DC circuit for a path indicated by the dashed line in Figure 11.8(a). The voltage drop across $R_{bias}$ and the forward-biased diode is typically a fraction of a volt. Therefore, the DC voltage at the RFC port is slightly below the logic high voltage. The same voltage potential (at the RFC) provides the required negative gate-drain voltage to turn $Q_2$ off. It is easy to see the voltage potentials on all three RF ports are essentially the same regardless of the control logic. Therefore, unlike the depletion mode GaAs FET in amplifier applications, the GaAs FET switch does not need an external negative voltage supply. However, it does require DC blocking capacitors on all RF ports because of the presence of the positive DC voltages. The discussion on the selection of DC blocking capacitors in Section 9.1 is completely applicable to the switch circuit.

From Figure 11.9(a), we can also see that for GaAs switches, the two undefined logics, H, H and L,L are equivalent because in neither case are the transistors biased. The RF characteristics in these conditions are not specified and are generally not recommended for circuit applications due to the potential inconsistency in production.

Section 11.3.2.1 considers a special circuit condition that is related to the DC blocking capacitors for an RF switch.

### 11.3.2.1  An Unsettling Condition Due to Transient External Voltage

From the above analysis, we can see that all transistors in a switch need to be properly DC-biased in order to perform the RF switching function properly. In a wireless system, multiple DC supplies are often present to support various functions. It is not uncommon that some of the DC supplies need to be regularly turned on and off, based on the system requirement. If one of these DC supplies is directly connected to the DC-blocking capacitor for the switch, special care needs to be taken; this is explained as follows.

Figure 11.9(a) shows an example, where the DC supply, $V_{PA}$, for the power amplifier (PA) is connected to the DC blocking capacitor $C_{blk}$. In the operation, $V_{PA}$ is turned on first, and subsequently the data transmission starts after a specific delay. The sudden change in the DC voltage on the PA side of $C_{blk}$ induces a transient response at the RFC port of the switch. This process can be modeled by a transient response at the switch port to a step function on the PA side as illustrated in Figure 11.9(b). During this transient process, the switch is not ready to function. If the settling time [$T_{set}$ in Figure 11.9(b)] is longer than the delay between the turn-on of the PA and the data transmission, a severe degradation may occur in the link quality for a packet-based communication system.

The settling time in this case is essentially a capacitance discharge process with an RC time constant. By examining Figure 11.8(b), we realize that the discharge paths from the RFC port to the ground are through reverse-biased diodes. Therefore, the equivalent resistance in the RC discharge circuit can be quite large and hence cause a long delay. This effect is usually more prominent in switches with very low control currents because of the large equivalent resistance. In most practical cases, the transient process is not long enough to cause any noticeable effect. Nevertheless, if a similar condition as that seen in Figure 11.9(a) is expected and the switching time is critical, the designer should be aware of this phenomena. A more detailed study on this issue is described in [7].

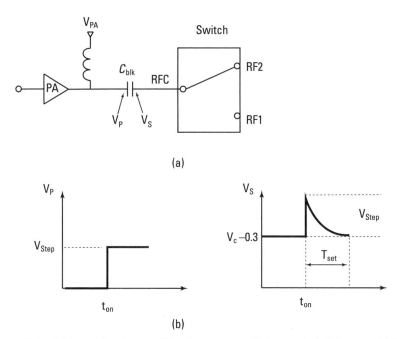

**Figure 11.9** (a) An application condition for a potentially long switch delay, and (b) the step response of the voltage at the RFC port.

## 11.4 RF Switch Selection in Applications

So far in this chapter we have considered three major types of RF switches in today's market. It is difficult to provide an accurate performance metric for these different switches, especially in a quantitative manner, because of the constantly changing nature of the technologies and market demands. Nevertheless, in the following we outline some considerations for RF switch selection in practice.

We start with a summary of key specifications to be considered in the switch selection:

- *Linear specifications:* frequency range, insertion loss, isolation, and return loss;
- *Nonlinearity (or power) specifications:* $P_{1dB}$ and/or $P_{0.1dB}$, IP3, harmonics, and EVM data;
- *DC control–related specifications:* Number of control voltages, switching speed, and control current.

In general, among the device technologies considered, the PIN diode switch offers a unique performance capability. However, because of the requirement of inductors for isolation, the PIN diode switch cannot be conveniently implemented in an IC, and it is limited in its operational frequency range. In practice, a PIN diode switch is usually selected only if no IC solutions are capable of the required performance specifications. Exceedingly higher power and faster switching speed are two common reasons for using the PIN diode. Additionally, it should be noted that the PIN diode is only one of the elements in a switch circuit. The bias circuit and condition also play major roles in determination of switch performance.

In comparison, FET-based RF switches are almost always offered as IC components. As such, the switch specifications are provided by the manufacturers. The remaining discussion focuses on RF switch ICs.

In the RF switch selection for a practical application, the effect of the switch on the system specification is generally well understood for the insertion loss, but it is not always obvious for the isolation. In actuality, it is rare that a system would fail to function due to a mediocre insertion loss. In contrast, however, insufficient isolation can render the system inoperable or cause it to fail the emission requirements. Take an antenna Tx/Rx switch as an example. For many wireless systems, the transmit power can be 1W or higher. If the switch has an isolation of 20 dB, which is generally considered adequate, the power coupled to the receiving path is 10 mW. This amount of RF power may be high enough to activate the front-end LNA, which is otherwise unpowered. (Recall that Chapter 7 shows that an RF signal can induce a DC bias.) This activated amplifier may radiate RF power at harmonic frequencies, causing a failure in spectrum compliance, or induce interferences to other functions of the system. In general, the isolation requirement of an RF switch in a specific application is not always easy to determine from the system specifications. There can be a number of factors that need to be considered. In some cases the isolation requirement has to be fully evaluated in the actual circuit environment.

Regarding the linearity specifications, a compression point, either by $P_{1dB}$ or $P_{0.1dB}$, is convenient for device comparison. However, in most practical cases, an RF switch operates at a power range that is significantly backed off from the compression point. This is certainly true when the switch is directly placed at the antenna port, because, as discussed in Section 11.3, the harmonic levels at the compression point are generally too high to pass the emission regulation. When the switch is used in the transmission path, another critical linearity specification is the signal integrity, usually specified by the EVM specifications.

A practical rule is that the switch should not degrade the EVM specifications of the RF power amplifier that drives the switch. This is because the switch is a relatively inexpensive component and should not be the limiting component in terms of the linearity specifications of the system. Generally, for the application of the antenna switch in a wireless system, the two most relevant nonlinearity specifications are the harmonics and EVM data. In the absence of these specs on the datasheet, the onset power for EVM degradation can be estimated to be in a range of 5–10 dB below the $P_{0.1dB}$ point, depending on the absolute EVM levels (e.g., it usually takes an increase of a few decibels in input power to degrade EVM from 0.5% to 2%). On the other hand, the harmonic performance is less consistent in relating to the compression point. Traditionally, GaAs technology is considered superior in linearity performance. Si devices, however, have been vastly improved in recent years.

In the case when it is desirable to use a switch below the lower frequency limit specified on the datasheet, the general rule is that the insertion loss and isolation can only be better but the power performance will be degraded.

Finally, we consider DC-related performance merits. Si devices have several advantages over their GaAs counterparts, including simplicity in DC bias, such as single-supply voltage, no need for DC blocking capacitors, and remarkably higher ESD ratings. GaAs devices, however, still hold the edge in switching speed, particularly when power specifications are included in the consideration. Again, the best practice in component selection is to survey the market to determine the best options for each specific application.

# References

[1] Bahl, I. J., *Control Components Using Si, GaAs, and GaN Technologies*, Norwood, MA: Artech House, 2014.

[2] Carverly, R., *Microwave and RF Semiconductor Control Device Modeling*, Norwood, MA: Artech House, 2016.

[3] Boles, T., "High Power mmW Switch Technologies," *2017 IEEE International Conference on Microwaves, Antennas, Communications and Electronic Systems*, 2017. Tel Aviv.

[4] Schwierz, F., and J. J. Liou, *Modern Microwave Transistors, Theory, Design and Performance*, John Wiley and Sons, 2003.

[5] Tiwari, S., *Compound Semiconductor Device Physics*, Academic Press, 1992.

[6] Jain, N., and R. J. Gutmann, "Modeling and Design of GaAs MESFET Control Devices for Broad-Band Applications," *IEEE Trans. Microwave Theory and Tech.*, Vol. 38, No. 2, 1990.

[7] Dong, M., "Analyze Transient Delays in GaAs MMIC Switches," *Microwave & RF*, February 2017.

# About the Author

Mouqun Dong received his Ph.D. in physics from Syracuse University. He has been an RF application engineer at California Eastern Laboratories for nearly 18 years. He has also worked on wireless product development at Qualcomm (formerly Atheros Communications). Before joining the industry, he was an academic scientist conducting research at Stanford University's Hansen Lab on a project concerning microwave frequency standards.

# Index

Adjacent channel power (ACP), 250
Admittance
    coordinates, 81, 82
    network, 82–83, 88
    normalized, 79
    parallel, 82
    Smith chart, 78
    source, noise generators and, 124
Amplifier design
    circuit specifications and, 141–52
    with constant-circle method, 169–75
    stability considerations in, 152–69
Amplifiers
    balanced, 175–77
    broadband, 177–78
    circuit specifications, 139
    configurations for, 175–81
    frequency characteristics of, 141–42
    fundamental function of, 142
    gain, 197
    negative feedback, 177–79
    power gain, 177
    return loss, 142
    RF circuit block diagram, 68
    RF performance of, 149
    S-parameters for, 32–33
    stability of, 44–55
    transistor, block diagram, 104
    in wireless transceiver ICs, 179
    *See also* Power amplifiers (PAs)
Amplitude modulation (AM), 308

Analog circuits, 1–2
Antenna noise temperature, 136–38
Attenuation
    constant, 11, 12, 13
    of EM wave, 11
    formulas, 13
    measure of, 104
    transmission line, 10, 12
    *See also* Power loss
Automatic test equipment (ATE), 318
Available gain
    constant circle for, 61
    defined, 40
    maximum (MAG), 57
    measurement, 41
Available power, 116

Balanced amplifiers
    about, 175–76
    feature and design considerations, 176
    illustrated, 176
    use of, 177
    *See also* Amplifiers
Bandwidth
    analysis, network for, 92
    expressed in $Q$, 89–95
    frequency response, 94
    limit, 95
    of matching networks, 86–96, 106–9
    reflection coefficient and, 95
    resolution (RBW), 302, 303–6

Bandwidth *(Cont.)*
  thermal noise and, 116
  video (VBW), 302, 306
  wireless communications systems, 237
Baseband, 235–40
Bias current, 128
Bias networks
  DC characteristics, 160
  for FET switches, 334–37
  impedance at low frequencies, 160–61
  RF isolation, 160
  unintended oscillations, 159–64
Bias-tees, 272–73
Bilinear transformation, 46
Bill of materials (BOM), 269
Binary sequence, 236
Binary states, 238
Bipolar junction transistors (BJTs), 144–45
Blackbody radiation, 136
Board loss, 173
Bode-Fano criterion, 95

Calibration
  checking, 300
  SOLT, 290
  VNA and, 287–91
Capacitance
  bypass, 164
  DC blocking, 106
  ideal, 265
  nominal, 265
Capacitors
  concerns of, 263
  DC-blocking, 187, 266–68, 272, 339
  equivalent circuit for, 262
  impedance curve, 264
  impedance plot, 263
  in matching networks, 263–66
  power dissipation, 268–69
  $Q$ factor for, 88, 261
  SRF of, 265
  usable frequency range, 264
Carrier-to-noise (CNR) ratio, 254
Cartesian coordinates, 78, 79
Cascaded noise figure, 242–43
Cavity effect, 168–69

Characteristic impedance
  about, 7
  effect of, 7–8
  electrical length and, 10
  formulas, 13
Circulators, 279–80
Class A amplifiers
  conduction angle, 206–10
  efficiency of, 210–14
  modes, 215–19
  operation mode, 192
Class AB amplifiers
  conduction angle, 206–10
  efficiency of, 210–14
Class B amplifiers
  conduction angle, 206–10
  efficiency of, 210–14
  modes, 215–20
Class C amplifiers
  conduction angle, 206–10
  efficiency of, 210–14
Class E amplifiers, 228–29
Class F amplifiers, 225–27
Clock signal, interference from, 255–56
Cold-source method, 317–18
Combiners, 282–83
Complementary metal-oxide-
    semiconductor (CMOS)
    technology, 143–44
Computerized Smith chart
  about, 101
  defined, 75–76
  guide for writing code, 101–2
  implementation, 103
  user interface concept, 101
  *See also* Smith chart
Conditionally stable condition, 48, 58
Conductance, source, 132, 134
Conduction angles
  about, 206
  class A and B amplifiers, 207
  class AB and C amplifiers, 207
  classes, 206
  dependence, 209
  efficiency and, 213
  illustrated, 207

output power and efficiency versus input power, 216
theory, 218
Conjugate matching
   complex, 24, 37
   at device output, 56
   noise figure and, 171
   simultaneous, 55–58, 169
Constant circles
   for available gain, 61
   linear circuit design using constant, 63
   in low-noise RF amplifier design, 169–75
   mapping of, 62
   mismatch factor estimation and, 64
   noise, 171
   on planes, 64, 65
   plots, design using, 66, 67
   plotting tools, 63
   for power gain, 61
   properties, 59, 63
Cooling effect, 204, 205, 229
Coupling factor, 277
Cryoelectronics, 139
Curvilinear coordinates, 76–77

DC bias circuits, matching networks and, 103–4
DC blocking capacitors
   effective source and, 267
   method in selecting, 267
   placement, 266
   sensitvity to, 268
DC blocking function, 104, 105, 106
DC resistance (DCR), 164, 272
Delay line parameters, 295, 296
Device temperature, 139
Digital signal processing (DSP), 247
Directional couplers
   directivity, 277–78, 279
   four-port, 276–77
   function of, 276
   isolation, 278
   specifications of, 277–78
   three-port, 279
Distortion-free condition, 190, 191–92

Dividers, 280–82
Doherty amplifier, 230–31
Double-pole double-throw (DPDT) switch, 288, 289
Double-sideband (DSB) modulation, 241
Drain biasing, 162
Drain current, 188, 189, 205, 212, 220, 225, 227
Drain efficiency, 204, 211, 227
Drain voltage, 188, 189, 212, 225, 227

Effective dielectric constant formulas, 13
Effective maximum voltage swing, 192
Efficiency
   class A, AB, B, and C amplifiers, 210–14
   conduction angle and, 213, 216
   drain, 211, 227
Electrical length
   about, 8
   characteristic impedance and, 10
   defined, 9
   parameter format, 9–10
Electromagnetic interference (EMI), 273
Electronic design automation (EDA)
   calculation tools, 18, 30, 57
   plotting constant circles, 63
   simulators, 127
   tuning process, 156
EM compatibility (EMC)
   control of wireless products, 258
   noise emission and, 113
   problem, 259
EM fields, 1, 5–6, 258
Equivalent circuits
   for capacitors, 262
   for ferrite beads, 274
   for FET in RF switch applications, 331
   for FETs, 146, 185
   for inductors, 270
   for noise figure, 131
   for ON state in linearity analysis, 333
Equivalent noise circuits
   concept, 119
   for noise figure, 131
   for noisy two-port network, 120
Equivalent noise conductance, 125

Equivalent noise resistance, 125
EVM
    about, 249
    impact of specification, 248
    at low power levels, 252
    measurement, 249–50
    numerically, 250
    specification, 252

Fast Fourier transform (FFT), 247
Ferrite beads, 273–75
FET switches
    about, 328–29
    DC bias networks for, 334–37
    as IC components, 338
    power-handling capability, 332
    RF performance of, 329–34
    transient external voltage and, 336–37
    use illustration, 329
    See also RF switches
Field-effect transistors (FETs)
    amplifier function, 185
    bias circuit for, 145
    depletion-mode, 167
    drain efficiency, 204
    equivalent circuit in RF switch
        applications, 331
    equivalent circuits for, 146, 185, 333
    gate bias circuit for, 158
    idealized characteristics, 187
    low-noise GaAs, 156
    maximum gate current, 151
    maximum voltage rating, 150
    source as ground, 150
    upper frequency limit, 144
    See also FET switches
Filtering functions, matching networks, 104–6
Frequency modulation (FM), 309
Frequency point, 146
Frequency response
    bandwidth, 94
    of matching networks, 91
    power gains, 177
    of reflection coefficient, 107
    transistors, 146
Frequency stability, 312

GaAs FET switch, 335
Gain
    amplifier, 197
    available, 40
    compression region, 197
    defined, 2
    frequency versus, 148
    measurement, 2
    transistor, 152
    See also Power gain
Grounding, in RF circuits, 258–60

Harmonic flattening
    about, 221–22
    concept, 222
    favorable phase condition for, 225
    second, 222–24, 225
    third, 224, 226
Harmonic trap, 210, 212, 216, 221

Impedance
    combination interpreted as network
        transformation, 84
    coordinates, 80, 81
    defined, 14
    load, 192, 193–94
    noise figure and, 129–35
    normalized, 78–79
    reference, 92, 99
    Smith chart, 78
    transmission line, 86, 99
Impedance matching, 91
Impedance switching, 330
Impedance transformation
    about, 17–18
    illustrated, 19
    network $Q$ factors and, 89
    ratio, 106–7
    Smith chart operation for, 84
    through a matching network, 74
    with transmission lines, 100
Inductance
    bias, 154, 163
    parallel, 100
    parasitic, 263
    as RF choke, 160
    series, 157

shunt, 162
source, 167
unit length, 7
Inductors
  in bias circuits and bias-tees, 272–73
  electrical characteristics of, 269
  equivalent circuit for, 270
  in matching networks, 271
  multilayer, 269
  power dissipation, 271
  Q factor for, 88, 261
  SMT, 269
  ultra-broadband, 272
Insertion loss
  defined, 30
  directional coupler, 278
  lossless networks, 91–92
  measurement, 30
  S-parameter simulation of, 328
Interference
  from clock signal, 255–56
  from intermodulation of external signals, 256–58
  in receivers, 255–58
  receiver sensitivity and, 257
Intermodulation
  frequencies, 199
  interference from, 256–58
  measurement, 320–21
  out-of-band, 257
  third-order, 199–200, 201
  two-tone, 320
Intermodulation distortion, 213
Inverse class F amplifiers, 225–27
$IP_3$, 245
I-Q diagram, 239
Isolation, 328
Isolators, 280

Johnson noise. *See* Thermal noise

K factor, 50, 51, 54–55
Kirchoff's current law (KCL), 83
Knee voltage, 216

Ladder network, 83
LC baluns, 179–81

Linearity specifications, 244–45
Load
  complex conjugate matched, 23
  defined, 15–16;
  impedance, 5, 192, 193–94
  maximum power delivered to, 20, 24
  power dissipation in, 22
  power transfer to, 37–44
  resistance, 212, 214
  transmission line termination with, 14
Load lines
  deriving, 210
  illustrated, 191
  modified, 211
  selection, 214–15, 221
Load resistance, 215, 216
Lossless networks
  general property for, 70
  insertion loss, 91–92
  invariance of mismatch factors of, 68–72
  invariance theorem of, 70
  matching, 69, 75–76
Lossy two-port networks, 30–32
Low-noise amplifier (LNA)
  attenuation and, 10
  cascaded design, 243
  front-end, 253
Low-passive-intermodulation components, 285
Lumped elements
  on Smith chart, 97–99
  transmission lines versus, 109–10
Lumped LC baluns, 180

M-ary modulation, 239
M-ary QAM, 239, 250
Matching circuits
  design on Smith chart, 162
  resistor allowed for, 74–76
Matching factor, 23
Matching networks
  bandwidth of, 86–96, 106–9
  capacitors in, 263–66
  circuit realizations of, 73–76, 108
  DC bias circuits and, 103–4
  DC blocking of, 104, 105

Matching networks *(Cont.)*
  design considerations of, 103–10
  design with the Smith chart, 97–100
  filtering functions of, 104–6
  frequency response of, 91
  impedance transformation through, 74
  inductors in, 271
  lossless, 75–76
  $Q$ factors of, 88–89, 106–9
  RF performance and, 103
  Smith chart operations of, 108
Maximum available gain (MAG), 57
Maximum RF power ratings, 147–51
Maximum stable gain (MSG), 57, 152
Maximum voltage, 147–51
Maxwell's equations, 5
Measurements
  noise figure, 312–18
  nonlinearity, 319–21
  phase noise and frequency stability, 306–12
  power, 318–19
  S-parameters, 287–301
  spectrum, 301–6
  VNA, 287, 299–301
Micro-electromechanical system (MEMS) devices, 323
Mismatch factors
  constant circles of gains and, 58–63
  conversion of return loss to, 69
  as in decibels, 69
  defined, 24
  estimated with plots of constant mismatch circles, 64
  input, 65, 66
  invariance of, 68–72
  load, 42
  lossless networks and, 72
  output, 66
  phase uncertainty and, 143
  source, 42
  target, 67
Modulation
  AM, 308
  FM, 309
  PM, 308–9
  by small perturbation signal, 310

types, 309–11
Multiple-section networks, design and evaluation of, 70

Negative feedback, RF amplifier with, 177–79
Negative-resistance method, 155, 159–69
Network admittance, 88
Network impedance, 82–83
Network $Q$ factor, 89–91, 94
90-degree hybrid couplers, 283
Noise
  analysis, 120, 130
  emission, 113
  $1/f$, 114–15
  parameters, 122, 127–28
  performance, 130
  phase, 248–49, 309–11
  as representation, 114
  shot, 114
  thermal, 114, 115–17
  white, 114, 115
Noise constant circles, 171
Noise current, 120
Noise figure
  antenna noise, 136–38
  antenna noise temperature and, 136–38
  cascaded, 242–43
  circuit model for calculation, 123
  defined, 122
  derivation of, 124
  device, effect on noise figure, 139
  device temperature and, 139
  equivalent circuit for, 131
  as figure of merit, 124
  for linear two-port networks, 122–29
  manufacturers and, 128–29
  minimum, 169
  noise parameters, 127–28
  in practical applications, 129–39
  receiver sensitivity and, 253–55
  return loss and, 173, 174–75
  SNRs for definition, 131
  source impedance and, 129–35
  standard temperature in, 136
  terms used for, 123
  Y-factor-based, 312

Noise figure measurement
   cold-source method, 317–18
   uncertainty and temperature correction, 316–17
   Y-factor method, 312–16
Noise power, 136
Noise power waves, 44–45
Noise sources
   categories of, 113
   diode-based, 314
   dominant, 120
Noise temperature
   about, 138
   antenna, 136–38
   equivalent input, 138
   as noise characteristic, 138
Noisy, 119–20
Nonlinearity
   analysis of, 195
   measurements, 319–21
   specifications, 194–202
Norton currents, 124, 133–34
NPN transistors, 144, 149
N-port networks, S-parameters, 33–34
Nyquist's law, 115, 118, 125

$1/f$ noise, 114–15
Optimal load resistance, 212
Orthogonal frequency division modulation (OFDM), 247
Oscillations
   about, 152
   bias networks, 159–64
   cavity effect, 168–69
   cavity-related, 168–69
   conditions for, 152–55
   inductance for DC biasing and, 162–63
   poor ground via design, 164–67
   unintended, 154–55, 159–69
Oscillator design, 154
Oscilloscopes, 2, 3
Out-of-band intermodulation, 257
Out-of-band rejection, 104

$P_{1dB}$, 244–45, 320
Parasitic resistance, 261, 263–64
Passband, 240–42

Passband of modulated RF carriers, 235–40
Passive components
   about, 261
   capacitors, 88, 261–62, 263–69
   circulators, 279–80
   combiners, 282–83
   directional couplers, 276–79
   inductors, 88, 269–75
   isolators, 280
   low-passive-intermodulation, 285
   90-degree hybrid couplers, 283
   power dividers, 280–82
   for power wave manipulations, 276–85
   resistors, 275
Peak-to-average power ratio (PAPR), 230, 246–48, 250
Phase modulation (PM), 308–9
Phase noise
   measurement, 309–11
   origin, 249
   RF transmitters, 248–49
   single-sideband, 311
Phase shift, 18
Pigtail technique, 299–301
PIN diodes
   defined, 326
   structure illustration, 327
   switches, 326–28
Planck distribution, 115–16
PNP transistors, 144
Polar chart, 77
Polar coordinates, 78, 79
Port extension technique, 296–301
Power-added efficiency (PAE), 204
Power amplifiers (PAs)
   about, 183–84
   in communication systems, 198
   conduction angles, 206–10
   cooling efficiencies of, 203–5
   discrete transistors, 184
   efficiency of, 210–14
   harmonics of, 196
   idealized equivalent circuit, 188
   load-line and matching for power, 187–94
   load line selection, 214–15

Power amplifiers (PAs) *(Cont.)*
  nonlinearity specifications, 194–202
  operation in overdriven conditions, 219–21
  performance, 187
  performance characteristics, 217
  performance parameters, 215
  resistance, 156
  technologies for, 184
  transistor DC characteristics and, 184
  *See also* Amplifiers; *specific classes of power amplifiers*
Power capacity, 184
Power dissipation
  calculation of, 32
  capacitors, 268–69
  inductors, 271
  in load, 22
  in-phase components contribution, 121
  resistors, 275
  zero, 229
Power dividers, 280–82
Power gain
  constant circle for, 61
  defined, 40–41
  formulas, 37–38
  frequency response, 177
  as ratio, 41, 42
  transducer, 37–38, 42, 43
  unilateral, 43
  unmatched, 43
Power loss, 11–12
Power measurements, 318–19
Power waves
  equivalent source, 39
  general description of, 14
  manipulations, devices for, 276–85
  noise, 44–45
  propagation between source and load, 25
  reflection, 15–16
  reflection coefficient in, 15
  in RF circuit theory, 13
  source, 21, 25
  two-port network, 27
Poynting vector, 14

Printed circuit board (PCB), 165, 166
Printed circuit board (PCB)-based transmission lines, 6, 8, 9
Propagating waves, 3, 25–27
Pulse-coded modulation (PCM), 236

$Q$ factors
  defined, 86
  as dimensionless ratio, 89
  essence of, 87–88
  of inductors and capacitors, 88, 261
  of matching networks, 88–89, 106–9
  network, 89–91, 94
  physical implications, 86–87
  of resonant circuits, 87–88
  in three circuits, 90
  transformation ratio and, 91
$Q$ lines, 95–96, 106
Quadrature amplitude modulation (QAM)
  M-ary, 239
  signal generation, 240
  signal variability, 248–49
Quarter-wavelength transmission lines, 19
Quiescent current, 192, 215–16, 219

Reference impedance
  defined, 17
  for S-parameters, 28–29
  using, 18
Reflection coefficients
  bandwidth and, 95
  boundary conditions, 22
  defined, 16
  frequency response, 107
  for one-port networks, 44, 47
  in power waves, 15
  as preferred parameter in RF circuits, 19
  transformations of, 17
  transmission line effect on, 18
  two-port networks, 28
Reflections, multiple, theory of, 27
Resistance
  adding for stability, 158–59
  DC, 272
  equivalent noise, 125
  load, 212, 214, 215, 216

parasitic, 261, 263–64
PAs, 156
shunt, 157
Resistive power divider, 281
Resistors, 275
Resolution bandwidth (RBW), 302, 303–6
Return loss
   amplifiers, 142
   of circuit input, 68
   conversion to mismatch factor, 69
   directional coupler, 277
   input, 169
   noise figure and, 173, 174–75
   output, 169
Reverse gain, 33
RF
   about, 1
   defined, 5
   source, 20–27
   term usage, this book, 5–6
RF amplifiers. *See* Amplifiers
RF choke (RFC), 103, 160, 162
RF circuits
   amplifier, block diagram, 68
   analog circuits and, 1–2
   design, with constant circle method, 63–68
   grounding in, 258–60
   introduction to, 1–6
   passive components in, 261–85
RF cooling effect, 204, 205, 229
RF isolation, 160
RF receiver specifications
   about, 252–53
   determination, 252
   interference, 255–58
   receiver sensitivity, 253–55
   *See also* Wireless communications systems
RF switches
   about, 323–24
   EVM specifications and, 339
   FET-based, 328–37
   GaAs, 335
   key specification in selection of, 337

PIN diode, 326–28
   selection in applications, 337–39
RF transmitter specifications
   EVM, 249–50
   peak-to-average power ratio (PAPR), 246–48
   phase noise, 248–49
   spectral regrowth, 250–52
   *See also* Wireless communications systems
*RLC* resonators, 87
Rollett factor, 51, 54

Scattering parameters. *See* S-parameters
Second harmonic flattening, 222–24, 225
Sensitivity, wireless receiver, 253–55, 257
Shot noise, 114
Signal flow graphs, 44
Signal-to-noise ratio (SNR), 131, 135
Single-dipole double-throw (SPDT) switch, 324, 325
Single-sideband phase noise, 311
Smith chart
   computerized, 75, 101–3
   constant $Q$ lines on, 96
   construction of, 76–82
   coordinate systems, 78
   curvilinear coordinates, 76–77
   destination point on, 106
   impedance and admittance and, 78–79
   for impedance transformation, 84
   of load impedance, 193
   lumped elements on, 97–99
   matching circuit design, 162
   matching network design, 97–110
   matching network operations, 108
   network performance and, 105–6
   original form, 76
   as overlay of coordinate systems, 81
   $Q$ lines, 95–96, 106
   representation of networks, 82–86
   term usage, 81–82
   transmission lines on, 99–100
SnP file format, 33–34
SOLT calibration, 290
Source conductance, 132, 134

Source impedance, noise figure and, 129–35
Sources
  defined, 20
  power transfer from, 37–44
  power waves, 21, 25
  representation of, 21
S-parameter measurement
  pigtail technique, 299–301
  port extension technique, 296–301
  techniques, 287–301
  TRL calibration, 291–96
  VNA and calibration, 287–91
S-parameters
  for amplifiers, 32–33
  circuit analysis block diagram, 38
  in constant-circle plots, 63
  as linear parameters, 29
  for lossy two-port networks, 30–32
  measurements of two-port network, 29
  network analysis using, 44
  N-port networks, 33–34
  reference impedance for, 28–29
  simulation of insertion loss and isolation, 328
  for transmission lines, 29–30
  two-port, 27–33
Spectral regrowth, 250–52
Spectrum analyzers, 301–2
Spectrum measurements
  about, 301
  resolution bandwidth (RBW), 302, 303–6
  spectrum analyzers and, 301–2
  video bandwidth (VBW), 302, 306
Square wave, 220
SRF effect, 265
Stability
  about, 44–45
  adding resistance for, 158–59
  in amplifier design, 152–69
  analysis, 45–47, 153, 155–59
  auxiliary conditions, 50, 52, 55
  bilinear transformation and, 46
  boundaries, 57, 63
  circuit, evaluation of, 162
  circuit, metal enclosure on, 168
  conditionally stable condition, 48, 58
  conditions, 49–50
  conditions for oscillation and, 152–55
  frequency, 312
  gain tradeoff, 64
  importance of, 52
  margin, 48
  overall characteristics, 161
  parameter criteria, 50–52
  problem, 45
  transition from conditional to unconditional, 156
  unconditional, 49–50, 52–54, 192
  unit circle mappings, 54
Stability circles
  about, 47–48
  defined, 48
  mapping, 53
  unit circle and, 49
  unit circle mappings and, 54
Stabilization process, 157
Stable region, 47, 48
Standing wave equation, 3–4
Steady-state solution, propagating waves, 25–27
Surface acoustic wave (SAW) filters, 105
Surface-mount technology (SMT), 179
Switches. See RF switches
Switching functions, realization of, 324–26
Switching-mode operation, 220

Temperature
  correction, 316–17
  effects on noise performance and characterization, 135–39
  Kelvin scale, 135
  standard, in noise figure, 136
  thermal noise and, 115, 116
Theory of multiple reflections, 27
Thermal noise
  about, 115
  available, 116
  bandwidth and, 116
  cause of, 114
  circuit models for, 117
  as ergodic, 118

floor, 118
  temperature and, 115, 116
Thermodynamic equilibrium condition, 117
Third harmonic flattening, 224, 226
Third-order intercept point, 200
Third-order intermodulation, 199–200, 201
Thru, reflect, and line (TRL) calibration
  about, 291
  delay parameter, 295, 296
  design process, 291–95
  standards, 291, 293, 295
Transducer gain, 42, 43, 55, 57
Transistors
  BJTs, 144–45, 149–50
  breakdown voltage, 149
  datasheet specifications for, 145–52
  DC characteristics, 184–87
  FETs, 144–45, 150–51
  frequency response, 146
  $f_T$ specification, 146
  gain specifications, 152
  gains versus frequency, 58
  high-frequency, 147
  maximum RF power ratings, 147–51
  maximum voltage, 147–51
  NPN, 144, 149
  PNP, 144
  for RF applications, 143–45
  transition frequency, 145–47
  usable frequency range, 145–47
Transition frequency, 145–47
Transmission lines
  attenuation, 10–13
  characteristic impedance, 7–8
  coaxial cables, 6, 7
  conductors, 6
  electrical length, 7, 8–10
  illustrated, 7
  impedance, 86, 99
  impedance transformation, 19, 100
  lumped elements versus, 109–10
  network interpretation, 110
  in parallel configuration, 86, 99
  parameters, 99
  PCB-based, 6

quarter-wavelength, 19
  on Smith chart, 99–100
  S-parameters for, 29–30
Two-port networks
  lossy, 30–32
  noise figure for, 122–29
  noisy, 119–20
  power waves, 27

Unconditional stability, 49–50, 52–54, 192
Unilateral power gain, 43
Unintended oscillations, 154, 155
Unmatched power gain, 43
Usable frequency range, 145–47, 184–85

Variable gain amplifier (VGA), 253
Vector network analyzers (VNAs)
  detectors, 287–88
  error model, 290
  implementations on standard modeling, 291
  measurement tips, 299
  measurement with pigtail, 300
  operation principle, 288
  port extension capability, 297
  in S-parameter measurement, 287–91
Vias, ground, 164–67
Video bandwidth (VBW), 302, 306
Voltage standing wave ratio (VSWR), 32

White noise, 114, 115
Wilkinson power divider, 281
Wireless communications systems
  bandwidth, 237
  baseband and M-ary scheme, 235–40
  basic function of, 234
  cascaded noise figure, 242–43
  grounding in RF circuits, 258–60
  introduction to, 233–35
  linearity specifications, 244–45
  modulation scheme, 236–37
  passband of modulated RF carriers, 240–42
  reliability, 235
  RF receiver specifications, 252–58
  RF transmitter specifications, 246–52

Wireless communications systems *(Cont.)*
  transceiver block diagram, 234
Wireless receivers
  block diagram, 253
  functional blocks, 252–53
  interference in, 255–58
  sensitivity, 253–55
Wireless transceiver, 234

Y-factor method
  about, 312
  functional blocks, 312–13
  graphic representation, 315
  Y-factor definition and, 314
  *See also* Noise figure measurement

Zero-power dissipation, 229

# Artech House Microwave Library

*Behavioral Modeling and Linearization of RF Power Amplifiers,* John Wood

*Chipless RFID Reader Architecture,* Nemai Chandra Karmakar, Prasanna Kalansuriya, Randika Koswatta, and Rubayet E-Azim

*Control Components Using Si, GaAs, and GaN Technologies,* Inder J. Bahl

*Design of Linear RF Outphasing Power Amplifiers,* Xuejun Zhang, Lawrence E. Larson, and Peter M. Asbeck

*Design Methodology for RF CMOS Phase Locked Loops,* Carlos Quemada, Guillermo Bistué, and Iñigo Adin

*Design of CMOS Operational Amplifiers,* Rasoul Dehghani

*Design of RF and Microwave Amplifiers and Oscillators, Second Edition,* Pieter L. D. Abrie

*Digital Filter Design Solutions,* Jolyon M. De Freitas

*Discrete Oscillator Design Linear, Nonlinear, Transient, and Noise Domains,* Randall W. Rhea

*Distortion in RF Power Amplifiers,* Joel Vuolevi and Timo Rahkonen

*Distributed Power Amplifiers for RF and Microwave Communications,* Narendra Kumar and Andrei Grebennikov

*Electric Circuits: A Primer,* J. C. Olivier

*Electronics for Microwave Backhaul,* Vittorio Camarchia, Roberto Quaglia, and Marco Pirola, editors

*EMPLAN: Electromagnetic Analysis of Printed Structures in Planarly Layered Media, Software and User's Manual,* Noyan Kinayman and M. I. Aksun

*An Engineer's Guide to Automated Testing of High-Speed Interfaces, Second Edition,* José Moreira and Hubert Werkmann

*Envelope Tracking Power Amplifiers for Wireless Communications,* Zhancang Wang

*Essentials of RF and Microwave Grounding,* Eric Holzman

*Frequency Measurement Technology,* Ignacio Llamas-Garro, Marcos Tavares de Melo, and Jung-Mu Kim

*FAST: Fast Amplifier Synthesis Tool—Software and User's Guide,* Dale D. Henkes

*Feedforward Linear Power Amplifiers,* Nick Pothecary

*Filter Synthesis Using Genesys S/Filter,* Randall W. Rhea

*Foundations of Oscillator Circuit Design,* Guillermo Gonzalez

*Frequency Synthesizers: Concept to Product,* Alexander Chenakin

*Fundamentals of Nonlinear Behavioral Modeling for RF and Microwave Design,* John Wood and David E. Root, editors

*Generalized Filter Design by Computer Optimization,* Djuradj Budimir

*Handbook of Dielectric and Thermal Properties of Materials at Microwave Frequencies,* Vyacheslav V. Komarov

*Handbook of RF, Microwave, and Millimeter-Wave Components,* Leonid A. Belov, Sergey M. Smolskiy, and Victor N. Kochemasov

*High-Efficiency Load Modulation Power Amplifiers for Wireless Communications,* Zhancang Wang

*High-Linearity RF Amplifier Design,* Peter B. Kenington

*High-Speed Circuit Board Signal Integrity, Second Edition,* Stephen C. Thierauf

*Integrated Microwave Front-Ends with Avionics Applications,* Leo G. Maloratsky

*Intermodulation Distortion in Microwave and Wireless Circuits,* José Carlos Pedro and Nuno Borges Carvalho

*Introduction to Modeling HBTs,* Matthias Rudolph

*An Introduction to Packet Microwave Systems and Technologies,* Paolo Volpato

*Introduction to RF Design Using EM Simulators,* Hiroaki Kogure, Yoshie Kogure, and James C. Rautio

*Introduction to RF and Microwave Passive Components,* Richard Wallace and Krister Andreasson

*Klystrons, Traveling Wave Tubes, Magnetrons, Crossed-Field Amplifiers, and Gyrotrons,* A. S. Gilmour, Jr.

*Lumped Elements for RF and Microwave Circuits,* Inder Bahl

*Lumped Element Quadrature Hybrids,* David Andrews

*Microstrip Lines and Slotlines, Third Edition,* Ramesh Garg, Inder Bahl, and Maurizio Bozzi

*Microwave Circuit Modeling Using Electromagnetic Field Simulation,* Daniel G. Swanson, Jr. and Wolfgang J. R. Hoefer

*Microwave Component Mechanics,* Harri Eskelinen and Pekka Eskelinen

*Microwave Differential Circuit Design Using Mixed-Mode S-Parameters,* William R. Eisenstadt, Robert Stengel, and Bruce M. Thompson

*Microwave Engineers' Handbook, Two Volumes,* Theodore Saad, editor

*Microwave Filters, Impedance-Matching Networks, and Coupling Structures,* George L. Matthaei, Leo Young, and E. M. T. Jones

*Microwave Imaging Methods and Applications,* Matteo Pastorino and Andrea Randazzo

*Microwave Material Applications: Device Miniaturization and Integration,* David B. Cruickshank

*Microwave Materials and Fabrication Techniques, Second Edition,* Thomas S. Laverghetta

*Microwave Materials for Wireless Applications,* David B. Cruickshank

*Microwave Mixer Technology and Applications,* Bert Henderson and Edmar Camargo

*Microwave Mixers, Second Edition,* Stephen A. Maas

*Microwave Network Design Using the Scattering Matrix,* Janusz A. Dobrowolski

*Microwave Power Amplifier Design with MMIC Modules,* Howard Hausman

*Microwave Radio Transmission Design Guide, Second Edition,* Trevor Manning

*Microwave and RF Semiconductor Control Device Modeling,* Robert H. Caverly

*Microwave Transmission Line Circuits,* William T. Joines, W. Devereux Palmer, and Jennifer T. Bernhard

*Microwaves and Wireless Simplified, Third Edition,* Thomas S. Laverghetta

*Modern Microwave Circuits,* Noyan Kinayman and M. I. Aksun

*Modern Microwave Measurements and Techniques, Second Edition,* Thomas S. Laverghetta

*Modern RF and Microwave Filter Design,* Protap Pramanick and Prakash Bhartia

*Neural Networks for RF and Microwave Design,* Q. J. Zhang and K. C. Gupta

*Noise in Linear and Nonlinear Circuits,* Stephen A. Maas

*Nonlinear Microwave and RF Circuits, Second Edition,* Stephen A. Maas

*On-Wafer Microwave Measurements and De-Embedding,* Errikos Lourandakis

*Parameter Extraction and Complex Nonlinear Transistor Models,* Günter Kompa

*Passive RF Component Technology: Materials, Techniques, and Applications,* Guoan Wang and Bo Pan, editors

*Practical Analog and Digital Filter Design,* Les Thede

*Practical Microstrip Design and Applications,* Günter Kompa

*Practical Microwave Circuits,* Stephen Maas

*Practical RF Circuit Design for Modern Wireless Systems, Volume I: Passive Circuits and Systems,* Les Besser and Rowan Gilmore

*Practical RF Circuit Design for Modern Wireless Systems, Volume II: Active Circuits and Systems,* Rowan Gilmore and Les Besser

*Principles of RF and Microwave Design,* Matthew A. Morgan

*Production Testing of RF and System-on-a-Chip Devices for Wireless Communications,* Keith B. Schaub and Joe Kelly

*Q Factor Measurements Using MATLAB*, Darko Kajfez

*QMATCH: Lumped-Element Impedance Matching, Software and User's Guide,* Pieter L. D. Abrie

*Radio Frequency Integrated Circuit Design, Second Edition*,
  John W. M. Rogers and Calvin Plett

*Reflectionless Filters,* Matthew A. Morgan

*RF Bulk Acoustic Wave Filters for Communications,*
  Ken-ya Hashimoto

*RF Circuits and Applications for Practicing Engineers*, Mouqun Dong

*RF Design Guide: Systems, Circuits, and Equations,* Peter Vizmuller

*RF Linear Accelerators for Medical and Industrial Applications,*
  Samy Hanna

*RF Measurements of Die and Packages,* Scott A. Wartenberg

*The RF and Microwave Circuit Design Handbook*, Stephen A. Maas

*RF and Microwave Coupled-Line Circuits*, Rajesh Mongia, Inder Bahl,
  and Prakash Bhartia

*RF and Microwave Oscillator Design*, Michal Odyniec, editor

*RF Power Amplifiers for Wireless Communications, Second Edition,*
  Steve C. Cripps

*RF Systems, Components, and Circuits Handbook,* Ferril A. Losee

*Scattering Parameters in RF and Microwave Circuit Analysis and
  Design,* Janusz A. Dobrowolski

*The Six-Port Technique with Microwave and Wireless Applications,*
  Fadhel M. Ghannouchi and Abbas Mohammadi

*Solid-State Microwave High-Power Amplifiers,* Franco Sechi and
  Marina Bujatti

*Spin Transfer Torque Based Devices, Circuits, and Memory,* Brajesh
  Kumar Kaushik and Shivam Verma

*Stability Analysis of Nonlinear Microwave Circuits,* Almudena Suárez
  and Raymond Quéré

*Substrate Noise Coupling in Analog/RF Circuits,* Stephane Bronckers,
  Geert Van der Plas, Gerd Vandersteen, and Yves Rolain

*System-in-Package RF Design and Applications,* Michael P. Gaynor

*Technologies for RF Systems,* Terry Edwards

*Terahertz Metrology,* Mira Naftaly, editor

*TRAVIS 2.0: Transmission Line Visualization Software and User's Guide, Version 2.0,* Robert G. Kaires and Barton T. Hickman

*Understanding Microwave Heating Cavities,* Tse V. Chow Ting Chan and Howard C. Reader

*Understanding Quartz Crystals and Oscillators,* Ramón M. Cerda

*Vertical GaN and SiC Power Devices,* Kazuhiro Mochizuki

*The VNA Applications Handbook,* Gregory Bonaguide and Neil Jarvis

For further information on these and other Artech House titles, including previously considered out-of-print books now available through our In-Print-Forever® (IPF®) program, contact:

Artech House Publishers
685 Canton Street
Norwood, MA 02062
Phone: 781-769-9750
Fax: 781-769-6334
e-mail: artech@artechhouse.com

Artech House Books
16 Sussex Street
London SW1V 4RW UK
Phone: +44 (0)20 7596 8750
Fax: +44 (0)20 7630 0166
e-mail: artech-uk@artechhouse.com

Find us on the World Wide Web at: www.artechhouse.com